T0172404

Circular and Linear Regression

Fitting Circles and Lines by Least Squares

Monographs on Statistics and Applied Probability 117

Circular and Linear Regression
Fitting Circles and Lines by Least Squares

Nikolai Chernov
University of Alabama at Birmingham
U.S.A.

CRC Press
Taylor & Francis Group
Boca Raton   London   New York

CRC Press is an imprint of the
Taylor & Francis Group, an **informa** business

A CHAPMAN & HALL BOOK

CRC Press
Taylor & Francis Group
6000 Broken Sound Parkway NW, Suite 300
Boca Raton, FL 33487-2742

First issued in paperback 2020

© 2019 by Taylor & Francis Group, LLC
CRC Press is an imprint of Taylor & Francis Group, an Informa business

No claim to original U.S. Government works

ISBN-13: 978-0-367-57717-9 (pbk)
ISBN-13: 978-1-4398-3590-6 (hbk)

This book contains information obtained from authentic and highly regarded sources. Reasonable efforts have been made to publish reliable data and information, but the author and publisher cannot assume responsibility for the validity of all materials or the consequences of their use. The authors and publishers have attempted to trace the copyright holders of all material reproduced in this publication and apologize to copyright holders if permission to publish in this form has not been obtained. If any copyright material has not been acknowledged please write and let us know so we may rectify in any future reprint.

Except as permitted under U.S. Copyright Law, no part of this book may be reprinted, reproduced, transmitted, or utilized in any form by any electronic, mechanical, or other means, now known or hereafter invented, including photocopying, microfilming, and recording, or in any information storage or retrieval system, without written permission from the publishers.

For permission to photocopy or use material electronically from this work, please access www copyright.com (http://www.copyright.com/) or contact the Copyright Clearance Center, Inc. (CCC), 222 Rosewood Drive, Danvers, MA 01923, 978-750-8400.CCC is a not-for-profit organization that provides licenses and registration for a variety of users. For organizations that have been granted a photocopy license by the CCC, a separate system of payment has been arranged.

Trademark Notice: Product or corporate names may be trademarks or registered trademarks, and are used only for identification and explanation without intent to infringe.

Visit the Taylor & Francis Web site at
http://www.taylorandfrancis.com

and the CRC Press Web site at
http://www.crcpress.com

Contents

Preface

This book is devoted to an active research topic in modern statistics—fitting geometric contours (lines and circles) to observed data, in particular, to digitized images. In such applications both coordinates of the observed points are measured imprecisely, i.e., both variables (x and y) are subject to random errors. Statisticians call this topic the Errors-In-Variables (EIV) model. It is radically different, and much more complex, than the classical regression where only one variable (usually, y) is random.

Fitting straight lines to observed data with errors in both variables is an old problem dating back to the 1870s [1, 2, 117], with applications in general statistics, sciences, econometrics, and image processing. Its studies have a colorful history (which we overview in Chapter 1) through the twentieth century, and its most active period perhaps lasted from 1975 to 1995. By the late 1990s all the major issues in the linear EIV problem appeared to be resolved, and now this topic is no longer an active research area.

For a detailed and complete account of the linear EIV regression studies, see surveys [8, 73, 126, 127, 132, 187] and books [40], [66], as well as Chapter 10 in [128] and Chapter 29 in [111]. We note that the linear EIV problem, despite its illusive simplicity, is deep and vast; entire books, such as [66] and [40], are devoted to this subject. We only overview it as much as it is related to our main theme — fitting circles and other curves.

Fitting *nonlinear* models to data with errors in both variables has been studied by statisticians since the 1930s [50, 55]. This topic can be divided into two parts. In one, the main goal is to describe observed data by a nonlinear function $y = g(x)$, such as a polynomial, or an exponential function, etc. In those applications the x and y variables usually have different natures, measured in different units, and errors in x and y may have different magnitude. Such applications are common in statistics and econometrics. A detailed presentation of this type of nonlinear model can be found [27]; see also the latest edition [28], updated and expanded.

The second type of nonlinear EIV problems is common in image processing applications. In those, data points come from a picture, photograph, map, etc. Both x and y variables measure length and are given in the same units; the choice of the coordinate system is often quite arbitrary, hence errors in x and y have the same magnitude, on average. Fitting explicit functions $y = g(x)$ to images is not the best idea: it inevitably forces a different treatment of the x and y variables, conflicting with the very nature of the problem. Instead, one fits geometric shapes that are to be found (or expected) on the given image. Those shapes are usually described geometrically: lines, rectangles and other polygons, circles, ovals (ellipses), etc. Analytically, the basic curved shapes—circles and ovals—are defined by implicit quadratic functions. More complicated curves may be approximated by cubic or quartic implicit polynomials [150, 176]; however, the latter are only used on special occasions and are rare in practice. Our book is devoted to fitting most basic geometric curves—circles and circular arcs—to observed data in image processing applications. This topic is different from the nonlinear EIV regression in other statistical applications mentioned above and covered in [27, 28]. Fitting ellipses to observed data is another important topic that deserves a separate book (the author plans to publish one in the future).

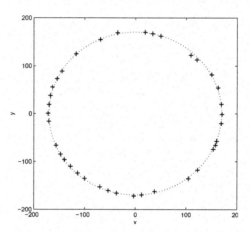

Figure 0.1 *The Brogar Ring on Orkney Islands [15, 178]. The stones are marked by pluses; the fitted circle is the dotted line.*

The problem of fitting circles and circular arcs to observed points in 2D images dates back to the 1950s. Its first instance was rather peculiar: English engineers and archaeologists examined megalithic sites (stone rings) in the British Isles trying to determine if ancient people who had built those mysterious structures used a common unit of length. This work started in the 1950s and continued for several decades [15, 65, 177, 178, 179]; see an example in Fig. 0.1, where the data are borrowed from [178].

In the 1960s the necessity of fitting circles emerged in geography [155]. In the 1970s circles were fitted to experimental observations in microwave engineering [54, 108]; see an example in Fig. 0.2, where the data are taken from

[15]. Since about 1980, fitting circles became an agenda in many areas of human practice. We just list some prominent cases below.

In medicine, one estimates the diameter of a human iris on a photograph [141], or designs a dental arch from an X-ray [21], or measures the size of a fetus on a picture produced by ultrasound. Archaeologists examine the circular shape of ancient Greek stadia [157], or determine the size of ancient pottery by analyzing potsherds found in field expeditions [44, 80, 81, 190]. In industry, quality control requires estimation of the radius and the center of manufactured mechanical parts [119]. In mobile robotics, one detects round objects (pillars, tree trunks) by analyzing range readings from a 2D laser range finder used by a robot [197].

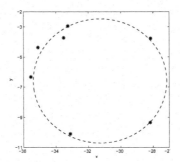

Figure 0.2 *Reflection coefficients in microwave engineering [15]. The observed values are marked by stars; the fitted circle is the dashed line.*

But perhaps the single largest field of applications where circles are fitted to data is nuclear physics. There one deals with elementary particles born in accelerators and colliders. The newborn particles move along circular arcs in a constant magnetic field; physicists determine the energy of the particle by measuring the radius of its trajectory; to this end they fit an arc to a string of mechanical or electrical signals the particle leaves in the detector [45, 53, 82, 106, 107, 136, 173, 174, 175]. Particles with high energy move along arcs with large radii (low curvature), thus fitting arcs to nearly straight-looking trajectories is quite common; this task requires very elaborate techniques to ensure accurate results.

We illustrate our discussion by a real–life example from archaeology. To estimate the diameter of a potsherd from a field expedition, the archaeologist traces the profile of a broken pot — such as the outer rim or base — with a pencil on a sheet of graph paper. Then he scans his drawing and transforms it into an array of pixels (data points). Lastly, he fits a circle to the digitized image by using a computer.

A typical digitized arc tracing a circular wheelmade antefix is shown in

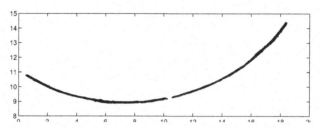

Figure 0.3 *A typical arc drawn by pencil with a profile gauge from a circular wheel-made antefix.*

Fig. 0.3 (this image contains 7452 pixels). The best fitting circle found by a standard least squares procedure has parameters

$$\text{center} = (7.4487, 22.7436), \qquad \text{radius} = 13.8251. \qquad (1)$$

This does not seem challenging, as the arc in Fig. 0.3 is clearly visible to the naked eye, so one can even reconstruct a circle manually.

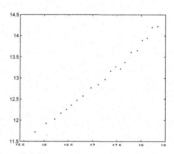

Figure 0.4: *A fragment of the arc shown in Fig. 0.3.*

Now suppose we can only see a small fragment of the above arc, with very few points on it. Fig. 0.4 shows a sample of merely 22 randomly chosen points from a tiny part of the original arc. Suppose we are to fit a circle to these points, without seeing the rest of the image. Is it possible?

Visually, the 22 points in Fig. 0.4 do not even form a clear circular arc, they rather look like a shapeless string. Reconstructing a circle manually from these 22 points appears an impossibility. However, the best computer algorithm returns the following parameters:

$$\text{center} = (7.3889, 22.6645), \qquad \text{radius} = 13.8111. \qquad (2)$$

Compare this to (1). The estimates are strikingly accurate!

The algorithm that produced the estimates (2) is the Levenberg-Marquard geometric fit (minimizing the geometric distances from the given points to the circle); it is described in Section 4.5. One may naturally want to estimate errors of the returned values of the circle parameters, but this is a difficult task for the EIV regression problems. In particular, under the standard statistical models described in Chapter 6, the estimates of the center and radius of the circle have infinite variances and infinite mean values! Thus, the conventional error estimates (based on the standard deviations) would be absurdly infinite. An approximate error analysis developed in Chapter 7 can be used to assess errors in a more realistic way; then the errors of the estimates (2) happen to be ≈ 0.1.

We see that the problem of fitting circles and circular arcs to images has a variety of applications. It has attracted the attention of scientists, engineers, statisticians, and computer programmers. Many good (and not-so-good) algorithms were proposed; some to be forgotten and later rediscovered. For example, the Kåsa algorithm, see our Chapter 5, was published independently at least 13 times, the first time in 1972 and the last (so far) in 2006, see references in Section 5.1. But, despite the popularity of circle fitting applications, until the 1990s publications were sporadic and lacked a unified approach.

An explosion of interest in the problem of fitting circles and other geometric shapes to observed points occurred in the 1990s when it became an agenda issue for the rapidly growing computer science community, because fitting simple contours (lines, circles, ellipses) to digitized images was one of the basic tasks in pattern recognition and computer vision. More general curves are often approximated by a sequence of segments of lines or circular arcs that are stitched together ("circular splines"); see [12, 145, 158, 164, 165].

Since the early 1990s, many new algorithms (some of them truly brilliant) for fitting circles and ellipses have been invented; among those are circle fits using the Riemann sphere [123, 175] and conformal maps of the complex plane [159], "direct ellipse fit" by Fitzgibbon et al. [61, 63, 147] and Taubin's eigenfit [176], a sophisticated renormalization procedure due to Kanatani [47, 94, 95] and a no less superb HEIV method due to Leedan and Meer [48, 49, 120], as well as the Fundamental Numerical Scheme by Chojnacki et al. [47, 48]. Chojnacki and his collaborators developed a unified approach to several popular algorithms [49] and did a remarkable job of explaining the underlying ideas.

Theoretical investigation also led to prominent accomplishments. These include consistent curve and surface fitting algorithms due to Kukush, Markovsky, and van Huffel [114, 130, 167], "hyperaccurate" ellipse fitting methods by Kanatani [102, 104], and a rather unconventional adaptation of the classical Cramer-Rao lower bound to general curve fitting problems [42, 96].

The progress made in the last 15 years is indeed spectacular, and the total output of all these studies is more than enough for a full size book on the subject. To the author's best knowledge, no such book exists yet. The last book on fitting geometric shapes to data was published by Kanatani [95] in 1996. A good (but rather limited) tutorial on fitting parametric curves, due to Zhang, appeared on the Internet in about the same year (and in print in 1997, see [198]). These two publications covered ellipse fitting methods existing in 1996, but not specifically circle fitting methods. The progress made after 1996 remains unaccounted for.

The goal of this book is to present the topic of fitting circles and circular arcs to observed points in full, especially accounting for all the recent achievements since the mid-1990s. I have tried to cover all aspects of this problem: geometrical, statistical, and computational. In particular, my purpose is to present numerical algorithms in relation to one another, with underlying ideas, to emphasize strong and weak points of each algorithm, and to indicate how to combine them to achieve the best performance. The book thoroughly addresses theoretical aspects of the fitting problem which are essential for understanding advantages and limitations of practical schemes. Lastly, an attempt was made to identify issues that remain obscure and may be subjects of future investigation.

At the same time the book is geared toward the end user. It is written for practitioners who want to learn the topic or need to select the right tool for their particular task. I have tried to avoid purely abstract issues detached from practice, and presented topics that were deemed most important for image processing applications.

I assume the reader has a good mathematical background (being at ease with calculus, geometry, linear algebra, probability and statistics) and some experience in numerical analysis and computer programming. I am not using any specific machine language in the book, though the MATLAB® code of all relevant algorithms may be found on our Web page [84].

The author is deeply indebted to his former supervisor at the Joint Institute for Nuclear Research (Russia), G. Ososkov, for his constant guidance in the studies on the circle fitting problem. The author is grateful to K. Kanatani for his strong support in research and especially in the design of this book. The author thanks his graduate students C. Lesort and A. Al-Sharadqah for their devotion to the subject and their help in preparing the manuscript and posting the computer code on the Web. Lastly, the author is partially supported by National Science Foundation, grant DMS-0652896.

The book is organized as follows, see diagram in Fig. 0.5. Chapter 1 is an introduction to the Errors-In-Variables regression analysis and gives its brief history (mostly in the context of the linear model). Chapter 2 summarizes the solution of the linear EIV problem and highlights its main properties (geometric and statistical). These two chapters do not deal with circles or arcs.

Chapter 3 gives the theory of fitting circles by least squares. It addresses the existence and uniqueness of the solution, describes various parametrization

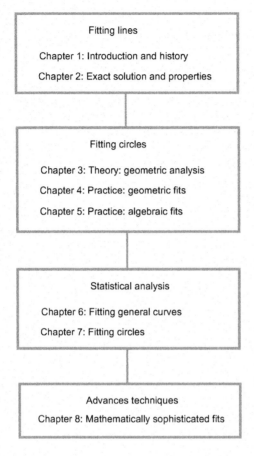

Figure 0.5: *The structure of the book.*

schemes for circles, and analyzes the shape of the objective function to be minimized (culminating in the important Two Valley Theorem).

Chapters 4 and 5 are devoted to practical circle fitting methods.

In Chapter 4 circles are fitted by minimizing geometric distances from observed points to the fitting circle, which is a classical (or geometric) fit. This is a nonlinear problem that has no closed form solution, so all algorithms are iterative, thus computationally intensive and subject to occasional divergence. We describe all popular schemes, in a historic perspective, emphasizing relations between one another, highlighting their advantages and drawbacks.

Chapter 5 deals with simplified circle fits, so called *algebraic fits*. They are fast, noniterative, and do not suffer from divergence. However, they are (in many cases) less accurate than the geometric fits of Chapter 4. Algebraic fits are often used in mass data processing (especially in nuclear physics), where speed is of paramount importance. Algebraic fits are also used for initializing iterative geometric fitting procedures.

The reader interested in only practical algorithms can find all the relevant information is Chapters 3–5.

Chapters 6 and 7 make a sharp turn and plunge into statistical analysis of curve fitting methods. This is theoretical material, but I have tried to relate it to practice and explain all constructions and conclusions in practical terms. Chapter 6 is devoted to general nonlinear EIV regression, i.e., it covers arbitrary curves. Chapter 7 focuses on the specific task of fitting circles and circular arcs.

Chapters 6 and 7 may be of interest to professional statisticians.

Lastly, Chapter 8 presents a sample of "exotic" circle fits, including some mathematically sophisticated procedures — they make use of complex numbers and conformal mappings of the complex plane. This chapter is best for scientists with a solid mathematical background. The ideas behind methods of Chapter 8 are quite intriguing and resulting fits look very promising. This may be a starting point for future development of this subject.

Most illustrations in the book have been prepared with MATLAB. MATLAB® is a registered trademark of The MathWorks, Inc. For product information, please contact:

The MathWorks, Inc.
3 Apple Hill Drive
Natick, MA 01760-2098 USA
Tel: 508 647 7000
Fax: 508-647-7001
E-mail: info@mathworks.com
Web: www.mathworks.com

Symbols and notation

Here we describe some notation used throughout the book. First we describe our notational system for matrices and vectors:

- $\mathbb{R}$ denotes the set of real numbers (real line), $\mathbb{R}^n$ the n-dimensional Euclidean space, and $\mathbb{C}$ the complex plane.

- Matrices are always denoted by capital letters typeset in bold face, such as $\mathbf{M}$ or $\mathbf{U}$. The identity matrix is denoted by $\mathbf{I}$.

- Vectors are denoted by letters in bold face, either capital or lower case, such as $\mathbf{A}$ or $\mathbf{a}$. By default, all vectors are assumed to be column vectors. Row vectors are obtained by transposition.

- The superscript T denotes the transpose of a vector or a matrix. For example, if $\mathbf{A}$ is a vector, then it is (by default) a column-vector, and the corresponding row-vector is denoted by $\mathbf{A}^T$.

- $\mathrm{diag}\{a_1, a_2, \ldots, a_n\}$ denotes a diagonal matrix of size $n \times n$ with diagonal entries $a_1, a_2, \ldots, a_n$.

- For any vector and matrix, $\|\mathbf{A}\|$ means its 2-norm, unless otherwise stated.

- $\varkappa(\mathbf{A}) = \|\mathbf{A}\|\,\|\mathbf{A}^{-1}\|$ denotes the condition number of a square matrix $\mathbf{A}$ (relative to the 2-norm).

- Equation $\mathbf{A}\mathbf{x} \approx \mathbf{b}$, where $\mathbf{A}$ is an $n \times m$ matrix, $\mathbf{x} \in \mathbb{R}^m$ is an unknown vector, and $\mathbf{b} \in \mathbb{R}^n$ is a known vector $(n > m)$, denotes the classical least square problem whose solution is $\mathbf{x} = \mathrm{argmin}\,\|\mathbf{A}\mathbf{x} - \mathbf{b}\|^2$.

- The singular value decomposition (SVD) of an $n \times m$ matrix $\mathbf{A}$ is denoted by $\mathbf{A} = \mathbf{U}\Sigma\mathbf{V}^T$, where $\mathbf{U}$ and $\mathbf{V}$ are orthogonal matrices of size $n \times n$ and $m \times m$, respectively, and Σ is a diagonal $n \times m$ matrix whose diagonal entries are real nonnegative and come in a decreasing order:

$$\sigma_1 \geq \sigma_2 \geq \cdots \geq \sigma_p, \qquad p = \min\{m, n\}.$$

If $n > m$, then a short SVD is given by $\mathbf{A} = \mathbf{U}'\Sigma'\mathbf{V}^T$, where $\mathbf{U}$ consists of the first (left) m columns of $\mathbf{U}$ and Σ' consists of the first (top) m rows of Σ.

- $\mathbf{A}^-$ denotes the Moore-Penrose pseudoinverse of a matrix $\mathbf{A}$. It is given by $\mathbf{A}^- = \mathbf{V}\Sigma^-\mathbf{U}^T$, where Σ^- is a diagonal $m \times n$ matrix whose diagonal entries are

$$\sigma_i^- = \begin{cases} 1/\sigma_i & \text{if } \sigma_i > 0 \\ 0 & \text{if } \sigma_i = 0 \end{cases}$$

For probability, we use the following notation:

- $\text{Prob}(A)$ denotes the probability of an event A.
- $\mathbb{E}(X)$ denotes the mean value of a random variable X.
- $\text{Var}(X)$ denotes the variance of a random variable X.
- $\text{Cov}(X,Y)$ denotes the covariance of random variables X and Y.
- $N(\mu, \sigma^2)$ denotes a normal random variable with mean μ and variance σ^2.
- $X_n \to_L X$ denotes the weak convergence of random variables, i.e., the convergence of the distribution functions of X_n to the distribution function of X at every point where the latter is continuous.
- $\mathcal{O}_P(\sigma^k)$ denotes a random variable, X, that may depend on σ and such that $\sigma^{-k}X$ is bounded in probability; i.e., such that for any $\varepsilon > 0$ there exists $A_\varepsilon > 0$ such that $\text{Prob}\{\sigma^{-k}X > A_\varepsilon\} < \varepsilon$ for all $\sigma > 0$.

For statistics, we use the following notation:

- Given a sample $x_1, \ldots, x_n$ we denote its sample mean by $\bar{x} = \frac{1}{n}\sum_{i=1}^n x_i$.
- We conveniently extend the above sample mean notation as follows:

$$\overline{xx} = \frac{1}{n}\sum x_i^2, \qquad \overline{xy} = \frac{1}{n}\sum x_i y_i, \qquad \text{etc.}$$

- Θ usually denotes the vector of unknown parameters, and $\theta_1, \theta_2, \ldots$ its components. For example, a and b are the parameters of an unknown line $y = a + bx$.
- We use tildas for the true values of the unknown parameters, i.e., we write $\tilde{\Theta} = (\tilde{\theta}_1, \tilde{\theta}_2, \ldots)$. For example, $\tilde{a}$ and $\tilde{b}$ are the true values of the parameters a and b.
- We use 'hats' for estimates of the unknown parameters, i.e., we write $\hat{\Theta} = (\hat{\theta}_1, \hat{\theta}_2, \ldots)$. For example, $\hat{a}$ and $\hat{b}$ denote estimates of the parameters a and b.
- MLE is an abbreviation for Maximum Likelihood Estimate. For example, we write $\hat{a}_{\text{MLE}}$ for the MLE of the parameter a.
- $\text{bias}(\hat{a}) = \mathbb{E}(\hat{a}) - \tilde{a}$ denotes the bias of an estimate $\hat{a}$. An estimate is unbiased if its bias is zero.
- MSE is the Mean Squared Error (of an estimate). For example,

$$\text{MSE}(\hat{a}) = \mathbb{E}\left[(\hat{a} - \tilde{a})^2\right] = \text{Var}(\hat{a}) + \left[\text{bias}(\hat{a})\right]^2.$$

- For a parameter vector Θ, the MSE is a matrix

$$
\begin{aligned}
\mathrm{MSE}(\hat{\Theta}) &= \mathbb{E}\left[(\hat{\Theta} - \tilde{\Theta})(\hat{\Theta} - \tilde{\Theta})^T\right] \\
&= \mathrm{Cov}(\hat{\Theta}) + \left[\mathrm{bias}(\hat{\Theta})\right]\left[\mathrm{bias}(\hat{\Theta})\right]^T,
\end{aligned}
$$

where $\mathrm{Cov}(\hat{\Theta})$ stands for the covariance matrix of the estimate $\hat{\Theta}$.

List of Figures

List of Tables

Chapter 1

Introduction and historic overview

1.1   Classical regression

In a classical regression problem, one deals with a functional relation $y = g(x)$ between two variables, x and y. As an archetype example, let x represent time and $y = g(x)$ a certain quantity observed at time x (say, the outside temperature or the stock market index), then one would like to model the evolution of g.

One records a number of observations $(x_1, y_1), \ldots, (x_n, y_n)$ and tries to approximate them by a relatively simple model function, such as linear $y = a + bx$ or quadratic $y = a + bx + cx^2$ or exponential $y = ae^{bx}$, etc., where $a, b, c, \ldots$ are the respective coefficients (or parameters of the model).

Generally, let us denote the model function by $y = g(x; \Theta)$, where $\Theta = (a, b, \ldots)$ is the vector of relevant parameters. The goal is to find a particular function $g(x; \hat{\Theta})$ in that class (i.e., choose a particular value $\hat{\Theta}$ of Θ) that approximates (fits) the observed data $(x_1, y_1), \ldots, (x_n, y_n)$ best. It is not necessary to achieve the exact relations $y_i = g(x_i; \hat{\Theta})$ for all (or any) i, because y_i's are regarded as imprecise (or noisy) observations of the functional values.

A standard assumption in statistics is that y_i's are small random perturbations of the true values $\tilde{y}_i = g(x_i; \tilde{\Theta})$, i.e.

$$y_i = g(x_i; \tilde{\Theta}) + \varepsilon_i, \qquad i = 1, \ldots, n$$

where $\tilde{\Theta}$ stands for the true (but unknown) value of Θ, and (small) errors ε_i are independent normally distributed random variables with zero mean and, in the simplest case, common variance σ^2. Then the joint probability density function is

$$f(y_1, \ldots, y_n) = \frac{1}{(2\pi\sigma^2)^{n/2}} \exp\left[-\frac{1}{2\sigma^2} \sum_{i=1}^{n} (y_i - g(x_i; \Theta))^2\right],$$

so the log-likelihood function is

$$\log L(\Theta, \sigma^2) = -\ln(2\pi\sigma^2)^{n/2} - \frac{1}{2\sigma^2} \sum_{i=1}^{n} [y_i - g(x_i; \Theta)]^2. \qquad (1.1)$$

Thus the maximum likelihood estimate $\hat{\Theta}$ of Θ is obtained by minimizing the sum of squares

$$\mathscr{F}(\Theta) = \sum_{i=1}^{n} [y_i - g(x_i; \Theta)]^2, \qquad (1.2)$$

which leads us to the classical least squares. This method for solving regression problems goes back to C.-F. Gauss [69] and A.-M. Legendre [121] in the early 1800s. It is now a part of every standard undergraduate statistics course.

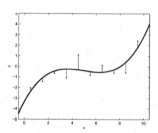

Figure 1.1 *Ordinary regression minimizes the sum of squares of vertical distances: a cubic polynomial fitted to 10 data points.*

We emphasize that the x and y variables play different roles: x is called a *control* variable (controlled by the experimenter), its values $x_1, \ldots, x_n$ are error-free, and y is called a *response* variable (observed as a response), its values $y_1, \ldots, y_n$ are imprecise (contaminated by noise). Geometrically, the regression procedure minimizes the sum of squares of *vertical* distances (measured along the y axis) from the data points (x_i, y_i) to the graph of the function $y = g(x; \Theta)$, see Fig. 1.1.

For example, if one deals with a linear relation $y = a + bx$, then the least squares estimates $\hat{a}$ and $\hat{b}$ minimize the function

$$\mathscr{F}(a,b) = \sum_{i=1}^{n}(y_i - a - bx_i)^2.$$

Solving equations $\partial\mathscr{F}/\partial a = 0$ and $\partial\mathscr{F}/\partial b = 0$ gives

$$\hat{a} = \bar{y} - \hat{b}\bar{x} \quad \text{and} \quad \hat{b} = s_{xy}/s_{xx}, \tag{1.3}$$

where $\bar{x}$ and $\bar{y}$ are the "sample means"

$$\bar{x} = \frac{1}{n}\sum_{i=1}^{n}x_i \quad \text{and} \quad \bar{y} = \frac{1}{n}\sum_{i=1}^{n}y_i \tag{1.4}$$

and

$$s_{xx} = \sum_{i=1}^{n}(x_i - \bar{x})^2$$

$$s_{yy} = \sum_{i=1}^{n}(y_i - \bar{y})^2$$

$$s_{xy} = \sum_{i=1}^{n}(x_i - \bar{x})(y_i - \bar{y}).$$

are the components of the so called "scatter matrix"

$$\mathbf{S} = \begin{bmatrix} s_{xx} & s_{xy} \\ s_{xy} & s_{yy} \end{bmatrix}, \tag{1.5}$$

which characterizes the "spread" of the data set about its centroid $(\bar{x}, \bar{y})$.

Remark. To estimate a and b, one does not need to know the variance σ^2. It can be estimated separately by maximizing the log-likelihood function (1.1) with respect to σ^2, which gives

$$\hat{\sigma}^2 = \frac{1}{n}\sum_{i=1}^{n}(y_i - \hat{a} - \hat{b}x_i)^2. \tag{1.6}$$

This estimate is slightly biased, as $\mathbb{E}(\hat{\sigma}^2) = \frac{n-2}{n}\sigma^2$. It is customary to replace n in the denominator with $n-2$, which gives an unbiased estimate of σ^2. Both versions of $\hat{\sigma}^2$ are strongly consistent, i.e., they converge to σ^2 with probability one.

The regression model has excellent statistical properties. The estimates $\hat{a}$ and $\hat{b}$ are strongly consistent, i.e., $\hat{a} \to a$ and $\hat{b} \to b$ as $n \to \infty$ (with probability one), and unbiased, i.e., $\mathbb{E}(\hat{a}) = a$ and $\mathbb{E}(\hat{b}) = b$. They have normal distributions with variances

$$\sigma_a^2 = \sigma^2 \left(\frac{\bar{x}^2}{s_{xx}} + \frac{1}{n} \right), \qquad \sigma_b^2 = \frac{\sigma^2}{s_{xx}}.$$

These variances are the smallest among the variances of unbiased estimators, i.e., they coincide with the Cramer-Rao lower bounds. Hence the estimates $\hat{a}$ and $\hat{b}$ are 100% efficient. All around, they are statistically optimal in every sense.

Remark. Suppose the errors ε_i are *heteroscedastic*, i.e., have different variances: $\varepsilon_i \sim N(0, \sigma_i^2)$. The maximum likelihood estimate of Θ is now obtained by the weighted least squares:

$$\mathscr{F}(\Theta) = \sum_{i=1}^{n} w_i \left[y_i - g(x_i; \Theta) \right]^2,$$

where the weights are set by $w_i = \sigma_i^{-2}$. In the linear case, $y = a + bx$, the estimates are still given by (1.3), but now the formulas for the sample mean and the scatter matrix should incorporate weights, e.g.,

$$\bar{x} = \frac{1}{n} \sum_{i=1}^{n} w_i x_i, \qquad s_{xx} = \sum_{i=1}^{n} w_i (x_i - \bar{x})^2, \quad \text{etc.} \tag{1.7}$$

Thus, heteroscedasticity only requires minor changes in the regression formulas.

1.2 Errors-in-variables (EIV) model

Recall that the classical regression problem was solved in the early 1800s. In the late nineteenth century statisticians encountered another problem, which looked very similar, but turned out to be substantially different and far more difficult. In fact the superficial similarity between the two caused a great deal of confusion and delayed the progress for several decades.

That new problem is reconstructing a functional relation $y = g(x)$ given observations $(x_1, y_1), \ldots, (x_n, y_n)$ in which *both* variables are subject to errors. We start with an example and describe a formal statistical model later.

Suppose (see Madansky [127]) we wish to determine ρ, the density of iron, by making use of the relation

$$\text{MASS} = \rho \times \text{VOLUME.} \tag{1.8}$$

We can pick n pieces of iron and measure their volumes $x_1, \ldots, x_n$ and masses $y_1, \ldots, y_n$. Given these data, we need to estimate the coefficient ρ in the functional relation $y = \rho x$. We cannot use the exact formula $y_i = \rho x_i$ for any i, because the measurements may be imprecise (our pieces of iron may be contaminated by other elements).

Similar problems commonly occur in economics (where, for instance, x may be the price of a certain good and y the demand, see Wald [187]) and in sociology. For a fascinating collection of other examples, including the studies of A-bomb survivors, see Chapter 1 in [28].

So how do we solve the iron density problem? For example, we can assume (or rather, pretend) that the volumes x_i's are measured precisely and apply the classical regression of y on x, i.e., determine $y = bx$ and set $\rho = b$. Alternatively, we can assume that our masses y_i's are error-free and do the regression of x on y, i.e., find $x = b'y$ and then set $\rho = 1/b'$.

This may sound like a good plan, but it gives us two different estimates, $\rho_1 = b$ and $\rho_2 = 1/b'$, which should make us at least suspicious. An objection was raised against this strategy as early as in 1901 by K. Pearson, see p. 559 in [144]: "we get one straight line or plane if we treat some one variable as independent, and a quite different one if we treat another variable as the independent variable." See Fig. 1.2.

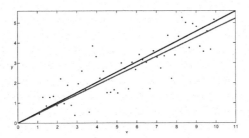

Figure 1.2 *50 data points (marked by dots) are fitted by two methods: the regression of y on x is the lower line and the regression of x on y is the upper line. Their slopes are 0.494 and 0.508, respectively.*

It was later determined that under natural statistical assumptions (to be described shortly) both estimates, ρ_1 and ρ_2, are inconsistent and may be heavily biased, see e.g., [8, 118, 142]; the consequences of this biasedness in econometrics are discussed in Chapter 10 of [128]. In fact, ρ_1 systematically underestimates the true density ρ, and ρ_2 systematically overestimates it.

Thus the new type of regression problem calls for nonclassical approaches. First we need to adopt an appropriate statistical model in which both x_i's and y_i's are subject to errors; it is called *errors-in-variables* (EIV) model[1]. It assumes that there are some 'true' values $\tilde{x}_i$ and $\tilde{y}_i$, that are linked by the (unknown) functional relation $\tilde{y}_i = g(\tilde{x}_i)$, and the experimenters observe their per-

---

[1] Another popular name is *measurement error* (ME) model, but we prefer EIV.

turbed values:

$$x_i = \tilde{x}_i + \delta_i, \qquad y_i = \tilde{y}_i + \varepsilon_i, \qquad i = 1, \ldots, n. \tag{1.9}$$

Here $\delta_1, \ldots, \delta_n, \varepsilon_1, \ldots, \varepsilon_n$ are $2n$ independent random variables with zero mean.

In the simplest case, one can assume that δ_i's have a common variance σ_x^2 and ε_i's have a common variance σ_y^2. Furthermore, it is common to assume that δ_i and ε_i are normally distributed, i.e.

$$\delta_i \sim N(0, \sigma_x^2) \qquad \text{and} \qquad \varepsilon_i \sim N(0, \sigma_y^2). \tag{1.10}$$

We also need to make some assumptions about the true values $\tilde{x}_i$'s and $\tilde{y}_i$'s, as they are neither random observations nor the model parameters (yet). There are two basic ways of treating these 'intermediate' objects.

First, the true values $\tilde{x}_i$'s and $\tilde{y}_i$'s may be regarded as fixed (nonrandom), then they have to be treated as additional parameters. They are sometimes referred to as "incidental" or "latent" parameters, or even "nuisance" parameters (as their values are normally of little interest). This interpretation of $\tilde{x}_i$'s and $\tilde{y}_i$'s is known as the *functional model*.

Alternatively, one can regard $\tilde{x}_i$'s and $\tilde{y}_i$'s as realizations of some underlying random variables that have their own distribution. It is common to assume that $\tilde{x}_i$'s are sampled from a normal population $N(\mu, \sigma^2)$, and then $\tilde{y}_i$'s are computed by $\tilde{y}_i = g(\tilde{x}_i)$. In that case δ_i and ε_i's are usually assumed to be independent of $\tilde{x}_i$'s and $\tilde{y}_i$'s. The mean μ and variance σ^2 of the normal population of $\tilde{x}_i$'s can be then estimated along with the parameters of the unknown function $g(x)$. This treatment of the true values is known as the *structural model*.

This terminology is not quite intuitive, but it is currently adopted in the statistics literature. It goes back to Kendall's works [109, 110] in the 1950s and became popular after the first publication of Kendall and Stuart's book [111]. Fuller [66] suggests a simple way of remembering it: the model is Functional (F) if the true points are Fixed; and the model is Structural (S) if the true points are Stochastic.

Before we turn to the solution of the EIV regression problem (which is typified by the iron density example), we describe a special version of the EIV model, which constitutes the main subject of this book.

1.3   Geometric fit

In the late 1800s statisticians encountered a special case of the EIV regression that arose in the analysis of images (photographs, drawings, maps). For example, given an imperfect line on an image, one wants to straighten it up, i.e., find an ideal line approximating the visible line contour. To this end, one can mark several points on the contour and try to fit a perfect straight line to the marked points.

More generally, one may want to approximate a round object on an image by a perfect circle, or an oval by a perfect ellipse, or a box by a perfect rectangle, etc. We call this task *geometric fitting problem*. It consists of approximating a visible contour on an image by a simple geometric figure (line, curve, polygon, etc). We discuss approximation by lines in this section.

In a coordinate system, the given points on the visible contour can be recorded as $(x_1, y_1), \ldots, (x_n, y_n)$, and one looks for the best fitting line in the form $y = a + bx$. Hence again the problem looks like a familiar regression. But a close look reveals that both x_i's and y_i's may be imprecise, hence we are in the framework of the EIV model.

Furthermore, there is a novel feature here: due to the geometric character of the problem, the errors in x and y directions should have the same magnitude, on average, hence we have a special case of the EIV model characterized by

$$\sigma_x^2 = \sigma_y^2. \tag{1.11}$$

In this case the "noise" vector $(\delta_i, \varepsilon_i)$ has a normal distribution with zero mean and a scalar covariance matrix, i.e., the random noise is *isotropic* in the xy plane. The isotropy means that the distribution of the noise vector is invariant under rotations. This property is natural in image processing applications, as the choice of coordinate axes on the image is often arbitrary, i.e., there should not be any differences between the x, or y, or any other directions.

Conversely, suppose that the random vector $(\delta_i, \varepsilon_i)$ has two basic properties (which naturally hold in image processing applications):

(a) it is isotropic, as described above,

(b) its components δ_i and ε_i are independent.

Then it necessarily has a normal distribution. This is a standard fact in probability theory, see e.g., [14] or Section III.4 of [60]. Thus the assumption about normal distribution (1.10) is not a luxury anymore, but a logical consequence of the more basic assumptions (a) and (b).

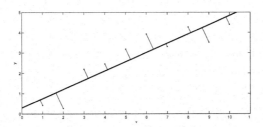

Figure 1.3 *Orthogonal regression minimizes the sum of squares of orthogonal distances.*

A practical solution to the special case $\sigma_x^2 = \sigma_y^2$ of the EIV model was proposed as early as in 1877 by Adcock [1] based on purely geometric (rather than statistical) considerations. He defines the fitting line $y = a + bx$ that is overall closest to the data points, i.e., the one which minimizes

$$\mathscr{F} = \sum_{i=1}^{n} d_i^2, \tag{1.12}$$

where d_i denotes the geometric (orthogonal) distance from the point (x_i, y_i) to the fitting line, see Fig. 1.3. By using elementary geometry, we obtain

$$\mathscr{F}(a,b) = \frac{1}{1+b^2} \sum_{i=1}^{n} (y_i - a - bx_i)^2. \tag{1.13}$$

Solving the equation $\partial \mathscr{F}/\partial a = 0$ yields

$$a = \bar{y} - b\bar{x}, \tag{1.14}$$

where $\bar{x}$ and $\bar{y}$ are the sample means, cf. Section 1.1. By the way, recall that (1.14) also holds in the classical case, cf. (1.3). Now eliminating a from (1.13) gives us a function of one variable

$$\mathscr{F}(b) = \frac{s_{yy} - 2bs_{xy} + s_{xx}b^2}{1+b^2},$$

where s_{xx}, s_{xy}, s_{yy} are the components of the scatter matrix, cf. Section 1.1. Next, the equation $\partial \mathscr{F}/\partial b = 0$ reduces the problem to a quadratic equation,

$$s_{xy}b^2 - (s_{yy} - s_{xx})b - s_{xy} = 0. \tag{1.15}$$

It has two roots, but a careful examination reveals that the minimum of $\mathscr{F}$ corresponds to the following one:

$$b = \frac{s_{yy} - s_{xx} + \sqrt{(s_{yy} - s_{xx})^2 + 4s_{xy}^2}}{2s_{xy}}. \tag{1.16}$$

This formula applies whenever $s_{xy} \neq 0$. In the case $s_{xy} = 0$, we need to set $b = 0$ if $s_{xx} > s_{yy}$ and $b = \infty$ if $s_{xx} < s_{yy}$. We encourage the reader to derive the formula (1.16) and carefully examine the special case $s_{xy} = 0$.

The above solution may be elementary, by our modern standards, but it has a history showing its nontrivial character. It was first obtained in 1878 by Adcock [2], who incidentally made a simple calculational error. Adcock's error was corrected the next year by Kummell [117], but in turn, one of Kummell's formulas involved a more subtle error. Kummell's error was copied by some other authors in the 1940s and 1950s (see [89, 126]). Finally it was corrected in 1959 by Madansky [127]. Madansky's work [127] is perhaps the most cited in the early studies on the EIV regression.

We call the fitting method based on minimization of the sum of squares of orthogonal (geometric) distances from the data points to the fitted contour *orthogonal fit* or *geometric fit*. Despite the natural appeal of the orthogonal fitting line, the early publications [1, 2, 117] in 1877–79 passed unnoticed. Twenty years later the orthogonal fitting line was independently proposed by Pearson [144], and another 20 years later, by Gini [72].

Pearson and Gini made another important observation: the line which minimizes (1.12) is the major axis of the scattering ellipse associated with the data set. The scattering ellipse is defined by equation

$$
\begin{bmatrix} x - \bar{x} \\ y - \bar{y} \end{bmatrix}^T \mathbf{S} \begin{bmatrix} x - \bar{x} \\ y - \bar{y} \end{bmatrix} = 1,
$$

its center is $(\bar{x}, \bar{y})$ and its axes are spanned by the eigenvectors of the scatter matrix $\mathbf{S}$. This fact establishes a link between the orthogonal fit and the principal component analysis of linear algebra.

Pearson [144] also estimated the angle $\theta = \tan^{-1} b$ which the fitting line made with the x axis and found a simple formula for it:

$$
\tan 2\theta = \frac{2s_{xy}}{s_{xx} - s_{yy}}. \tag{1.17}
$$

We leave its verification to the reader as an exercise.

Adcock and Pearson were motivated by geometric considerations and did not use probabilities. Only in the 1930s their method was incorporated into the formal statistical analysis. Koopmans [113] (see also Lindley [126]) determined that the orthogonal fit provided the maximum likelihood estimate under the assumptions (1.9)–(1.11). Recall that the classical least squares fit (1.2) also maximizes the likelihood in the ordinary regression model (1.1). Thus there is a deep analogy between the two regression models.

The geometric nature of the orthogonal fit makes the resulting line independent of the choice of the coordinate system on the image. In other words, the geometric fit is invariant under orthogonal transformations (rotations and translations) of the coordinate frame.

The invariance under certain transformations is very important. We say that a fitting line is invariant under translations if changing the data coordinates by

$$
T_{c,d} : (x, y) \mapsto (x + c, y + d) \tag{1.18}
$$

will leave the line unchanged, i.e., its equation in the new coordinate system will be $y + d = a + b(x + c)$. Similarly we define invariance under rotations

$$
R_\theta : (x, y) \mapsto (x \cos \theta + y \sin \theta, -x \sin \theta + y \cos \theta) \tag{1.19}
$$

and under scaling of variables

$$
S_{\alpha, \beta} : (x, y) \mapsto (\alpha x, \beta y). \tag{1.20}
$$

An important special case of a scaling transformation is $\alpha = \beta$; it is called a similarity (or sometimes a dilation; in formal mathematics it is known as a homothety). We will denote it by

$$S_\alpha = S_{\alpha,\alpha}: (x,y) \mapsto (\alpha x, \alpha y). \qquad (1.21)$$

It takes little effort to verify that the orthogonal fitting line is invariant under $T_{c,d}$ and R_θ, as well as S_α, but *not* invariant under general scaling transformations $S_{\alpha,\beta}$ with $\alpha \neq \beta$. We leave the verification of these facts to the reader.

The orthogonal fit has a clear appeal when applied to regular geometric patterns. Fig. 1.4 shows four data points placed at vertices of a rectangle. While classical regression lines are skewed upward or downward (the first and second panels of Fig. 1.4), the orthogonal regression line cuts right through the middle of the rectangle and lies on its axis of symmetry. Arguably, the orthogonal fitting line would "please the eye" more than any other line.

However, the orthogonal fit leads to an inconsistency if one applies it to a more general EIV model, where $\sigma_x^2 \neq \sigma_y^2$. This inconsistency stems from the noninvariance of the orthogonal fitting line under scaling transformations $S_{\alpha,\beta}$.

For example, let us again consider the task of determining the iron density by using (1.8) and measuring volumes x_i's and masses y_i's of some iron pieces, cf. the previous section. If we employ the orthogonal fit to the measurements $(x_1,y_1),\ldots,(x_n,y_n)$, then the fitting line $y = bx$, and the resulting estimate of the iron density $\rho = b$, would depend on the choice of units in which the measurements x_i's and y_i's are recorded. That is, if we rescale the variables by $(x,y) \mapsto (\alpha x, \beta y)$, the equation of the orthogonal fitting line in the new coordinate system would be $\beta y = b'(\alpha x)$, where $b' \neq b$. In other words, a different density would be obtained if we change pounds to kilograms or tons, and similarly liters to bushels or cubic meters.

This objection was raised in 1937 by Roos [156] and further discussed in the statistics literature in the 1940s [89, 187]. Thus the orthogonal fit has its limitations, it is essentially restricted to the special case $\sigma_x^2 = \sigma_y^2$ of the EIV model. Some modern books, see e.g., [28], strongly warn against the use of orthogonal fitting line in EIV applications with $\sigma_x^2 \neq \sigma_y^2$, and more generally, against the use of other techniques that are based on any unreliable assumptions about σ_x^2 and σ_y^2.

We briefly overview basic features of the general EIV model in the next section (though not attempting anything close to a comprehensive coverage).

1.4 Solving a general EIV problem

Let us turn back to the EIV model (1.9)–(1.10) without assuming (1.11), i.e., leaving σ_x^2 and σ_y^2 unconstrained.

Kummell [117] was perhaps the first who examined, in 1879, the task of

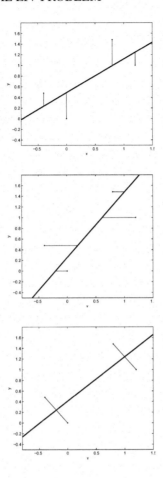

Figure 1.4 *The regression of y on x minimizes the sum of squares of vertical distances (top); the regression of x on y does the same with horizontal distances (middle); the orthogonal regression minimizes the sum of squares of orthogonal distances (bottom).*

determining the underlying functional relation $y = g(x)$ in the EIV context, and realized that this could not be done in any reasonable sense (!), unless one makes an extra assumption on the relation between σ_x^2 and σ_y^2. Even in the simplest, linear case $y = a + bx$, there is no sensible way to estimate the parameters a and b without extra assumptions. The problem is just *unsolvable*, however simple it may appear!

Many other researchers arrived at the same conclusion in the early twentieth century. The realization of this stunning fact produced a long turmoil in the community lasting until about the 1950s and marked by confusion and controversy. A. Madansky, for example, devotes a few pages of his 1959 paper [127] describing the shock of an average physicist who would learn about the unsolvability of the "simple" regression problem, and how statisticians could explain it to him.

Later the insolvability of this problem was proved in mathematical terms. First, it was established in 1956 by Anderson and Rubin [9] (see also [73]) that even in the linear case $y = a + bx$ the likelihood function was unbounded (its supremum was infinite), thus maximum likelihood estimates could not be determined. Interestingly, the likelihood function has critical points, which have been occasionally mistaken for maxima; only in 1969 the issue was resolved: M. Solari [168] proved that all critical points were just saddle points.

Second (and more importantly), it was shown in 1977 by Nussbaum [139] (see also page 7 in [40]) that no statistical procedure could produce strongly consistent estimates $\hat{a}$ and $\hat{b}$ (which would converge to the true values of a and b as $n \to \infty$). See also the discussion of identifiability in the book [40] by Cheng and Van Ness.

To make the EIV regression model solvable, Kummel [117] assumed that

$$\text{the ratio} \qquad \kappa = \sigma_x / \sigma_y \qquad \text{is known.} \qquad (1.22)$$

He justified his assumption by arguing that experimenters "usually know this ratio from experience." Later this assumption was commonly adopted in the statistics literature. Recently Fuller [66] called the EIV model satisfying the assumptions (1.9), (1.10), and (1.22) the "classical EIV model."

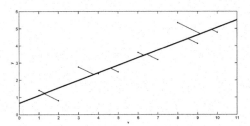

Figure 1.5 *The EIV fit minimizes the sum of squares of "skewed" distances from the data points to the line. Here* $\kappa = 2$.

Now the EIV regression problem has a well defined solution. In 1879 Kummell [117] gave formulas for the best fitting line that involved κ. His line $y = a + bx$ minimizes

$$\mathscr{F} = \frac{1}{1 + \kappa^2 b^2} \sum_{i=1}^{n} (y_i - a - bx_i)^2 \qquad (1.23)$$

and its slope is estimated by

$$b = \frac{\kappa^2 s_{yy} - s_{xx} + \sqrt{(\kappa^2 s_{yy} - s_{xx})^2 + 4\kappa^2 s_{xy}^2}}{2\kappa^2 s_{xy}}, \qquad (1.24)$$

compare this to (1.16). The intercept is again $a = \bar{y} - b\bar{x}$, as in (1.14).

This line minimizes the sum of squares of the distances to the data points (x_i, y_i) measured along the vector $(\kappa b, -1)$, see Fig. 1.5. Kummell arrived at his formula rather intuitively, but later it was determined that he actually found the maximum likelihood solution, cf. [113, 126].

In the special case $\kappa = 1$, i.e., $\sigma_x^2 = \sigma_y^2$, the vector $(\kappa b, -1) = (b, -1)$ is normal to the line $y = a + bx$, thus we arrive at the familiar orthogonal fit. Hence, the EIV linear regression (1.24) includes the orthogonal fit as a particular case.

The slope b given by (1.24) is monotonically increasing with κ (this follows from the standard fact $s_{xy}^2 \leq s_{xx} s_{yy}$ by some algebraic manipulations, which we leave to the reader as an exercise). In the limit $\kappa \to 0$, the EIV regression line converges to the classical regression of y on x with the slope $b = s_{xy}/s_{xx}$, cf. (1.3). Similarly, in the limit $\kappa \to \infty$, the EIV regression line converges to the classical regression of x on y with the slope $b = s_{yy}/s_{xy}$. Thus the classical regressions (of y on x and of x on y) are the extreme cases of the EIV regression.

The EIV line minimizing (1.23) can be made invariant under rescaling of coordinates $x \mapsto \alpha x$ and $y \mapsto \beta y$, as the scaling factors α and β can be incorporated into the ratio κ by the obvious rule $\kappa \mapsto \kappa \alpha/\beta$. This fact was pointed out in 1947 by Lindley [126], who concluded that the estimate (1.24) thus conformed to the basic requirement of the EIV model: it does not depend on the units in which the measurements $x_1, \ldots, x_n$ and $y_1, \ldots, y_n$ are recorded.

Actually, if one rescales the coordinates by $x \mapsto x$ and $y \mapsto \kappa y$, then in the new variables we have $\kappa = 1$. Thus the EIV regression line can be transformed to the orthogonal fitting line. Therefore, the EIV linear regression model (with the known ratio $\kappa = \sigma_x/\sigma_y$) can be converted to the orthogonal regression model by a simple rescaling of the coordinates, and vice versa.

But we emphasize that the general EIV regression and the orthogonal fit must conform to different requirements:

- The EIV regression must be invariant under scaling of the variables x and y (resulting from the change of units in which these variables are measured).

- The orthogonal fit (due to its geometric nature) must be invariant under rotations and translations of the coordinate frame on the xy plane, as well as under similarities resulting from a change of unit of length.

This difference has important consequences for nonlinear regression discussed in Section 1.9.

1.5 Nonlinear nature of the "linear" EIV

It may be enlightening to interpret the orthogonal regression problem geometrically in the space $\mathbb{R}^{2n}$ with coordinates $x_1, y_1, \ldots, x_n, y_n$. We follow Malinvaud (Chapter 10 of [128]). Our observations $(x_1, y_1), \ldots, (x_n, y_n)$ are represented by one point (we denote it by $\mathscr{X}$) in this multidimensional space. To understand the construction of the orthogonal fitting line, consider the subset $\mathbb{P} \subset \mathbb{R}^{2n}$ defined by

$$(x_1, y_1, \ldots, x_n, y_n) \in \mathbb{P} \iff \exists a, b \colon y_i = a + b x_i \ \forall i,$$

i.e., $\mathbb{P}$ consists of all $(x_1, y_1, \ldots, x_n, y_n) \in \mathbb{R}^{2n}$ such that the n planar points $(x_1, y_1), \ldots, (x_n, y_n)$ are collinear. Note that the true values $\tilde{x}_1, \tilde{y}_1, \ldots, \tilde{x}_n, \tilde{y}_n$ are represented by one point (we denote it by $\tilde{\mathscr{X}}$) in $\mathbb{P}$, i.e., $\tilde{\mathscr{X}} \in \mathbb{P}$.

The orthogonal fitting line minimizes the sum

$$\sum_{i=1}^{n} (x_i - \tilde{x}_i)^2 + (y_i - \tilde{y}_i)^2,$$

which is the square of the distance (in the Euclidean metric) between the points $\mathscr{X}$ and $\tilde{\mathscr{X}}$. Thus, the orthogonal fitting procedure corresponds to choosing a point $\tilde{\mathscr{X}} \in \mathbb{P}$ closest to the point $\mathscr{X} \in \mathbb{R}^{2n}$ representing the data. In other words, we simply project the given point $\mathscr{X}$ onto $\mathbb{P}$ orthogonally. Or is it that simple?

It takes a little effort to verify that $\mathbb{P}$ is a *nonlinear* submanifold ('surface') in $\mathbb{R}^{2n}$. Indeed, it is specified by $n - 2$ independent relations

$$\frac{y_{i+1} - y_i}{x_{i+1} - x_i} = \frac{y_{i+2} - y_i}{x_{i+2} - x_i}, \qquad i = 1, \ldots, n-2, \tag{1.25}$$

each of which means the collinearity of the three planar points (x_i, y_i), (x_{i+1}, y_{i+1}), and (x_{i+2}, y_{i+2}). The relations (1.25) are obviously quadratic, hence $\mathbb{P}$ is an $(n+2)$-dimensional *quadratic* surface (variety) in $\mathbb{R}^{2n}$.

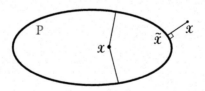

Figure 1.6: *Projection of the point $\mathscr{X}$ onto the quadratic manifold $\mathbb{P}$.*

Projecting a point $\mathscr{X}$ onto a quadratic surface is not a trivial, and definitely not a linear, problem. This geometric interpretation should dispel our illusion (if we still have any) that we deal with a linear problem, it unmasks its truly nonlinear character.

Imagine, for example, the task of projecting a point $\mathcal{X} \in \mathbb{R}^2$ onto a quadric, say an ellipse $a^2x^2 + b^2y^2 = 1$. This is not a simple problem, its exact solution involves finding roots of a 4th degree polynomial [162]. In a sense, we are lucky that the projection of our data point $\mathcal{X} \in \mathbb{R}^{2n}$ onto $\mathbb{P}$ reduces to just a quadratic equation (1.15).

Besides, the projection may not be unique (for example when $\mathcal{X}$ lies on the major axis of the ellipse near the center, see Fig. 1.6). We will actually see that the orthogonal fitting line may not be unique either, cf. Section 2.3.

To further emphasize the nonlinear nature of the EIV regression, suppose for a moment that the errors are heteroscedastic, i.e.

$$\delta_i \sim N(0, \sigma_{x,i}^2) \qquad \text{and} \qquad \varepsilon_i \sim N(0, \sigma_{y,i}^2),$$

where the ratio of variances is known, but it differs from point to point, i.e., we assume that

$$\kappa_i = \sigma_{x,i}/\sigma_{y,i}$$

is known for every $i = 1, \ldots, n$. Recall that in the classical regression the heteroscedasticity of errors does not affect the linear nature of the problem. Now, in the EIV model, the best fitting line should minimize

$$\mathcal{F}(a,b) = \sum_{i=1}^{n} \frac{(y_i - a - bx_i)^2}{1 + \kappa_i^2 b^2}.$$

Despite its resemblance to (1.23), the minimization of this $\mathcal{F}$ cannot be reduced to a quadratic (or any finite degree) polynomial equation. Here "finite degree" means a degree independent of the sample size n. This is a hard-core nonlinear problem that has no closed form solution; its numerical solution requires iterative algorithms.

In other words, the hidden nonlinear nature of the "linear" EIV fit may come in different ways at different stages. The more general assumptions on errors one makes the more serious difficulties one faces.

Yet another explanation why the linear EIV regression has an essentially nonlinear character was given by Boggs et al., see [23].

Overall, the linear EIV model, though superficially resembling the classical linear regression, turns out to be dissimilar in many crucial ways. The sharp contrast between these two models is now recognized by many authors. As the textbook [29] puts it, "Regression with errors in variables (EIV) ... is so fundamentally different from the simple linear regression ... that it is probably best thought of as a completely different topic."

1.6   Statistical properties of the orthogonal fit

Our book is devoted to the orthogonal fitting problem, and from now on we adopt the statistical model assumptions (1.9), (1.10), and (1.11). Under these assumptions the orthogonal fit maximizes the likelihood function, i.e., provides the Maximum Likelihood Estimate (a formal proof of this fact will be given in Section 6.3).

In this section we touch upon the basic statistical properties of the linear orthogonal fit, i.e., the behavior of the estimates $\hat{\alpha}$ and $\hat{\beta}$ of the parameters of the fitting line $y = \alpha + \beta x$ (we use α and β here, instead of the previous a and b, to be consistent with the notation in the papers we will refer to).

Our discussion will shed more light on a stark dissimilarity between the orthogonal fit and the classical regression, whose nice features we mentioned in Section 1.1. Even a quick look reveals a totally different (and somewhat shocking) picture.

To begin with, the distribution of the estimates $\hat{\alpha}$ and $\hat{\beta}$ is not normal and does not belong to any standard family of probability distributions. Only in 1976, explicit formulas for their density functions were found by Anderson and others [7, 10]; see Section 2.4. Those expressions are overly complicated, involve double-infinite series, and Anderson [7] promptly conceded that they are not very useful for practical purposes. Instead, he and Kunitomo [118] derived various approximations to the distribution functions of $\hat{\alpha}$ and $\hat{\beta}$, which turned out to be practically accurate.

Second, and worse, the estimates $\hat{\alpha}$ and $\hat{\beta}$ do not have finite moments, i.e.,

$$\mathbb{E}(|\hat{\alpha}|) = \infty \qquad \text{and} \qquad \mathbb{E}(|\hat{\beta}|) = \infty.$$

Thus they also have infinite mean squared errors:

$$\mathbb{E}([\hat{\alpha} - \alpha]^2) = \infty \qquad \text{and} \qquad \mathbb{E}([\hat{\beta} - \beta]^2) = \infty.$$

These stunning facts were also revealed in 1976 by Anderson [7]. Intuitively, one can see why this happens from (1.16), where the denominator can take value zero, and its probability density does not vanish at zero. We encourage the reader to closely examine this observation.

Until Anderson's discovery, researchers "approximated" the moments of the estimates $\hat{\alpha}$ and $\hat{\beta}$ as follows. They employed Taylor expansion, dropped higher order terms, and obtained some "approximate" formulas for the moments of $\hat{\alpha}$ and $\hat{\beta}$ (including their means and variances). Anderson demonstrated that all those formulas were fundamentally flawed, as the actual moments did not exist. Anderson said that those formulas should be regarded as "moments of some approximations," rather than "approximate moments."

Once Anderson made his discovery, it immediately lead to fundamental methodological questions: how can one trust a statistical estimate that has an infinite mean squared error (not to mention infinite bias)? Should these facts imply that the estimate is totally unreliable? Why did not anybody notice these bad features in practice? Can an estimate with infinite moments be practically better than others which have finite moments? These questions lead to further studies, see next.

In the late 1970s, Anderson [7, 8], Patefield [142], and Kunitomo [118] compared the slope $\hat{\beta}$ of the orthogonal fitting line, given by (1.16), with the slope $\hat{\beta}$ of the classical regression line, given by (1.3) (of course, both estimates were used in the framework of the same model (1.9), (1.10), and (1.11)). They denote the former by $\hat{\beta}_M$ (Maximum likelihood) and the latter by $\hat{\beta}_L$ (Least squares). Their results can be summarized in two seemingly conflicting verdicts:

(a) The mean squared error of $\hat{\beta}_M$ is infinite, and that of $\hat{\beta}_L$ is finite (whenever $n \geq 4$), thus $\hat{\beta}_L$ appears (infinitely!) more accurate;

(b) The estimate $\hat{\beta}_M$ is consistent and asymptotically unbiased, while $\hat{\beta}_L$ is inconsistent and asymptotically biased (it is consistent and unbiased only in the special case $\beta = 0$), thus $\hat{\beta}_M$ appears more appropriate.

Going further, Anderson shows that

$$\text{Prob}\{|\hat{\beta}_M - \beta| > t\} < \text{Prob}\{|\hat{\beta}_L - \beta| > t\}$$

for all $t > 0$ that are not too large, i.e., for all $t > 0$ of practical interest. In other words, the accuracy of $\hat{\beta}_M$ *dominates* that of $\hat{\beta}_L$ everywhere, except for very large deviations (large t). It is the heavy tails of $\hat{\beta}_M$ that make its mean squared error infinite, otherwise it tends to be closer to β than its rival $\hat{\beta}_L$.

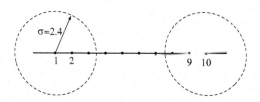

Figure 1.7: *The true points location and the noise level in our experiment.*

Furthermore, when one observes values of $\hat{\beta}_M$ in practice, or in simulated experiments, nothing indicates that $\hat{\beta}_M$ has infinite moments; its values group around a certain center and have a seemingly normal distribution. Large deviations occur so rarely that they usually pass unregistered. However, those large deviations are, ultimately, responsible for the lack of moments. In order to make them visible, i.e., have them appear at a noticeable rate in computer experiments, one needs to increase the noise level $\sigma = \sigma_x = \sigma_y$ way above what it normally is in image processing applications, see next.

For example, we generated 10^6 random samples of $n = 10$ points on the line $y = x$ whose true positions were equally spaced on a stretch of length 10, with $\sigma = 2.4$ (note how high the noise is: its standard deviation is a quarter of the length of the interval where the data are observed; see Fig. 1.7). Fig. 1.8 plots the average estimate $\hat{\beta}_M$ over k samples, as k runs from 1 to 10^6. It behaves very much like the sample mean of the Cauchy random variable (whose moments do not exist either). Thus one can see, indeed, that the estimate $\hat{\beta}_M$ has infinite moments. But if one decreases the noise level to $\sigma = 2$ or less, then the erratic behavior disappears, and the solid line in Fig. 1.8 turns just flat, as it is for the finite moment estimate $\hat{\beta}_L$.

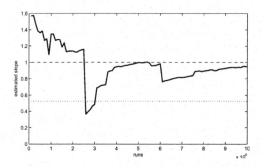

Figure 1.8 *The average estimate $\hat{\beta}_M$ over k randomly generated samples (solid line), as k runs from 1 to 10^6. The true slope $\beta = 1$ is marked by the dashed line. The average estimate $\hat{\beta}_L$ is the dotted line, it remains stable at level 0.52, systematically underestimating β.*

Now which estimate, $\hat{\beta}_M$ or $\hat{\beta}_L$, should we prefer? This may be quite a dilemma for a practitioner who is used to trusting the mean squared error as an absolute and ultimate criterion. Anderson argues that in this situation one has to make an exception and choose $\hat{\beta}_M$ over $\hat{\beta}_L$, despite its infinite mean squared error.

In the early 1980s, as if responding to Anderson's appeal, several statisticians (most notably, Gleser [73, 74, 75], Malinvaud [128], and Patefield [143]) independently established strong asymptotic properties of the orthogonal fitting line (and more generally, the classical EIV fitting line):

(a) the estimates $\hat{\alpha}$ and $\hat{\beta}$ are strongly consistent[2] and asymptotically normal;

---

[2]However, the maximum likelihood estimates of σ_x^2 and σ_y^2 are not consistent, in fact

$$\hat{\sigma}_x^2 \to \tfrac{1}{2}\sigma_x^2 \qquad \text{and} \qquad \hat{\sigma}_y^2 \to \tfrac{1}{2}\sigma_y^2$$

as $n \to \infty$, in the functional model. This odd feature was noticed and explained in 1947 by Lindley

(b) in a certain sense, these estimates are efficient.

They also constructed confidence regions for α and β. More details can be found in [66], [40], [128], and our Chapter 2.

These results assert very firmly that the maximum likelihood estimate $\hat{\beta}_M$ is the best possible. Certain formal statements to this extent were published by Gleser [74], see also [39, 40, 128] and our Chapter 2.

After all these magnificent achievements, the studies of the linear EIV regression seem to have subsided in the late 1990s; perhaps the topic exhausted itself. The statistical community turned its attention to nonlinear EIV models.

For further reading on the linear EIV regression, see excellent surveys in [8, 73, 126, 127, 132, 187], and books [66], [40], [128] (Chapter 10), and [111] (Chapter 29). We give a summary of the orthogonal line fitting in Chapter 2.

1.7   Relation to total least squares (TLS)

The EIV linear regression is often associated with the so-called *total least squares* (TLS) techniques in computational linear algebra. The latter solve an overdetermined linear system

$$\mathbf{A}\mathbf{x} \approx \mathbf{b}, \qquad \mathbf{x} \in \mathbb{R}^m, \ \mathbf{b} \in \mathbb{R}^n, \ n > m, \qquad (1.26)$$

where not only the vector $\mathbf{b}$, but also the matrix $\mathbf{A}$ (or at least some of its columns) are assumed to be contaminated by errors. If only $\mathbf{b}$ is corrupted by noise, the solution of (1.26) is given by the ordinary least squares

$$\mathbf{x} = \operatorname{argmin} \|\mathbf{A}\mathbf{x} - \mathbf{b}\|^2,$$

where $\|\cdot\|$ denotes the 2-norm. Equivalently, it can be paraphrased by

$$\mathbf{x} = \operatorname{argmin} \|\Delta \mathbf{b}\|^2 \qquad \text{subject to} \qquad \mathbf{A}\mathbf{x} = \mathbf{b} + \Delta \mathbf{b}. \qquad (1.27)$$

If $\mathbf{A}$ has full rank, the (unique) explicit solution is

$$\mathbf{x} = (\mathbf{A}^T \mathbf{A})^{-1} \mathbf{A}^T \mathbf{b}.$$

If A is rank deficient, the solution is not unique anymore, and one usually picks the minimum-norm solution

$$\mathbf{x} = \mathbf{A}^- \mathbf{b},$$

where $\mathbf{A}^-$ denotes the Moore-Penrose pseudoinverse.

If both $\mathbf{b}$ and $\mathbf{A}$ are corrupted by noise, the solution of (1.26) is more complicated, and it is the subject of the TLS techniques. In the simplest case, where

---

[126]; the factor 1/2 here is related to the degrees of freedom: we deal with $2n$ random observations and $n+2$ parameters of the model, thus the correct number of degrees of freedom is $n-2$, rather than $2n-2$.

all errors in $\mathbf{A}$ and $\mathbf{b}$ are independent and have the same order of magnitude, the solution is given by

$$\mathbf{x} = \text{argmin} \left\| [\Delta \mathbf{A} \ \Delta \mathbf{b}] \right\|_F^2 \qquad \text{subject to} \qquad (\mathbf{A} + \Delta \mathbf{A})\mathbf{x} = \mathbf{b} + \Delta \mathbf{b}, \qquad (1.28)$$

where $[\Delta \mathbf{A} \ \Delta \mathbf{b}]$ denotes the "augmented" $n \times (m+1)$ matrix and $\| \cdot \|_F$ stands for the Frobenius norm (the "length" of the $[n(m+1)]$-dimensional vector). Note the similarities between (1.27) and (1.28).

To compute $\mathbf{x}$ from (1.28), one uses the singular values (and vectors) of the augmented matrix $[\mathbf{A} \ \mathbf{b}]$. In the basic case, see Chapter 2 of [185], it is given by

$$\mathbf{x} = (\mathbf{A}^T \mathbf{A} - \sigma_{m+1}^2 \mathbf{I})^{-1} \mathbf{A}^T \mathbf{b},$$

where σ_{m+1} is the smallest singular value of $[\mathbf{A} \ \mathbf{b}]$, and $\mathbf{I}$ denotes the identity matrix. This is the TLS in the "nutshell;" we refer to [77, 185, 160, 161] for an extensive treatment.

To see how the EIV and TLS models are related, consider an EIV problem of fitting a line $y = a + bx$ to data points (x_i, y_i), $i = 1, \dots, n$. This problem is equivalent to (1.26) with

$$\mathbf{A} = \begin{bmatrix} 1 & x_1 \\ \vdots & \vdots \\ 1 & x_n \end{bmatrix}, \qquad \mathbf{x} = \begin{bmatrix} a \\ b \end{bmatrix}, \qquad \mathbf{b} = \begin{bmatrix} y_1 \\ \vdots \\ y_n \end{bmatrix}.$$

We see that the vector $\mathbf{b}$ and the second column of $\mathbf{A}$ are corrupted by noise, thus we arrive at a particular TLS problem. If the errors in x_i's and y_i's are independent and have the same variance, then we can solve it by (1.28), and this solution is equivalent to the orthogonal least squares fit.

The link between the EIV regression models and the TLS techniques of computational linear algebra is very helpful. Many efficient tools of the TLS (especially, the SVD) can be employed to solve linear (or nonlinear but linearized) EIV problems, see [185, 160, 161].

1.8   Nonlinear models: General overview

> *...the errors in variables are bad enough in linear models.*
> *They are likely to be disastrous to any attempts to estimate*
> *additional nonlinearity or curvature parameters...*
> Z. Griliches and V. Ringstad; see [79]

Fitting a straight line to observed points may appear as a "linear" regression problem, but it has a truly nonlinear character (Section 1.5). Its solution is given by an irrational formula (1.16), and it may not be unique (Section 2.2). The probability distributions of the resulting estimates do not belong to any standard family and are described by overly complicated expressions (Section 2.4). The estimates do not have moments, i.e., their bias is indeterminate and their mean square errors are infinite. One might just wonder if things could get any worse.

Sadly, things do get worse when one has to fit *nonlinear* functions to data with errors in variables. We only overview some new troubles here. First of all, the nonlinear fitting problem may not even have a solution. More precisely, if one fits a curve of a certain type (say, a circle) by minimizing the orthogonal distances to the data points, then such a curve may not exist; we will see examples in Section 3.3. The nonexistence is a phenomenon specific to nonlinear problems only. Next, even if the best fitting curve exists, it may not be unique, there may be multiple solutions, all of which are "equally good;" see examples in Section 3.5. This leads to confusion in theoretical analysis.

Furthermore, even when the best fit exists and is unique, nothing is known about the distribution of the resulting parameter estimates; there are no explicit formulas for their densities or moments. In fact, theoretical moments quite often fail to exist. This happens even in the linear case, see Section 1.6. For the problem of fitting circles, see Section 6.4. The nonexistence of moments appears to be a common feature of the EIV regression and orthogonal fitting problems. To resolve these difficulties, statisticians have developed a nontraditional error analysis based on approximating distributions. We devote almost the entire Chapter 6 to those new statistical theories.

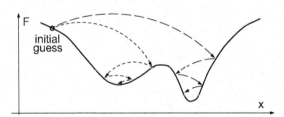

Figure 1.9 *Two algorithms minimizing a function $F(x)$. One makes shorter steps and converges to a local minimum. The other makes longer steps and converges to the global minimum.*

Down to more practical issues, the estimates of parameters in nonlinear EIV regression cannot be found in closed form, by explicit formulas like (1.16). They can only be computed by numerical algorithms, i.e., approximately. Numerical schemes, at best, converge to the desired estimate iteratively. However, in practice, the iterations may very well diverge, and even if they do converge, one never knows if they arrive at the desired estimate (the procedure may just terminate at a local minimum of the objective function; see Fig. 1.9).

Quite often, different numerical algorithms return different estimates. Fig. 1.9 shows an example where an iterative procedure is trapped by a local minimum. Another example is shown in Fig. 1.10: there is no local minima, but the second (slow) algorithm takes a large number of steps to reach the area near the minimum. In computer programs the number of iterations is always limited (usually, the limit is set to 50 or 100), thus the returned estimate may be still far from the actual minimum.

The estimates returned by different algorithms may even have seemingly different statistical characteristics (bias, variance, etc.). Thus the choice of the algorithm becomes a critical factor in practical applications, as well as in many theoretical studies. There is simply little point of studying an abstract "solution" that is not accessible in practice, while practical solutions heavily depend on the particular algorithm used to compute them.

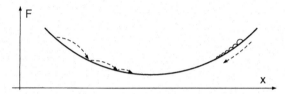

Figure 1.10 *Two algorithms minimizing a function $F(x)$ with a unique minimum. One approaches it fast (from the left) and arrives in a vicinity of the minimum in 5–10 steps. The other moves very slowly (from the right); it may take 100 or 1000 iterations to get near the minimum.*

Thus the analysis of numerical schemes becomes an integral part of the research. Sizable portions of published articles and books are now devoted to computer algorithms, their underlying ideas, performance, limitations, numerical stability, etc. This is all unavoidable, due to the nature of the subject.

1.9 Nonlinear models: EIV versus orthogonal fit

So far we have discussed two large topics—the orthogonal (geometric) fit and the EIV regression (with the known ratio of variances)—in parallel. In the linear context, these models can be transformed to one another by a simple scaling the variables x and y (Section 1.4), and both models have very similar properties.

In the nonlinear context, a strong link between these two models is lost. They can no longer be transformed to one another. The crucial disparity derives from the different requirements stated at the end of Section 1.4: the EIV regression must be invariant under scaling of the variables x and y, and the orthogonal (geometric) fit — under rotations and translations on the xy plane.

These requirements affect the very classes of nonlinear models used in each case.

For example, one may fit polynomials

$$y = a_0 + a_1 x + \cdots + a_k x^k \tag{1.29}$$

to observed points, which is common in the EIV context [79, 140]. Scaling of variables $S_{\alpha,\beta}$, cf. Section 1.3, transforms one polynomial to another, so the class of polynomial remains conveniently invariant.

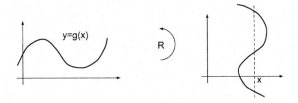

Figure 1.11 *The graph of an explicit nonlinear function $y = g(x)$ (left). After rotation, the same curve (right) does not represent any explicit function.*

However, a rotation R_θ of the coordinate plane transforms a polynomial to a different function; it becomes an implicit polynomial curve, which may not even allow an explicit representation $y = g(x)$. Thus explicit polynomials are *not* suited for orthogonal (geometric) fitting. The same applies to any other class of nonlinear explicit functions $y = g(x)$: the graph of a nonlinear function can always be rotated so that the resulting curve does not represent any explicit functional relation; see Fig. 1.11.

A natural class of models that remain invariant under rotations and translations consists of implicit polynomials of a certain degree $k \geq 1$. Polynomials of degree $k = 1$ are given by equation

$$Ax + By + C = 0, \tag{1.30}$$

which represents all straight lines on the plane (including vertical and horizontal lines). Polynomials of degree $k = 2$ are given by equation

$$Ax^2 + By^2 + Cxy + Dx + Ey + F = 0, \tag{1.31}$$

which represents all conic sections: ellipses, hyperbolas, and parabolas, in addition to straight lines. This class is large enough to cover a vast majority of the existing applications in computer vision and pattern recognition.

Implicit polynomials of higher degree $k \geq 3$ are occasionally used to describe more complex objects, see examples in [150, 176] where polynomials of degree $k = 3, 4$, and even $k = 6$, are mentioned. But the use of polynomials of degree $k \geq 3$ remains extremely rare, as most practitioners prefer to divide complex shapes into small segments that can be well approximated by lines and arcs of conics, or even arcs of circles. What one gets in the end is a sequence of circular arcs stitched together ('circular splines'); see [12, 145, 158, 164, 165]. Some authors plainly assert that "most of the objects in the world are made up of circular arc segments and straight lines;" see [146, 195].

Thus fitting circles and conics to observed data is practically the most important task in image processing applications, besides fitting lines.

We note that often one deals with objects in images that have rectangular or other polygonal shape, see e.g., [186, 189]. In that case a polygon of the right shape can be fit to data. Polygon consists of segments of straight lines, so that general line fitting algorithms can be used, but there are also vertices and corners that may require a special treatment. Such problems are not discussed in this book.

To summarize, we see that in the nonlinear context, the two large topics, (i) the EIV regression used in general statistics and (ii) the geometric fit used in image processing, go separate ways and become very different. There is another significant distinction here: these topics adopt different asymptotic models. In the general EIV regression, it is common to study properties of estimators as the sample size grows, i.e., as $n \to \infty$ (at the same time the noise level σ remains constant). In the image processing applications, the sample size is usually very limited, but the noise is quite small, hence a more appropriate asymptotic model is $\sigma \to 0$ while n is fixed. This issue will be discussed at length in Section 2.5.

We reiterate that our main subject is geometric curve fitting in image processing, i.e., the topic (ii) above. For a comprehensive presentation of the topic (i), i.e., the general nonlinear EIV regression, see a recent book [27] and its second edition [28], updated and expanded.

Chapter 2

Fitting lines

The problem of fitting a straight line to observed points by minimizing orthogonal distances has been around since late 1800s, and now all its aspects are well understood and documented in the statistics literature. We gave a historic overview in the previous chapter. Here we give a brief summary of the solution to this problem and its main features.

2.1 Parametrization

First we need to describe a line in the xy plane by an equation, and there are a few choices here.

Two parameters (slope and intercept). Until late 1980s, nearly every paper and book dealing with the problem of fitting lines used the equation $y = a + bx$. It is simple enough, and it represents all possible linear relations between x and y, which is indeed the goal in many applications. But this equation fails to describe all geometric lines in the plane: vertical lines ($x = \mathrm{const}$) are left unaccounted for.

Besides, lines which are nearly vertical involve arbitrarily large values of b, and this is hazardous in numerical computation. (The issue of numerical stability may not have been noteworthy before 1980s, but it is definitely vital in today's computerized world.)

Lastly, as we mentioned in Section 1.6, the estimates of a and b have infinite mean values and infinite variances; which makes it difficult to analyze their statistical behavior.

Three algebraic parameters. These are compelling reasons to use an alternative equation of a line:

$$Ax + By + C = 0. \tag{2.1}$$

An additional restriction $A^2 + B^2 > 0$ must be applied, as otherwise $A = B = 0$ and (2.1) would describe an empty set if $C \neq 0$, or the entire plane if $C = 0$.

Equation (2.1) represents all geometric lines in the plane, including vertical ($B = 0$) and horizontal ($A = 0$). Also, it allows us to keep the values A, B, C bounded, as we will prove in the next section. This also ensures finiteness of the moments of the estimates of the parameters, and helps to secure numerical stability in practical computations.

Constraints. A little problem is a multiple representation: each line corresponds to infinitely many parameter vectors (A, B, C), which are, obviously, all proportional to each other. This multiplicity can be eliminated by imposing a constraint

$$A^2 + B^2 + C^2 = 1, \tag{2.2}$$

or another constraint

$$A^2 + B^2 = 1. \tag{2.3}$$

The former is used, e.g., in [93, 95], and the latter in [83, 143]. The constraint (2.3) is a bit better, as it automatically enforces the restriction $A^2 + B^2 > 0$, see above. Besides, it allows clear geometric interpretation of the parameter values, see next. So we use (2.3) in what follows.

Parameter space. Due to (2.3), A and B can be replaced by a single parameter φ so that $A = \cos \varphi$ and $B = \sin \varphi$, with $0 \leq \varphi < 2\pi$. Note that φ is a cyclic parameter, i.e., its domain is the unit circle $\mathbb{S}^1$. Now a line can be described by

$$x \cos \varphi + y \sin \varphi + C = 0 \tag{2.4}$$

with two parameters, $\varphi \in \mathbb{S}^1$ and $C \in \mathbb{R}^1$. Observe that φ is the direction of the normal vector to the line. The parameter space $\{(\varphi, C)\}$ is an infinite cylinder, $\mathbb{S}^1 \times \mathbb{R}^1$.

An attentive reader shall notice that each geometric line is still represented by more than one parameter vector: there are exactly two vectors (φ, C) for every line; one is obtained from the other by transforming $C \mapsto -C$ and $\varphi \mapsto \varphi + \pi \pmod{2\pi}$. This ambiguity can be eliminated by requiring $C \geq 0$; then C will be the distance from the origin to the line, see Fig. 2.1. The parameter

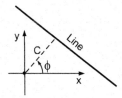

Figure 2.1: *Parameters φ and C of a straight line.*

space will be now a half-infinite cylinder, $\mathbb{S}^1 \times \mathbb{R}^1_+$. In addition, two points $(\varphi, 0)$ and $(\varphi + \pi, 0)$ for every $\varphi \in [0, \pi)$ represent the same line, so they need to be be identified as well.

Objective function. Now the orthogonal fitting line can be found by minimizing the objective function

$$\mathscr{F} = \frac{1}{A^2 + B^2} \sum_{i=1}^{n} (Ax_i + By_i + C)^2 \tag{2.5}$$

in the parametrization (2.1), without imposing the constraint (2.3), or

$$\mathscr{F} = \sum_{i=1}^{n} (x_i \cos \varphi + y_i \sin \varphi + C)^2 \tag{2.6}$$

in the parametrization (2.4).

2.2   Existence and uniqueness

Before we solve the above minimization problem in the new parameters, let us divert our attention to two fundamental theoretical questions: Does the solution always exist? Is it always unique?

Noncompactness of the parameter space. Our analysis here will involve mathematical concepts of continuity and compactness, which we will engage also in the next chapter to treat circles. The reader who is not so familiar with topology may go straight to the next section for a practical solution of the least squares problem, where the existence and uniqueness issues will be also resolved, in a different way.

The function $\mathscr{F}$ given by (2.6) is obviously continuous on the corresponding parameter space. It is known that continuous functions on compact spaces attain their minimum (and maximum) values. If our parameter space *were* compact, this would immediately guarantee the existence of the orthogonal fit. But the parameter space is *not* compact, as C may take arbitrarily large values.

Direct geometric argument. We can also demonstrate the noncompactness of the space of straight lines in a geometric manner, without referring to any parametrization.

Let us denote by $\mathfrak{L}$ the space of all lines in $\mathbb{R}^2$. First we need to introduce the right topology on $\mathfrak{L}$. We say that a sequence of lines $\mathbb{L}_i$ converges to a line $\mathbb{L}$ in the plane $\mathbb{R}^2$ if for any closed disk $\mathbb{D} \subset \mathbb{R}^2$ we have

$$\text{dist}_H(\mathbb{L}_i \cap \mathbb{D}, \mathbb{L} \cap \mathbb{D}) \to 0 \quad \text{as } i \to \infty \tag{2.7}$$

where $\text{dist}_H(\cdot, \cdot)$ is the so-called Hausdorff distance between compact subsets of $\mathbb{R}^2$. The latter is computed by the rule

$$\text{dist}_H(E, F) = \max\left\{\max_{x \in E} \text{dist}(x, F), \max_{y \in F} \text{dist}(y, E)\right\}, \tag{2.8}$$

where $E, F \subset \mathbb{R}^2$ are two compact sets.

Alternatively, one can compactify the plane $\mathbb{R}^2$ by mapping it onto the Riemann sphere (see details in Section 8.1), so that every line is transformed to a circle on the sphere (passing through the north pole), and then use the Hausdorff distance between the respective circles. Both constructions yield the same topology, which coincides with the topology induced by the parameters φ, C (i.e., a sequence of lines $\mathbb{L}_i = (\varphi_i, C_i)$ converges to a line $\mathbb{L} = (\varphi, C)$ if and only if $\varphi_i \to \varphi$ and $C_i \to C$).

Now compact spaces are characterized by the property that every sequence of its points contains a convergent subsequence. Hence one can demonstrate the noncompactness of the space of lines by producing a sequence of lines which has no convergent subsequences. To this end we can take, for example, lines $\mathbb{L}_i = \{x = i\}$, where $i = 1, 2, \ldots$.

Existence of a minimum. So everything tells us that our parameter space is not compact; hence an additional argument is needed to show the existence of the minimum of the objective function. We do this in two steps: first we reduce the space $\mathfrak{L}$ to a smaller space $\mathfrak{L}_0$, which is compact; second, we show that

$$\inf_{\mathfrak{L}_0} \mathscr{F} \leq \inf_{\mathfrak{L}} \mathscr{F},$$

thus we can restrict our search of the minimum of $\mathscr{F}$ to the subspace $\mathfrak{L}_0$.

Since the set of data points (x_i, y_i) is finite, we can enclose them in some rectangle $\mathbb{B}$ (a "bounding box"). In many experiments, all data points are naturally confined to a window corresponding to the physical size of the measuring device, so the box $\mathbb{B}$ may be known in advance, before the data are collected.

Now if a line $\mathbb{L}$ does not cross the box $\mathbb{B}$, it cannot provide a minimum of $\mathscr{F}$, see Fig. 2.2. Indeed, one can just move $\mathbb{L}$ closer to the box and thus reduce all the distances from the data points to the line. We denote by $\mathfrak{L}_0$ the space of all lines crossing the box $\mathbb{B}$ and restrict our search of the minimum of $\mathscr{F}$ to $\mathfrak{L}_0$.

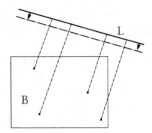

Figure 2.2: *A line* $\mathbb{L}$ *not crossing the bounding box* $\mathbb{B}$.

Recall that C represents the distance from the line to the origin. Since every line $\mathbb{L} \in \mathfrak{L}_0$ crosses the box $\mathbb{B}$, we have a restriction $C \leq C_{\max}$ where $C_{\max}$ is the distance from the origin to the most remote point of $\mathbb{B}$. Thus the reduced space $\mathfrak{L}_0$ is indeed compact.

Theorem 1 *The objective function* $\mathscr{F}$ *does attain its minimum value. Hence the orthogonal fitting line always exists.*

Remark. Our analysis also demonstrates that if one works with parameters A, B, C subject to the constraint $A^2 + B^2 = 1$, then $|A|, |B| \leq 1$ and $|C| \leq C_{\max}$. Thus our parameters never have to exceed the larger of $C_{\max}$ and 1, i.e., the parameters remain bounded, and there is no danger of running into arbitrarily large parameter values during computations; this is a necessary condition for the numerical stability of any algorithm.

Nonuniqueness. Lastly we address the uniqueness issue. While for typical data sets (see precise statement in Section 2.3), the orthogonal regression line is unique, there are exceptions. To give a simple example, suppose that $n > 2$ data points are placed at the vertices of a regular n-gon, $\mathbb{P}$. We show that there are more than one line that minimizes $\mathscr{F}$.

Suppose $\mathbb{L}$ is (one of) the orthogonal regression line(s). Let $R = R_{2\pi/n}$ denote the rotation of the plane through the angle $2\pi/n$ about the center of the polygon $\mathbb{P}$. Since R transforms $\mathbb{P}$ into itself, the line $R(\mathbb{L})$ provides exactly the same value of $\mathscr{F}$ as the line $\mathbb{L}$, hence $R(\mathbb{L})$ is another least squares line, different from $\mathbb{L}$.

In Section 2.3 we present exact condition under which the orthogonal regression line is not unique; we will also see that whenever that line is not unique, there are actually infinitely many of them.

Practical implications. One last remark for an impatient reader who might see little value in our theoretical discussion: the existence and uniqueness of a solution are practically relevant. For example, knowing under what conditions the problem does not have a solution might help understand why the computer program occasionally returns nonsense or crashes altogether.

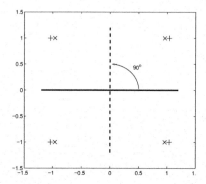

Figure 2.3 *A horizontal solid line fitted to four points (pluses) and a vertical dashed line fitted to four other points (crosses).*

To illustrate the practical role of the nonuniqueness consider an example in Fig. 2.3. The best line fitted to four points $\left(\pm(1+\delta),\pm1\right)$, marked by pluses, is a horizontal line (the x axis). On the other hand, the best line fitted to four nearby points $\left(\pm(1-\delta),\pm1\right)$, marked by crosses, is a vertical line (the y axis). Here δ may be arbitrarily small.

We see that a slight perturbation of the data points may result in a dramatic alteration of the best fitting line (here it is rotated by 90°). This occurs exactly because our data lie very close to an exceptional configuration (the vertices of a perfect square) that corresponds to a multiple solution of the fitting problem.

2.3 Matrix solution

Here we find the minimum of the objective function $\mathscr{F}$ given by (2.5) in terms of the three parameters A, B, C. We use methods of linear algebra; our formulas may look more complicated than the elementary solution (1.16), but they have several advantages: (i) they are more stable numerically; (ii) they work in any case (with no exceptions); (iii) they are convenient for the subsequent error analysis.

Elimination of C. Recall that we are minimizing the function

$$\mathscr{F}(A,B,C) = \sum_{i=1}^{n}(Ax_i + By_i + C)^2 \qquad (2.9)$$

subject to the constraint $A^2 + B^2 = 1$, cf. (2.3). Since the parameter C is unconstrained, we can eliminate it by minimizing (2.9) with respect to C while holding A and B fixed. Solving the equation $\partial \mathscr{F} / \partial C = 0$ gives us

$$C = -A\bar{x} - B\bar{y}, \tag{2.10}$$

where $\bar{x}$ and $\bar{y}$ are the sample means introduced in (1.4). In particular, we see that the orthogonal fitting line always passes through the centroid $(\bar{x}, \bar{y})$ of the data set.

Eigenvalue problem. Eliminating C from (2.9) gives

$$\mathscr{F}(A,B) = \sum_{i=1}^{n} \left[A(x_i - \bar{x}) + B(y_i - \bar{y}) \right]^2$$
$$= s_{xx}A^2 + 2s_{xy}AB + s_{yy}B^2, \tag{2.11}$$

where s_{xx}, s_{xy}, and s_{yy} are the components of the scatter matrix $\mathbf{S}$ introduced in (1.5). In matrix form,

$$\mathscr{F}(\mathbf{A}) = \mathbf{A}^T \mathbf{S} \mathbf{A}, \tag{2.12}$$

where $\mathbf{A} = (A,B)^T$ denotes the parameter vector. Minimizing (2.12) subject to the constraint $\|\mathbf{A}\| = 1$ is a simple problem of the matrix algebra: its solution is the eigenvector of the scatter matrix $\mathbf{S}$ corresponding to the smaller eigenvalue.

Observe that the parameter vector $\mathbf{A}$ is orthogonal to the line (2.1), thus the line itself is parallel to the other eigenvector. In addition, it passes through the centroid, hence it is the major axis of the scattering ellipse.

Practical algorithms. If one uses software with built-in matrix algebra operations, one can just assemble the scatter matrix $\mathbf{S}$ and call a routine returning its eigenvalues and eigenvectors: this gives $\mathbf{A}$; then one computes C via (2.10). The scatter matrix $\mathbf{S}$ may be found as the product $\mathbf{S} = \mathbf{X}^T \mathbf{X}$, where

$$\mathbf{X} = \begin{bmatrix} x_1 - \bar{x} & y_1 & \bar{y} \\ \vdots & & \vdots \\ x_n - \bar{x} & y_n - \bar{y} \end{bmatrix} \tag{2.13}$$

is the $n \times 2$ data matrix.

Alternatively, note that the eigenvectors of $\mathbf{S}$ coincide with the right singular vectors of $\mathbf{X}$. Thus one can use the singular value decomposition (SVD) of $\mathbf{X}$ to find $\mathbf{A}$; this procedure bypasses the evaluation of the scatter matrix $\mathbf{S}$ altogether and makes computations more stable numerically.

Other parameters. Another parameter of interest is φ, the angle between the normal to the fitting line and the x axis, cf. (2.4); it can be computed by

$$\varphi = \tan^{-1}(B/A).$$

The directional angle θ of the line itself is obtained by

$$\theta = \varphi + \pi/2 \qquad (\mathrm{mod}\ \pi).$$

Exceptional cases. For typical data sets, the above procedure leads to a unique orthogonal fitting line. But there are some exceptions.

If the two eigenvalues of $\mathbf{S}$ coincide, then every vector $\mathbf{A} \in \mathbb{R}^2$ is its eigenvector and the function $\mathscr{F}(A,B)$ is actually constant. In that case all the lines passing through the centroid of the data minimize $\mathscr{F}$; hence the problem has multiple (infinitely many) solutions. This happens exactly if $\mathbf{S}$ is a scalar matrix, i.e.

$$s_{xx} = s_{yy} \quad \text{and} \quad s_{xy} = 0. \tag{2.14}$$

One example, mentioned in the previous section, is the set of vertices of a regular polygon.

Thus the orthogonal regression line is not unique if and only if both equations in (2.14) hold. Obviously, this is a very unlikely event. If data points are sampled randomly from a continuous probability distribution, then indeed (2.14) occurs with probability zero. However, if the data points are obtained from a digital image (say, they are pixels on a computer screen), then the chance of having (2.14) may no longer be negligible and may have to be reckoned with. For instance, a simple configuration of 4 pixels making a 2×2 square satisfies (2.14), and thus the orthogonal fitting line is not uniquely defined.

Another interesting exception occurs when the matrix $\mathbf{S}$ is singular. In that case 0 is its eigenvalue, and so $\mathscr{F}(A,B) = 0$, which means a "perfect fit," or "interpolation" (the data points are collinear). Thus we see that the criterion of collinearity is

$$\det \mathbf{S} = s_{xx}s_{yy} - s_{xy}^2 = 0.$$

2.4   Error analysis: Exact results

No statistical estimation is complete without error analysis, whose main purpose is to determine the probability distribution and basic characteristics (such as the mean value and variance) of the respective estimators.

Linear functional model. In order to analyze our estimators $\hat{A}$, $\hat{B}$, $\hat{C}$, and $\hat{\theta}$, we need to adopt a statistical model. We recall the one described in Section 1.2. Namely, we suppose there is a true (but unknown) line

$$\tilde{A}x + \tilde{B}y + \tilde{C} = 0$$

with $\tilde{A}^2 + \tilde{B}^2 = 1$ and n true (but unknown) points $(\tilde{x}_1, \tilde{y}_1), \ldots (\tilde{x}_n, \tilde{y}_n)$ on it. Since the true points lie on the true line, we have

$$\tilde{A}\tilde{x}_i + \tilde{B}\tilde{y}_i + \tilde{C} = 0, \qquad i = 1, \ldots, n. \tag{2.15}$$

Each observed point (x_i, y_i) is as a random perturbation of the true point $(\tilde{x}_i, \tilde{y}_i)$ by a random isotropic Gaussian noise, i.e.,

$$x_i = \tilde{x}_i + \delta_i, \qquad y_i = \tilde{y}_i + \varepsilon_i,$$

where δ_i and ε_i are independent random variables with normal distribution $N(0, \sigma^2)$. The value of σ^2 is not needed for the estimation of A, B, C, and θ.

The true values $\tilde{x}_i, \tilde{y}_i$ are regarded as fixed (nonrandom), and therefore must be treated as additional model parameters. They are called "incidental" or "latent" parameters (or "nuisance" parameters, as their values are normally of little interest). This constitutes the functional model, cf. Section 1.2.

Distribution of the scatter matrix. These are, perhaps, the simplest possible assumptions in the context of the orthogonal fit. Still, under these assumptions the distribution of the estimates of the regression parameters is overly complicated. Without going into the depth of statistical analysis, we will only take a glimpse at some formulas.

For example, the components s_{xx}, s_{yy}, and s_{xy} of the scatter matrix $\mathbf{S}$ have the noncentral Wishart distribution with joint density function

$$\frac{e^{-S/2} e^{-(s_{xx}+s_{yy})/2} (s_{xx} s_{yy} - s_{xy}^2)^{(n-3)/2}}{2^n \pi^{1/2} \Gamma\left(\frac{n-1}{2}\right)} \sum_{j=0}^{\infty} \frac{\left[\left(\frac{S}{2}\right)^2 s_{xx}\right]^j}{j! \, \Gamma\left(\frac{n}{2} + j\right)}, \tag{2.16}$$

in the region $s_{xx} > 0$, $s_{yy} > 0$, and $s_{xy}^2 < s_{xx} s_{yy}$; see, e.g., [10]. In this expression, it is assumed, for simplicity, that the true line is horizontal and passes through the origin (i.e., $\tilde{A} = \tilde{C} = 0$) and $\sigma^2 = 1$. The density for general lines and $\sigma > 0$ can be obtained by a standard coordinate transformation.

The parameter S of the above distribution is defined by

$$S = s_{\tilde{x}\tilde{x}} + s_{\tilde{y}\tilde{y}}$$
$$= \sum_{i=1}^{n} \tilde{x}_i^2 - \frac{1}{n}\left[\sum_{i=1}^{n} \tilde{x}_i\right]^2 + \sum_{i=1}^{n} \tilde{y}_i^2 - \frac{1}{n}\left[\sum_{i=1}^{n} \tilde{y}_i\right]^2. \tag{2.17}$$

It characterizes the spread of the true points $(\tilde{x}_i, \tilde{y}_i)$ about their centroid. In fact, S is the sum of squares of the distances from the true points to their centroid. Note that S remains invariant under translations and rotations of the coordinate frame on the plane.

Estimation of θ. Next we turn our attention to the estimator $\hat{\theta}$ and make two important observations, following Anderson [7]. First, the distribution of $\hat{\theta} - \tilde{\theta}$ does not depend on $\tilde{\theta}$ (the true value of θ). Indeed, the model and the orthogonal fitting line are invariant under rotations, so we can rotate the line to achieve $\tilde{\theta} = 0$.

The other important observation is that the difference $\hat{\theta} - \tilde{\theta}$ has distribution symmetric about zero, i.e.,

$$\text{Prob}\{\hat{\theta} - \tilde{\theta} > t\} = \text{Prob}\{\hat{\theta} - \tilde{\theta} < t\} \qquad (2.18)$$

for every $t \in (0, \pi/2)$ (recall that θ is a cyclic parameter). Indeed, we can assume that $\tilde{\theta} = 0$, see above. Now positive and negative values of the errors δ_i are equally likely, hence $\hat{\theta}$ is symmetric about zero.

Remark. As an immediate consequence, we see that the estimator $\hat{\theta}$ is unbiased, i.e., $\mathbb{E}(\hat{\theta}) = \tilde{\theta}$.

Now the density function of the estimator $\hat{\theta}$ can be obtained by using (2.16) via the transformation (1.17). The final expression for that density is

$$\frac{e^{-S/2}}{2^{n-1}\sqrt{\pi}} \sum_{j=0}^{\infty} \frac{\Gamma(n+j)}{\Gamma(\frac{n}{2}+j)} \left(\frac{S}{4}\right)^j \sum_{i=0}^{j} \frac{\Gamma[\frac{i}{2}+1]\cos^i 2(\hat{\theta}-\tilde{\theta})}{\Gamma[\frac{n+1+i}{2}]i!(j-i)!},$$

where it is again assumed, for simplicity, that $\sigma^2 = 1$. This density formula for $\hat{\theta}$ is obtained in the 1970s by Anderson and others [7, 10].

Other estimates. Now one can easily derive the densities of the estimates $\hat{A} = \cos\hat{\theta}$ and $\hat{B} = \sin\hat{\theta}$ and the slope $\beta = -A/B$, they will be given by similar expressions. We do not provide them here because (as Anderson [7] remarked) they are not very useful for practical purposes.

Instead, Anderson [7] and Kunitomo [118] derive various asymptotic formulas for the distribution functions of the estimates $\hat{\theta}$ and $\hat{\beta}$. Those asymptotic expressions were accurate enough to allow Anderson and others to obtain (rigorously) asymptotic formulas for the variance of $\hat{\theta}$ (recall that $\hat{\beta}$ has infinite moments, cf. Section 1.6). Also, Patefield [142] verified numerically how good those approximative formulas are, and under what conditions they cease to be valid.

We present those asymptotic results in Section 2.6. We start by introducing various asymptotic models popular in the EIV regression analysis.

2.5   Asymptotic models: Large n versus small σ

The choice of an asymptotic model is an interesting issue itself in regression analysis, especially in the context of the functional model.

Traditional (large sample) approach. It is common in statistics to study asymptotic properties of estimators as the sample size grows, i.e., as $n \to \infty$. In the functional model, however, there is an obvious difficulty with this approach: increasing n requires introduction of more and more "true" points $(\tilde{x}_i, \tilde{y}_i)$, thus increasing the number of parameters of the model (recall that the true points are regarded as "latent" parameters).

The asymptotic behavior of the estimates might then depend on where those new points are placed, so certain assumptions on the asymptotic structure of the true points are necessary. For example, if the true points tend to cluster near a certain center, parameter estimates may have very poor asymptotic properties as $n \to \infty$.

Furthermore, general statistical theorems that guarantee good asymptotical properties of maximum likelihood estimators (MLE), as $n \to \infty$, will not apply in this case, because these theories are not valid when the number of parameters changes with n. Therefore, the properties of the MLE must be investigated by nontraditional methods.

Nontraditional (small noise) approach. Some statisticians prefer a different asymptotic model in regression analysis with latent parameters. They keep the number of true points, n, fixed but consider the limit $\sigma \to 0$. This is called the small-sigma (or small-disturbance) approach introduced in the early 1970s by Kadane [91, 92] and later used by Anderson [7] and others[1]. This model does not require extra assumptions, though the asymptotic properties of the estimators may again depend on the (fixed) locations of the true points.

Comparative analysis. Each asymptotic model has advantages and disadvantages. The classical model, where $n \to \infty$ but σ is fixed, is perhaps more appropriate for applications in econometrics and sociology, where the very nature of statistical analysis consists of collecting and processing large samples of data, and to increase the accuracy one collects more data. At the same time each individual observation (such as the income of a family or the price of a house) may not fit the ideal model very well, and there is no way to make all (or most of) deviations from the model small.

The other assumption, where n is fixed but $\sigma \to 0$, is more appropriate for image processing applications. Given an image, there is a limited number of points that can be read (scanned) from it. One might suggest that using higher resolution produces more data points, but this is not true: increasing resolution would just refine the existing data points, rather than produce new ones.

On the other hand, when points are marked on (or scanned from) an imperfect line or oval in an image, they are usually quite close to that line (or oval). Deviations are normally caused by technical reasons (an imperfect photograph, imprecise scanning), and improving technology allows one to reduce errors substantially, so the limiting assumption $\sigma \to 0$ is quite reasonable. Large deviations are regarded as abnormalities and often are eliminated by various methods (including manual preprocessing).

---

[1] Anderson [7] remarks that his small-sigma model is equivalent to the one where both n and $\sigma > 0$ are held fixed, but instead one homotetically expands the set of the true points on the line. While not quite realistic, this picture caught imagination and was termed "a fascinating new concept of asymptotic theory," see the discussion after [7], as well as [118].

An ardent advocate of the small-sigma model for image processing applications is Kanatani; we refer the reader to his well written articles [97, 98, 99].

A hybrid model. Some authors combine the above two models and assume that

$$n \to \infty \quad \text{and} \quad \sigma \to 0 \tag{2.19}$$

simultaneously. This is in fact a very strong assumption which allows one to derive the desired asymptotic properties more easily. Such hybrid models have been studied by Amemiya, Fuller and Wolter [5, 192] in the context of nonlinear regression. In the linear case, we can do without them. We note that a somewhat simplified version of the above hybrid model will be involved in our statistical error analysis of Chapter 6.

Terminology. Since our primary motivation is geometric fitting in image processing applications, the small-sigma model has a higher priority in our book. We call it the Type A model. The classical (large n) asymptotics is called the Type B model (this terminology was also used by Patefield [142]):

Type A model: The true points are fixed and $\sigma \to 0$. No additional assumptions are necessary.

Type B model: The variance $\sigma^2 > 0$ is fixed, but $n \to \infty$. In addition we assume that

$$\lim_{n \to \infty} \frac{1}{n} S = s_* > 0,$$

where S is the spread of the true points defined by (2.17). This assumption ensures that the line is identifiable.

If $s_* = 0$, then the true points may "cluster" too tightly near a common center, and the parameter estimates may not even be consistent (see [76]). In the extreme case $S = 0$, for example, the estimate of the slope is obviously independent of the true slope.

2.6 Asymptotic properties of estimators

Theory. First we state two theorems describing the asymptotics of the estimate $\hat{\theta}$ in the Type A and Type B models.

Theorem 2 ([7]) *For the Type A model, the estimate $\hat{\theta}$ is asymptotically normal, and its limit distribution is described by*

$$\sigma^{-1}(\hat{\theta} - \tilde{\theta}) \to_L N(0, S^{-1}). \tag{2.20}$$

Its variance satisfies

$$\text{Var}(\hat{\theta}) = \frac{\sigma^2}{S} \left(1 + \frac{(n-1)\sigma^2}{S} \right) + \mathscr{O}(\sigma^6). \tag{2.21}$$

Theorem 3 ([73, 118, 143]) *For the Type B model, the estimate $\hat{\theta}$ is asymptotically normal, and its limit distribution is described by*

$$\sqrt{n}(\hat{\theta} - \tilde{\theta}) \to_L N\left(0, \frac{\sigma^2}{s_*}\left(1 + \frac{\sigma^2}{s_*}\right)\right). \tag{2.22}$$

Its variance satisfies

$$\operatorname{Var}(\hat{\theta}) = \frac{\sigma^2}{ns_*}\left(1 + \frac{\sigma^2}{s_*}\right) + \mathcal{O}(n^{-3/2}). \tag{2.23}$$

Recall that the estimate $\hat{\theta}$ is unbiased, i.e., $\mathbb{E}(\hat{\theta}) = \tilde{\theta}$, cf. Section 2.4.

Remark. In the Type A model, the asymptotic normality (2.20) is ensured by the assumed normal distribution of errors δ_i and ε_i. If errors have a different distribution, then the asymptotic normality of $\hat{\theta}$ may be lost. In that case one can still derive an asymptotic formula for $\operatorname{Var}(\hat{\theta})$ similar to (2.21) assuming that the errors δ_i and ε_i have finite moments of order four.

Remark. In the Type B model, the asymptotic normality (2.22) follows, as usual, from the central limit theorem, and one does not need to assume any specific distribution of errors [73]. The formula (2.23) is derived under the assumption of normally distributed errors, but perhaps that assumption can be relaxed.

Practical remark. A more pragmatic question is: which approximation to $\operatorname{Var}(\hat{\theta})$, (2.21) or (2.23), should one use in practice, where $\sigma > 0$ and n is finite? Well, note that the leading terms in these formulas are identical and give a *first order approximation*:

$$\operatorname{Var}_1(\hat{\theta}) = \frac{\sigma^2}{S}, \tag{2.24}$$

as we should obviously set $s_* = S/n$. The second order terms are almost identical; they differ only by the factor $n - 1$ in (2.21) versus n in (2.23). Thus both formulas give essentially the same result in practice, we record it as

$$\operatorname{Var}_2(\hat{\theta}) = \frac{\sigma^2}{S}\left(1 + \frac{(n-1)\sigma^2}{S}\right). \tag{2.25}$$

Numerical test. We verified the accuracy of the approximations (2.24) and (2.25) to $\operatorname{Var}(\hat{\theta})$ numerically. We generated 10^6 random samples of n points whose true positions on a line were equally spaced on a stretch of length L, i.e., the distances between the points were $L/(n-1)$. The noise level σ changed[2]

---

[2]Because of invariance under rotations and similarities, the results of our experiment do not depend on the position of the true line in the plane or on the value of L.

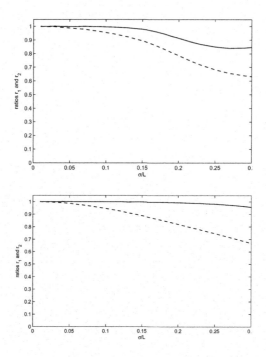

Figure 2.4 *The ratios r_1 (dashed line) and r_2 (solid line), as functions of σ/L. Here $n = 10$ (top) and $n = 100$ (bottom).*

from 0 to 0.3L. We note that $\sigma = 0.3L$ is a fairly large noise, beyond that value there is little sense in fitting a line at all. Fig. 2.4 shows how the ratios

$$r_1 = \frac{\mathsf{Var}_1(\hat{\theta})}{\mathsf{Var}(\hat{\theta})} \quad \text{and} \quad r_2 = \frac{\mathsf{Var}_2(\hat{\theta})}{\mathsf{Var}(\hat{\theta})}$$

change with σ/L.

The top portion of Fig. 2.4 is for $n = 10$ and the bottom for $n = 100$. We see that, as n increases, the accuracy of the approximation (2.24) remains about the same, but that of (2.25) noticeably improves, i.e., the ratio r_2 gets closer to one as n grows. No doubt, the factor $n - 1$ in the second order term is the big helper here.

Also note that the ratios r_1 and r_2 are always below 1, i.e., both approximations (2.24) and (2.25) underestimate the true values of $\mathsf{Var}(\hat{\theta})$. Perhaps the third order expansion would do even better, but no formula for it is published in the literature.

Taylor expansions. The asymptotic properties of the estimates $\hat{A}, \hat{B}, \hat{C}$, to the leading order, can now be obtained easily from those of $\hat{\theta}$ by using a Taylor approximation. We demonstrate this in the context of the Type A model.

First, $A = \sin\theta$ and $B = -\cos\theta$, hence using Taylor expansion up to the third order terms, we get

$$A = \tilde{A} - \tilde{B}\,\delta\theta - \tfrac{1}{2}\tilde{A}(\delta\theta)^2 + \frac{1}{6}\tilde{B}(\delta\theta)^3 + \mathscr{O}\big((\delta\theta)^4\big)$$

$$B = \tilde{B} + \tilde{A}\,\delta\theta - \tfrac{1}{2}\tilde{B}(\delta\theta)^2 - \frac{1}{6}\tilde{A}(\delta\theta)^3 + \mathscr{O}\big((\delta\theta)^4\big).$$

Means and variances of $\hat{A}$ and $\hat{B}$. Taking mean values (recall that $\mathbb{E}(\delta\theta) = \mathbb{E}(\delta\theta)^3 = 0$) gives

$$\mathbb{E}(\hat{A}) = \tilde{A}\left[1 - \frac{\sigma^2}{2S}\right] + \mathscr{O}(\sigma^4),$$

$$\mathbb{E}(\hat{B}) = \tilde{B}\left[1 - \frac{\sigma^2}{2S}\right] + \mathscr{O}(\sigma^4),$$

i.e., the estimates $\hat{A}$ and $\hat{B}$ are biased toward zero. Taking variances gives

$$\mathsf{Var}\,\hat{A} = \frac{\sigma^2\tilde{B}^2}{S} + \mathscr{O}(\sigma^4),$$

$$\mathsf{Var}\,\hat{B} = \frac{\sigma^2\tilde{A}^2}{S} + \mathscr{O}(\sigma^4),$$

$$\mathsf{Cov}(\hat{A},\hat{B}) = -\frac{\sigma^2\tilde{A}\tilde{B}}{S} + \mathscr{O}(\sigma^4).$$

Note that the standard deviation of every estimate is $\mathscr{O}(\sigma)$, while its bias is $\mathscr{O}(\sigma^2)$; hence the bias is of higher order of smallness. In other words, the bias is negligible when we, for instance, assess the mean squared error of an estimate, say

$$\mathsf{MSE}(\hat{A}) = \mathbb{E}\big[(\hat{A} - \tilde{A})^2\big] = \mathsf{Var}(\hat{A}) + \big[\mathbb{E}(\hat{A}) - \tilde{A}\big]^2 = \frac{\sigma^2\tilde{B}^2}{S} + \mathscr{O}(\sigma^4),$$

so the bias is simply absorbed by the $\mathscr{O}(\sigma^4)$ term.

Mean and variance of $\hat{C}$. Lastly, by using $\hat{C} = -\hat{A}\bar{x} - \hat{B}\bar{y}$ one obtains

$$\mathbb{E}(\hat{C}) = \tilde{C}\left[1 - \frac{\sigma^2}{2S}\right] + \mathscr{O}(\sigma^4),$$

and

$$\mathsf{Var}\,\hat{C} = \sigma^2\left[\frac{1}{n} + \frac{(\tilde{A}\bar{\bar{y}} - \tilde{B}\bar{\bar{x}})^2}{S}\right] + \mathscr{O}(\sigma^4). \tag{2.26}$$

where

$$\bar{\bar{x}} = \sum_{i=1}^{n} \tilde{x}_i \qquad \text{and} \qquad \bar{\bar{y}} = \sum_{i=1}^{n} \tilde{y}_i$$

are the sample means of the true values. The extra term $\frac{1}{n}$ in (2.26) comes from the fact that $\bar{x}$ and $\bar{y}$ (the sample means of the observations) are random variables correlated with $\hat{A}$ and $\hat{B}$.

Covariances. For the sake of completeness, we include three more covariances

$$\text{Cov}(\hat{A}, \hat{C}) = \frac{\sigma^2 \tilde{B}(\tilde{A}\bar{\bar{y}} - \tilde{B}\bar{\bar{x}})}{S} + \mathcal{O}(\sigma^4)$$

$$\text{Cov}(\hat{B}, \hat{C}) = -\frac{\sigma^2 \tilde{A}(\tilde{A}\bar{\bar{y}} - \tilde{B}\bar{\bar{x}})}{S} + \mathcal{O}(\sigma^4)$$

$$\text{Cov}(\hat{\theta}, \hat{C}) = -\frac{\sigma^2 (\tilde{A}\bar{\bar{y}} - \tilde{B}\bar{\bar{x}})}{S} + \mathcal{O}(\sigma^4).$$

Practical remarks. In practice, the true values that appear in the above formulas are not available. Normally, one makes further approximation replacing the true points with the observed ones and the true parameter values with their estimates (this does not alter the above expressions, the resulting errors are absorbed by the $\mathcal{O}(\sigma^4)$ terms). In the same fashion, one estimates S by

$$\hat{S} = s_{xx} + s_{yy}$$

and σ^2 by

$$\hat{\sigma}^2 = \frac{1}{n} \sum_{i=1}^{n} (\hat{A}x_i + \hat{B}y_i + \hat{C})^2. \tag{2.27}$$

Gleser [73] and Patefield [143] show that all these estimates are strongly consistent, thus their use in the construction of confidence intervals is justified.

2.7   Approximative analysis

Taylor approximation versus rigorous proofs. In the previous section, we used Taylor expansion to derive formulas for the variances and covariances of all important estimates based on the second order approximation (2.25) to the variance $\text{Var}(\hat{\theta})$, whose proof was left "behind the scene" (we simply referred to the exact results by Anderson [7] and Kunitomo [118]).

In this section we derive (2.25) itself by Taylor approximation without using the exact results. Thus we demonstrate that Taylor expansion, if properly used, is a powerful (albeit not rigorous) tool, which allows one to obtain correct approximations even including higher order terms. This will help us in the studies of nonlinear regression later, in particular circles and ellipses.

Matrix approximations. We use notation of Section 2.3 and focus on perturbations $\delta \mathbf{A} = \mathbf{A} - \tilde{\mathbf{A}}$, $\delta \mathbf{S} = \mathbf{S} - \tilde{\mathbf{S}}$, etc., where the tilde is used to denote the true values. The orthogonal fit minimizes

$$\mathscr{F}(\mathbf{A}) = \mathbf{A}^T \mathbf{S} \mathbf{A} = \mathbf{A}^T \mathbf{X}^T \mathbf{X} \mathbf{A} = (\tilde{\mathbf{A}} + \delta \mathbf{A})^T (\tilde{\mathbf{X}} + \delta \mathbf{X})^T (\tilde{\mathbf{X}} + \delta \mathbf{X})(\tilde{\mathbf{A}} + \delta \mathbf{A}).$$

Expanding, using the fact $\tilde{\mathbf{X}} \tilde{\mathbf{A}} = 0$, which follows from (2.15), and keeping only statistically significant terms[3] gives

$$\mathscr{F}(\mathbf{A}) = \tilde{\mathbf{A}}^T \delta \mathbf{X}^T \delta \mathbf{X} \tilde{\mathbf{A}} + \delta \mathbf{A}^T \tilde{\mathbf{X}}^T \tilde{\mathbf{X}} \delta \mathbf{A} + 2 \tilde{\mathbf{A}}^T \delta \mathbf{X}^T \tilde{\mathbf{X}} \delta \mathbf{A} + 2 \tilde{\mathbf{A}}^T \delta \mathbf{X}^T \delta \mathbf{X} \delta \mathbf{A}.$$

Now the minimum of $\mathscr{F}$ is attained at $\partial \mathscr{F} / \partial \mathbf{A} = 0$, i.e., at

$$\tilde{\mathbf{X}}^T \tilde{\mathbf{X}} \delta \mathbf{A} + (\tilde{\mathbf{X}}^T \delta \mathbf{X} + \delta \mathbf{X}^T \delta \mathbf{X}) \tilde{\mathbf{A}} = 0,$$

hence

$$\delta \mathbf{A} = -(\tilde{\mathbf{X}}^T \tilde{\mathbf{X}})^- (\tilde{\mathbf{X}}^T \delta \mathbf{X} + \delta \mathbf{X}^T \delta \mathbf{X}) \tilde{\mathbf{A}}, \tag{2.28}$$

where $(\tilde{\mathbf{X}}^T \tilde{\mathbf{X}})^-$ denotes the Moore-Penrose generalized inverse of the matrix $\tilde{\mathbf{X}}^T \tilde{\mathbf{X}}$ (which is rank deficient because $\tilde{\mathbf{X}} \tilde{\mathbf{A}} = 0$).

A convenient assumption. Since the distribution of $\delta \theta = \hat{\theta} - \tilde{\theta}$ does not depend on $\tilde{\theta}$ (Section 2.4), we can simplify our calculation by assuming that the line is horizontal, i.e., $\tilde{\mathbf{A}} = (0,1)^T$. Then $\delta \mathbf{A} = (\delta \theta, 0)^T$, to the leading order, and

$$\tilde{\mathbf{X}}^T \tilde{\mathbf{X}} = \begin{bmatrix} S & 0 \\ 0 & 0 \end{bmatrix}, \quad \text{hence} \quad (\tilde{\mathbf{X}}^T \tilde{\mathbf{X}})^- = \begin{bmatrix} \frac{1}{S} & 0 \\ 0 & 0 \end{bmatrix}.$$

The matrix $\delta \mathbf{X}$ is

$$\begin{bmatrix} \delta_1 - \bar{\delta} & \varepsilon_1 - \bar{\varepsilon} \\ \vdots & \vdots \\ \delta_n - \bar{\delta} & \varepsilon_n - \bar{\varepsilon} \end{bmatrix},$$

where we continue using the sample mean notation, i.e., $\bar{\delta} = \sum_{i=1}^n \delta_i$, etc. Now (2.28) takes form

$$\delta \theta = -\frac{1}{S} \sum_{i=1}^n (\tilde{x}_i - \bar{\tilde{x}})(\varepsilon_i - \bar{\varepsilon}) - \frac{1}{S} \sum_{i=1}^n (\delta_i - \bar{\delta})(\varepsilon_i - \bar{\varepsilon}).$$

Using our assumptions on the errors δ_i and ε_i we easily compute that $\mathbb{E}(\delta \theta) = 0$ and

$$\mathbb{E} \left[\sum_{i=1}^n (\tilde{x}_i - \bar{\tilde{x}})(\varepsilon_i - \bar{\varepsilon}) \right]^2 = \sigma^2 \sum_{i=1}^n (\tilde{x}_i - \bar{\tilde{x}})^2 = \sigma^2 S,$$

---

[3]We drop the term $\delta \mathbf{A}^T \delta \mathbf{X}^T \delta \mathbf{X} \delta \mathbf{A}$, which is of order σ^4 and $2\delta \mathbf{A}^T \tilde{\mathbf{X}}^T \delta \mathbf{X} \delta \mathbf{A}$, which is of order $\sigma^2 / \sqrt{n}$. We keep all the terms of order σ^2 or $\sigma^3 \sqrt{n}$ or higher.

$$\mathbb{E}\left[\left(\sum_{i=1}^{n}(\tilde{x}_i - \bar{\tilde{x}})(\varepsilon_i - \bar{\varepsilon})\right)\left(\sum_{i=1}^{n}(\delta_i - \bar{\delta})(\varepsilon_i - \bar{\varepsilon})\right)\right] = 0,$$

and

$$\mathbb{E}\left[\sum_{i=1}^{n}(\delta_i - \bar{\delta})(\varepsilon_i - \bar{\varepsilon})\right]^2 = (n-1)\sigma^4.$$

Summing up we arrive at (2.25).

Observe that our approximate analysis gives the correct asymptotic variance of $\hat{\theta}$ for both Type A model, cf. (2.21), and Type B model, cf. (2.23).

Final remarks. In the above calculations we only used the following assumptions on the errors δ_i and ε_i: they have mean zero, a common variance σ^2, and they are uncorrelated.

Our approach combines (and improves upon) the calculations made by Malinvaud, see pp. 399–402 in [128], who treated Type B model only, and Kanatani [93, 105], who treated Type A model only.

2.8 Finite-size efficiency

In this last section we discuss the efficiency of the parameter estimators. There are various approaches to this issue in the context of orthogonal (and more generally, EIV) regression.

Two approaches. First, one can directly compute the classical Cramer-Rao lower bound and compare it to the actual variances of the parameter estimators. We do this below. As it happens, the maximum likelihood estimates are not exactly optimal, but their efficiency is close to 100% in various senses.

Second, one may try to prove that the estimators are asymptotically efficient, in the context of the Type A model or the Type B model. There are, indeed, general results in this direction; we discuss them as well.

Cramer-Rao for finite samples. We start with the classical Cramer-Rao lower bound; our analysis is an adaptation of [128] (pp. 402–403). We work in the context of the functional model (Section 2.4) with $n + 2$ independent parameters: two principal parameters (θ and C) for the line and one "latent" parameter per each true point. We can describe the line by parametric equations

$$x = -C\sin\theta + t\cos\theta, \qquad y = C\cos\theta + t\sin\theta,$$

where t is a scalar parameter, and specify the coordinates of the true points by

$$\tilde{x}_i = -C\sin\theta + t_i\cos\theta, \qquad \tilde{y}_i = C\cos\theta + t_i\sin\theta.$$

Thus $t_1, \ldots, t_n$ play the role of the (latent) parameters of the model.

The log-likelihood function is

$$\ln L = \text{const} - \frac{1}{2\sigma^2} \sum_{i=1}^{n} (x_i - \tilde{x}_i)^2 + (y_i - \tilde{y}_i)^2. \tag{2.29}$$

According to the classical Cramer-Rao theorem, the covariance matrix of the parameter estimators is bounded below[4] by

$$\text{Cov}(\hat{\theta}, \hat{C}, \hat{t}_1, \ldots, \hat{t}_n) \geq \mathbf{F}^{-1},$$

where $\mathbf{F}$ is the Fisher information matrix

$$\mathbf{F} = -\mathbb{E}(\mathbf{H})$$

and $\mathbf{H}$ denotes the Hessian matrix consisting of the second order partial derivatives of $\ln L$ with respect to the parameters.

Computing the Cramer-Rao bound. Computing the second order partial derivatives of (2.29) with respect to $\theta, C, t_1, \ldots, t_n$ (which is a routine exercise) and taking their expected values (by using the obvious rules $\mathbb{E}(x_i) = \tilde{x}_i$ and $\mathbb{E}(y_i) = \tilde{y}_i$) gives the following results (where we omit the common factor σ^{-2} for brevity)

$$\mathbb{E}\left[\frac{\partial^2 \ln L}{\partial \theta^2}\right] = -\sum t_i^2 - nC^2, \qquad \mathbb{E}\left[\frac{\partial^2 \ln L}{\partial \theta \partial C}\right] = -\sum t_i, \qquad \mathbb{E}\left[\frac{\partial^2 \ln L}{\partial \theta \partial t_i}\right] = C,$$

$$\mathbb{E}\left[\frac{\partial^2 \ln L}{\partial C \partial \theta}\right] = -\sum t_i, \qquad \mathbb{E}\left[\frac{\partial^2 \ln L}{\partial C^2}\right] = -n, \qquad \mathbb{E}\left[\frac{\partial^2 \ln L}{\partial C \partial t_i}\right] = 0,$$

$$\mathbb{E}\left[\frac{\partial^2 \ln L}{\partial t_i \partial \theta}\right] = C, \qquad \mathbb{E}\left[\frac{\partial^2 \ln L}{\partial t_i \partial C}\right] = 0, \qquad \mathbb{E}\left[\frac{\partial^2 \ln L}{\partial t_i \partial t_j}\right] = -\delta_{ij},$$

where δ_{ij} denotes the Kronecker delta symbol. The Fisher information matrix now has structure

$$\mathbf{F} = \sigma^{-2} \begin{bmatrix} \mathbf{E} & \mathbf{G}^T \\ \mathbf{G} & \mathbf{I}_n \end{bmatrix},$$

where

$$\mathbf{E} = \begin{bmatrix} \sum t_i^2 + nC^2 & \sum t_i \\ \sum t_i & n \end{bmatrix}$$

and

$$\mathbf{G} = \begin{bmatrix} -C & 0 \\ \vdots & \vdots \\ -C & 0 \end{bmatrix}$$

---

[4]The inequality $\mathbf{A} \geq \mathbf{B}$ between two symmetric matrices is understood in the sense that $\mathbf{A} - \mathbf{B}$ is positive semi-definite, i.e., $\mathbf{x}^T \mathbf{A} \mathbf{x} \geq \mathbf{x}^T \mathbf{B} \mathbf{x}$ for every vector $\mathbf{x}$.

and $\mathbf{I}_n$ denotes the identity matrix of order n. By using the block matrix inversion lemma (see, e.g., p. 26 in [128]) we obtain

$$\begin{bmatrix} \mathbf{E} & \mathbf{G}^T \\ \mathbf{G} & \mathbf{I}_n \end{bmatrix}^{-1} = \begin{bmatrix} (\mathbf{E}-\mathbf{G}^T\mathbf{G})^{-1} & -(\mathbf{E}-\mathbf{G}^T\mathbf{G})^{-1}\mathbf{G}^T \\ -\mathbf{G}(\mathbf{E}-\mathbf{G}^T\mathbf{G})^{-1} & (\mathbf{I}_n-\mathbf{G}\mathbf{E}^{-1}\mathbf{G}^T)^{-1} \end{bmatrix}.$$

(One can easily verify this formula by direct multiplication.)

The 2×2 top left block of this matrix is the most interesting to us as it corresponds to the principal parameters θ and C. This block is

$$(\mathbf{E}-\mathbf{G}^T\mathbf{G})^{-1} = \begin{bmatrix} \sum t_i^2 & \sum t_i \\ \sum t_i & n \end{bmatrix}^{-1} = \frac{1}{nS}\begin{bmatrix} n & -\sum t_i \\ -\sum t_i & \sum t_i^2 \end{bmatrix},$$

because the determinant here is

$$n\sum t_i^2 - \left(\sum t_i\right)^2 = nS.$$

Therefore, the Cramer-Rao lower bound on the covariance matrix of the estimates of the principal parameters is

$$\mathsf{Cov}(\hat{\theta},\hat{C}) \geq \frac{\sigma^2}{S}\begin{bmatrix} 1 & -\frac{1}{n}\sum t_i \\ -\frac{1}{n}\sum t_i & \frac{1}{n}\sum t_i^2 \end{bmatrix}. \tag{2.30}$$

In particular,

$$\mathsf{Var}(\hat{\theta}) \geq \frac{\sigma^2}{S}$$

which is exactly our first order approximation (2.24) to the actual variance of $\hat{\theta}$. In fact, all the components of the matrix (2.30) are equal to the leading terms of the actual variances and covariances obtained in Section 2.6, because

$$\frac{1}{n}\sum t_i = \frac{1}{n}\left(\tilde{A}\sum \tilde{y}_i - \tilde{B}\sum \tilde{x}_i\right) = \tilde{A}\bar{\tilde{y}} - \tilde{B}\bar{\tilde{x}}$$

and

$$\frac{1}{n}\sum t_i^2 = \frac{S}{n} + \frac{1}{n^2}\left(\sum t_i\right)^2 = \frac{S}{n} + (\tilde{A}\bar{\tilde{y}} - \tilde{B}\bar{\tilde{x}})^2.$$

Concluding remarks. Our analysis shows that, to the leading order, the estimators of θ and C are optimal. However, the more accurate second order approximation (2.25) indicates that our estimates are not exactly optimal; in particular the (approximate) efficiency of $\hat{\theta}$ is

$$\left(1 + \frac{(n-1)\sigma^2}{S}\right)^{-1}. \tag{2.31}$$

This is close to one when the ratio $(n-1)\sigma^2/S$ is small, i.e., when the deviations of the data points *from* the line are small compared to their spread *along* the line.

2.9 Asymptotic efficiency

Here we discuss the asymptotic efficiency of our estimates, in the context of Type A model or Type B model.

In Type A model, as n is fixed and $\sigma^2 \to 0$, the efficiency (2.31) converges to one. Thus our estimates are asymptotically efficient (optimal). In fact, the maximum likelihood estimates of regression parameters (linear or nonlinear) are always asymptotically efficient in the context of Type A model, see Section 6.5.

In Type B model, where σ^2 is fixed and $n \to \infty$, our estimators are not asymptotically efficient. For example, the efficiency of $\hat{\theta}$ converges to

$$\left(1 + \frac{\sigma^2}{s_*}\right)^{-1} < 1.$$

On the other hand, in 1954 Wolfowitz proved [191], see also p. 403 in [128], that in the context of Type B model there exists no estimator which has asymptotic efficiency equal to 1. We do not know if there exists any estimator with asymptotic efficiency greater than that of $\hat{\theta}$.

Remark. There are general theorems in statistics stating that maximum likelihood estimators are asymptotically efficient, as $n \to \infty$. However, those theorems do not apply to models where the number of parameters grows with n, and this is exactly the case with the Type B model here.

Statistical optimality by Gleser. Nonetheless, the maximum likelihood estimators $\hat{A}$, $\hat{B}$, and $\hat{C}$ of the parameters of the line are asymptotically optimal, as $n \to \infty$, in the following restricted sense.

Consider an infinite sequence of true points $(\tilde{x}_i, \tilde{y}_i)$ lying on the (unknown) line $\tilde{A}x + \tilde{B}y + \tilde{C} = 0$ such that

$$\bar{\tilde{x}}_n = \frac{1}{n}\sum_{i=1}^n \tilde{x}_i \to \bar{x}_* \qquad \text{and} \qquad \bar{\tilde{y}}_n = \frac{1}{n}\sum_{i=1}^n \tilde{y}_i \to \bar{y}_*$$

in the manner satisfying

$$|\bar{\tilde{x}}_n - \bar{x}_*| = o(n^{-1/2}) \qquad \text{and} \qquad |\bar{\tilde{y}}_n - \bar{y}_*| = o(n^{-1/2})$$

and

$$\frac{1}{n}S_n = \frac{1}{n}\sum_{i=1}^n (\tilde{x}_i - \bar{\tilde{x}}_n)^2 + \frac{1}{n}\sum_{i=1}^n (\tilde{y}_i - \bar{\tilde{y}}_n)^2 \to s_* > 0$$

in the manner satisfying

$$\left|\tfrac{1}{n}S_n - s_*\right| = o(n^{-1/2}).$$

Now suppose we are estimating A, B, C from the noisy observation of the first n

points, and the noise level σ^2 is unknown and we are estimating it, too. Denote by $\hat{A}_n$, $\hat{B}_n$, $\hat{C}_n$, $\hat{\sigma}_n^2$ the estimators of these parameters.

Let $\mathfrak{E}$ denote the class of all estimators $\hat{A}_n$, $\hat{B}_n$, $\hat{C}_n$, $\hat{\sigma}_n^2$ which are consistent and asymptotically normal, with the asymptotic covariance matrix depending on the sequence of the true points $(\tilde{x}_i, \tilde{y}_i)$ only through their limit values $\bar{x}_*$, $\bar{y}_*$, and s_*. Then within $\mathfrak{E}$, the maximum likelihood estimators of A, B, C and the estimator (2.27) of σ^2 are the best estimators, i.e., they have the asymptotically smallest possible covariance matrix.

This remarkable result is due to Gleser [74, 75], see also Cheng and Van Ness [38, 39, 40].

Final remarks. This concludes our brief summary of the orthogonal line fitting. We have covered only selected aspects of this subject, which seemed to be most relevant to our main objectives — fitting circles and ellipses. We left out nonorthogonal EIV linear regression, multidimensional linear regression, structural and ultra-structural models, and many other interesting but not quite relevant topics.

Chapter 3

Fitting circles: Theory

3.1   Introduction

Variety of applications. The need of fitting circles or circular arcs to observed points arises in many areas. In medicine, one estimates the diameter of a human iris on a photograph [141] or designs a dental arch from an X-ray [21]. Archaeologists examine the circular shape of ancient Greek stadia [157] and mysterious megalithic sites (stone rings) in Britain [65, 177]; in other studies they determine the size of ancient pottery by analyzing potsherds found in field expeditions [44, 80, 81, 190]. In industry, quality control requires estimation of the radius and the center of manufactured mechanical parts [119] (other interesting applications involve microwave engineering [54, 169] and the lumber industry [13]).

In nuclear physics, one deals with elementary particles born in an accelerator — they move along circular arcs in a constant magnetic field; physicists determine the energy of the particle by measuring the radius of its trajectory; to this end experimenters fit an arc to the trace of the electrical signals the particle leaves in special detectors [53, 45, 106]. In mobile robotics, one detects round objects (pillars, tree trunks) by analyzing range readings from a 2D laser range finder used by a robot [197]. In computer vision, one uses a sequence of arcs "stitched together" (a "circular spline") to approximate more complex curved shapes [12, 145, 158, 164, 165].

Different practical requirements. All these applications involve fitting circles to planar images, but the character of data and the requirements may differ widely. In some cases one deals with data points sampled along a full circle (like in a human iris image or a log in lumber industry). In other cases the data are badly incomplete — one observes just a small arc of a big circle.

In some applications (e.g., in medicine) the fit must be very accurate and the computational cost is not an issue. Others are characterized by mass data processing (for instance, in high energy physics, millions of particle tracks per day can be produced by an accelerator), and the processing speed is of paramount importance. These variations certainly dictate different approaches to the fitting problem, as we will see below.

3.2   Parametrization

Standard parameters. To fit a circle to observed points $\{(x_i, y_i)\}$ by orthogonal least squares one minimizes the sum of squares

$$\mathscr{F} = \sum_{i=1}^{n} d_i^2, \tag{3.1}$$

where d_i is the (geometric) distance from the data point (x_i, y_i) to the hypothetical circle. The canonical equation of a circle is

$$(x-a)^2 + (y-b)^2 = R^2, \tag{3.2}$$

where (a,b) is its center and R its radius; then the (signed) distance is given by

$$d_i = \sqrt{(x_i - a)^2 + (y_i - b)^2} - R. \tag{3.3}$$

Note that $d_i > 0$ for points outside the circle and $d_i < 0$ inside it. Hence

$$\mathscr{F}(a,b,R) = \sum_{i=1}^{n} \left[\sqrt{(x_i - a)^2 + (y_i - b)^2} - R \right]^2. \tag{3.4}$$

Elimination of R. This function is just a quadratic polynomial in R, hence $\mathscr{F}$ has a unique global (conditional) minimum in R, when the other two variables a and b are kept fixed. That conditional minimum can be easily found. If we denote

$$r_i = r_i(a,b) = \sqrt{(x_i - a)^2 + (y_i - b)^2}, \tag{3.5}$$

then the minimum of $\mathscr{F}$ with respect to R is attained at

$$\hat{R} = \bar{r} = \frac{1}{n} \sum_{i=1}^{n} r_i. \tag{3.6}$$

This allows us to eliminate R and express $\mathscr{F}$ as a function of a and b only:

$$\mathscr{F}(a,b) = \sum_{i=1}^{n} \left[r_i - \bar{r} \right]^2$$

$$= \sum_{i=1}^{n} \left[\sqrt{(x_i - a)^2 + (y_i - b)^2} - \frac{1}{n} \sum_{j=1}^{n} \sqrt{(x_j - a)^2 + (y_j - b)^2} \right]^2 \tag{3.7}$$

This is still a complicated expression, and it cannot be simplified any further. The minimization of (3.7) is a nonlinear problem that has no closed form solution.

The little advantage of (3.7) over (3.4) is that a function of two variables, $\mathscr{F}(a,b)$, can be easily graphed and examined visually (which we will do below), while for a function $\mathscr{F}(a,b,R)$ such a visual inspection is difficult.

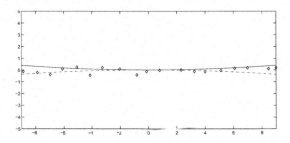

Figure 3.1 *Data points (diamonds) are sampled along a very small arc of a big circle. The correct fit (the solid line) and the wrong fit (the dashed line) have centers on the opposite sides of the data set.*

In this chapter we discuss theoretical properties of the objective function $\mathscr{F}$ and its minimum; in the next two chapters we present practical methods of computing the minimum of $\mathscr{F}$. Our first theme is the analysis of various parametrization schemes used in circle fitting applications.

Drawbacks of the standard parameters. The geometric parameters (a,b,R) of a circle are standard; they are simple and describe all circles in the plane. There is a problem, however, if one fits circles of very large radii. This may happen when the data are sampled along a small arc (of say less than 5 degrees), see Fig. 3.1; then the radius of the fitted circle can potentially take arbitrarily large values. In that case a small perturbation of data points or a small inaccuracy of the fitting procedure may result in the circle center being on the "wrong side" (see the dashed arc in Fig. 3.1).

In addition, numerical computations become unreliable: a catastrophic loss of accuracy may occur when two large nearly equal quantities are subtracted in (3.3). We call this a *problematic*, or a *singular case* of a circle fit. We will see shortly that in this case even more serious problems arise than catastrophic cancelations.

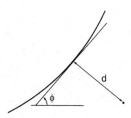

Figure 3.2: *Karimäki's parameters d and ϕ.*

Karimäki's parameters. If experiments where incomplete arcs (partially occluded circles) are frequent, statisticians often adopt different parametrization schemes. For example, in nuclear physics Karimäki [106] proposes to replace R with a signed curvature $\rho = \pm R^{-1}$ and the center (a,b) with the distance of the closest approach (d) to the origin, and the direction of propagation (φ) at the point of closest approach, see Fig. 3.2. We will describe these parameters more precisely in Section 8.8. The three parameters (ρ, d, ϕ) completely (and uniquely) describe the fitting circular arc, and they never have to take dangerously large values (unless $R \approx 0$, hence $\rho \approx \infty$, which is actually not a bothersome event).

Karimäki's parameters have a clear geometric meaning, but expressing the objective function $\mathscr{F}$ in terms of (ρ, d, ϕ) leads to cumbersome formulas; see [106] and our Section 8.8. Another pitfall in Karimäki's scheme is that for circles centered on the origin, the direction ϕ becomes indeterminate. If the center of the fitting circle gets close to the origin, numerical algorithms cannot handle the ϕ parameter adequately and may get stuck.

Algebraic parameters. A more elegant scheme was proposed by Pratt [150] and others [67], which describes circles by an algebraic equation,

$$A(x^2 + y^2) + Bx + Cy + D = 0 \qquad (3.8)$$

with an obvious constraint $A \neq 0$ (otherwise this equation describes a line) and a less obvious constraint

$$B^2 + C^2 - 4AD > 0. \qquad (3.9)$$

The necessity of the latter can be seen if one rewrites equation (3.8) as

$$\left(x+\frac{B}{2A}\right)^2 + \left(y+\frac{C}{2A}\right)^2 - \frac{B^2+C^2-4AD}{4A^2} = 0.$$

It is clear now that if $B^2+C^2-4AD < 0$, then (3.8) defines an empty set, and if $B^2+C^2-4AD = 0$, then (3.8) specifies a single point (a singleton).

Constraint. Since the parameters only need to be determined up to a scalar multiple, we can impose some constraints. The constraint $A = 1$ brings us back to the scheme (3.2). Some researchers try the constraint $D = 1$ or $A^2+B^2+C^2+D^2 = 1$ (see [67]) or $A^2+B^2+C^2 = 1$ (see [137]). However, the best choice is the constraint

$$B^2+C^2-4AD = 1, \tag{3.10}$$

because it automatically ensures (3.9). The constraint (3.10) was first proposed by Pratt [150], who clearly described its advantages.

The constraint eliminates the multiplicity of parametrization (A,B,C,D), but not entirely: each circle is now represented by exactly two parameter vectors (A,B,C,D), one being the negative of the other. If we additionally require that $A > 0$, then every circle will correspond to a unique quadruple (A,B,C,D), and vice versa.

Conversion between algebraic and geometric parameters. The conversion formulas between the natural geometric parameters (a,b,R) and the algebraic parameters (A,B,C,D) are

$$a = -\frac{B}{2A}, \qquad b = -\frac{C}{2A}, \qquad R^2 = \frac{B^2+C^2-4AD}{4A^2}, \tag{3.11}$$

or by using the constraint (3.10), $R^2 = (4A^2)^{-1}$. Conversely,

$$A = \pm\frac{1}{2R}, \qquad B = -2Aa, \qquad C = -2Ab, \qquad D = \frac{B^2+C^2-1}{4A}. \tag{3.12}$$

The distance from a data point (x_i, y_i) to the circle can be expressed, after some algebraic manipulations, by

$$d_i = \frac{2P_i}{1+\sqrt{1+4AP_i}}, \tag{3.13}$$

where

$$P_i = A(x_i^2+y_i^2) + Bx_i + Cy_i + D. \tag{3.14}$$

One can check that

$$1+4AP_i = \frac{(x_i-a)^2+(y_i-b)^2}{R^2}, \tag{3.15}$$

hence $1 + 4AP_i \geq 0$ for all i, so that (3.13) is always computable. The denominator is ≥ 1, thus the division is always safe (computationally).

The formula (3.13) is somewhat more complicated than (3.3), but it is numerically stable as it conveniently avoids catastrophic cancelations.

Natural bounds on algebraic parameters. Lastly, we show that (A, B, C, D), just as Karimäki parameters (ρ, d, φ), never have to take arbitrarily large values (under certain natural conditions). As in Section 2.2, let all the data points (x_i, y_i) be enclosed in a rectangle $\mathbb{B}$ (a "bounding box"). In many experiments, all data points are naturally confined to a window corresponding to the physical size of the measuring device, so the box $\mathbb{B}$ may be known in advance, before the data are collected. Also let

$$d_{\max} = \max_{i,j} \sqrt{(x_i - x_j)^2 + (y_i - y_j)^2}$$

denote the maximal distance between the data points.

Theorem 4 *The best fitting circle satisfies bounds*

$$|A| < A_{\max}, |B| < B_{\max}, |C| < C_{\max}, |D| < D_{\max} \qquad (3.16)$$

where $A_{\max}$, $B_{\max}$, $C_{\max}$, $D_{\max}$ are determined by the size and location of the bounding box $\mathbb{B}$ and by $d_{\max}$.

Thus whatever the configuration of the data points, the parameters (A, B, C, D) of the fitted circle stay bounded.

Proof. Due to (3.6), the best fitting circle has radius $R \geq d_{\max}/n$, hence

$$|A| \leq A_{\max} = n/2d_{\max}.$$

Let L denote the distance from the origin to the most remote point of $\mathbb{B}$. Just like in Section 2.2, it is easy to see that the the best fitting line or circle must cross $\mathbb{B}$, thus the distance from it to the origin is less than L. Substituting $x = y = 0$ into (3.13)–(3.14) gives

$$\frac{2|D|}{1 + \sqrt{1 + 4AD}} \leq L.$$

Solving this inequality for D gives

$$|D| \leq D_{\max} = 10A_{\max}L^2.$$

Lastly, (3.10) gives bounds on B and C with

$$B_{\max} = C_{\max} = \sqrt{1 + 4A_{\max}D_{\max}}.$$

The theorem is proved. $\qquad \qquad \square$

Remark. The restriction on d_{max} here is a just technicality, it is not of a critical importance. If d_{max} is small, the data points cluster together, and in the limit $d_{max} \to 0$ they merge into a single point. In this case the best fitting circle may have an arbitrarily small radius, so that A is unbounded, but practically such situations rarely occur and do not cause serious trouble.

3.3 (Non)existence

As in the previous chapter, we turn to two fundamental theoretical questions: Does the best fitting circle always exist? Is it always unique? Strangely enough, these issues have not been discussed in the literature until recently. Only in the early 2000s have they been resolved, independently, in [43, 138, 196]. The answers to the above question happen to be far less straightforward than those we found in the case of fitting lines.

Collinear data case. We begin with a rather unexpected fact: there are data sets for which the best fitting circle does not exist! The simplest example is a set of $n \geq 3$ collinear data points. No circle can interpolate more than two collinear points, so the function $\mathscr{F}$ in (3.1) is strictly positive for any (a,b,R). On the other hand, one can approximate the collinear points by a circular arc of very large radius and make $\mathscr{F}$ arbitrarily small. Thus the function $\mathscr{F}$ does not attain its minimum value. In practical terms, this means that for any circular arc fitted to a set of collinear points one can find another arc that fits them even better, and that process never stops.

Clearly, the best fit here is achieved by the line passing through all the data points (for that line $\mathscr{F} = 0$). This example suggests that if we want to fit circles to observed points we should try lines as well! Then the best fit should be chosen between all circles and lines, and the orthogonal least squares method may return either a circle or a line depending on the data set (this statement is made precise below).

Another example. Collinear points are not the only example where lines "beat" circles. Consider four points: $(X,0), (-X,0), (0,1), (0,-1)$, with some large $X \gg 1$. In this case the best fitting line (the x axis) provides $\mathscr{F} = 2$, and it can be easily seen that for any circular arc $\mathscr{F} > 2$.

Practical remarks. Admittedly, such examples are quite rare. More precisely, if the data are sampled randomly from a continuous distribution, then the probability that the orthogonal least squares method returns a line, rather than a circle, is zero. This is a mathematical fact, it will be proved in Section 3.9, see Theorem 8.

This fact explains why lines are often ignored in practice and one works with circles only. On the other hand, if the data points are obtained from a digital image (say, one fits circles to pixels on a computer screen), then lines may appear with a positive probability and have really to be reckoned with.

Lines – into the model. To add lines to our search for the best fitting circle we need to incorporate them into the parameter scheme. The natural circle parameters (a, b, R), unfortunately do not include lines. Karimäki's parameters (ρ, d, φ) incorporate lines if one allows $\rho = 0$; then d and φ are the same as our C and φ in equation (2.4). The algebraic parameter scheme (3.8) easily integrates lines, too, by setting $A = 0$, then we recover the linear equation (2.1).

Noncompactness of the parameter space. Here we provide a rigorous analysis involving the notion of compactness already employed in Section 2.2. We follow [43], and a similar argument is given by Nievergelt, see page 260 in [138].

The function $\mathscr{F}$ is obviously continuous in the circle parameters a, b, R, but the parameter space is not compact as a, b, R may take arbitrarily large values. What is worse, the "reduced" space of circles intersecting any given rectangle $\mathbb{B}$ ("bounding box") is not compact either, so the remedy we used in the case of lines (Section 2.2) will not save us anymore.

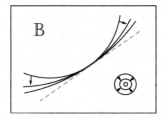

Figure 3.3: *Straightening a circular arc and shrinking a circle to a singleton.*

Direct geometric argument. The noncompactness can be again demonstrated geometrically if we introduce an appropriate topology on the space of circles. Following the pattern of (2.7)–(2.8) we say that a sequence of circles $\mathbb{S}_i$ converges to a circle $\mathbb{S}$ (or more generally, to a closed set $\mathbb{S}$) on the plane $\mathbb{R}^2$ if for any closed disk $\mathbb{D} \subset \mathbb{R}^2$ we have

$$\text{dist}_H(\mathbb{S}_i \cap \mathbb{D}, \mathbb{S} \cap \mathbb{D}) \to 0 \qquad \text{as } i \to \infty \tag{3.17}$$

where $\text{dist}_H(\cdot, \cdot)$ is again the Hausdorff distance between compact subsets of $\mathbb{R}^2$, see (2.8). Now let us take a circular arc crossing the given bounding box $\mathbb{B}$ and straighten (flatten) it by fixing one of its points and the tangent line at that point, and then increasing the radius steadily. The arc will converge to a straight line, see Fig. 3.3. Also, if one takes a sequence of concentric circles with decreasing radii, they will shrink to a point. These constructions show that if we want the space of circles crossing the box $\mathbb{B}$ be compact, we *must* include lines and singletons in it.

Existence of a minimum. From now on we work with the enlarged space containing all circles, lines, and singletons. On this space we define a topology by the same rule (3.17), in which $\mathbb{S}_i$ and $\mathbb{S}$ may denote either circles or lines or singletons.

Theorem 5 *Let $\mathbb{B}$ be a given bounding box containing all the data points. Then the 'enlarged' space of circles, lines, and singletons intersecting $\mathbb{B}$ is compact.*

Proof. Let $\{\mathbb{S}_i\}$ be a sequence of objects (circles, lines, singletons) that intersect $\mathbb{B}$. If there are infinitely many singletons in $\{\mathbb{S}_i\}$, then a subsequence of those converges to a point in $\mathbb{B}$ because the latter is compact. If there are infinitely many nonsingletons, then each of them is represented by a vector (A_i, B_i, C_i, D_i) with $A_i \geq 0$. The sequence $\{A_i\}$ contains a subsequence that either converges to a limit $\bar{A} < \infty$ or diverges to infinity. In the latter case we have a sequence of circles in $\mathbb{B}$ whose radii converge to zero, then they have a limit point in $\mathbb{B}$ just as singletons do, see above. If $\bar{A} = 0$, then we have a sequence of arcs whose radii grow to infinity, and it contains a subsequence converging to a line because the space of lines crossing $\mathbb{B}$ is compact (Section 2.2). Lastly, if $0 < \bar{A} < \infty$, then we have a sequence of circles crossing $\mathbb{B}$ whose radii converge to $\bar{R} = 1/(2\bar{A})$. Then their centers must stay within distance $2\bar{R}$ from the box $\mathbb{B}$, hence there is a subsequence of these circles whose centers converge to a limit point (a, b). Therefore our subsequence converges to the circle $(a, b, \bar{R})$. The theorem is proved. □

We remark that singletons need not really be involved, as they never provide the minimum of the objective function $\mathscr{F}$, unless all the data points collapse, but even in that case any line or circle through that collapsed data point would provide the best fit $\mathscr{F} = 0$ anyway.

Next, again as in Section 2.2 suppose a circle $\mathbb{S}$ (or a line $\mathbb{L}$) does not cross the box $\mathbb{B}$. Then it cannot provide a minimum of $\mathscr{F}$, as one can just move $\mathbb{S}$ (or $\mathbb{L}$) closer to the box and thus reduce all the distances from the data points to $\mathbb{S}$ (or $\mathbb{L}$). So we can restrict our search of the minimum of $\mathscr{F}$ to the space of circles and lines that intersect the bounding box $\mathbb{B}$. Since it is compact we obtain:

Theorem 6 *The objective function $\mathscr{F}$ always attains its minimum, though it may be attained either on a circle or on a line.*

This resolves the existence issue in a satisfactory way.

3.4 Multivariate interpretation of circle fit

The conclusions of the previous section can be illustrated by a multidimensional geometric construction similar to the one described in Section 1.5.

'Megaspace'. Given n data points $(x_1, y_1), \ldots, (x_n, y_n)$, we can represent them by one point ('megapoint') $\mathscr{X}$ in the $2n$-dimensional space $\mathbb{R}^{2n}$ with co-ordinates $x_1, y_1, \ldots, x_n, y_n$. Recall that the orthogonal line fit corresponds to projecting the point $\mathscr{X}$ onto an $(n+2)$-dimensional quadratic surface (manifold) $\mathbb{P} \subset \mathbb{R}^{2n}$.

We now interpret the orthogonal circle fit in a similar manner. Consider the subset $\mathbb{Q} \subset \mathbb{R}^{2n}$ defined by

$$(x_1, y_1, \ldots, x_n, y_n) \in \mathbb{Q} \iff \exists a, b, R \colon (x_i - a)^2 + (y_i - b)^2 = R^2 \ \forall i,$$

i.e., $\mathbb{Q}$ consists of all $(x_1, y_1, \ldots, x_n, y_n) \in \mathbb{R}^{2n}$ such that the n planar points $(x_1, y_1), \ldots, (x_n, y_n)$ lie on a circle (we call such points "co-circular," which is perhaps an unconventional term).

Projection in the megaspace. Now the best fitting circle minimizes the sum

$$\sum_{i=1}^{n} (x_i - \tilde{x}_i)^2 + (y_i - \tilde{y}_i)^2.$$

where the points $(\tilde{x}_1, \tilde{y}_1), \ldots, (\tilde{x}_n, \tilde{y}_n)$ are constrained to lie on one circle. Therefore, the true values $\tilde{x}_1, \tilde{y}_1, \ldots, \tilde{x}_n, \tilde{y}_n$ are represented by one megapoint (we denote it by $\tilde{\mathscr{X}}$) in $\mathbb{Q}$. The orthogonal circle fitting procedure corresponds to choosing a megapoint $\tilde{\mathscr{X}} \in \mathbb{Q}$ closest to the megapoint $\mathscr{X} \in \mathbb{R}^{2n}$ representing the data. In other words, we just need to project the point $\mathscr{X}$ onto $\mathbb{Q}$ orthogonally.

Dimensionality of $\mathbb{Q}$. It takes a little effort to verify that $\mathbb{Q}$ is specified by $n - 3$ independent relations

$$\det \begin{bmatrix} x_i - x_{i+1} & y_i - y_{y+1} & x_i^2 - x_{i+1}^2 + y_i^2 - y_{y+1}^2 \\ x_i - x_{i+2} & y_i - y_{y+2} & x_i^2 - x_{i+2}^2 + y_i^2 - y_{y+2}^2 \\ x_i - x_{i+3} & y_i - y_{y+3} & x_i^2 - x_{i+3}^2 + y_i^2 - y_{y+3}^2 \end{bmatrix} = 0 \qquad (3.18)$$

for $i = 1, \ldots, n - 3$, each of which means the co-circularity of the four planar points (x_i, y_i), (x_{i+1}, y_{i+1}), (x_{i+2}, y_{i+2}), and (x_{i+3}, y_{i+3}). The relations (3.18) are polynomials of fourth degree, hence $\mathbb{Q}$ is an $(n+3)$-dimensional algebraic manifold in $\mathbb{R}^{2n}$ defined by *quartic* polynomial equations.

Relation between $\mathbb{P}$ and $\mathbb{Q}$. Note that the dimension of $\mathbb{Q}$ is one higher than that of $\mathbb{P}$, i.e., $\dim \mathbb{Q} = \dim \mathbb{P} + 1$. A closer examination shows that $\mathbb{P}$ plays the role of the boundary of $\mathbb{Q}$, i.e., $\mathbb{Q}$ terminates on $\mathbb{P}$. Indeed, imagine n co-circular points lying on a small arc of a very large circle. Now let us straighten (flatten) that arc as illustrated in Fig. 3.3, and move our n points with the arc. These points are represented by one megapoint $\mathscr{X} \in \mathbb{Q}$, and when our arc is flattening, the point $\mathscr{X}$ is moving (sliding) on the surface $\mathbb{Q}$.

As the arc transforms into a straight line (the dashed line in Fig. 3.3), our moving point $\mathscr{X} \in \mathbb{Q}$ leaves the surface $\mathbb{Q}$ and instantaneously hits the manifold $\mathbb{P}$ representing all sets of n collinear points. Thus $\mathbb{Q}$ borders on $\mathbb{P}$, or $\mathbb{P}$ serves as the boundary (frontier) of $\mathbb{Q}$.

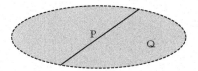

Figure 3.4: *A grey disk, $\mathbb{Q}$, is cut in half by a line, $\mathbb{P}$.*

Moreover, in the above construction we can approach the limit straight line from each of its two sides. Only a one-sided approach is shown in Fig. 3.3, but we can reflect all of the arcs across the dashed line and get another sequence of arcs converging to the same line from the other side. That shows that there are two parts of the surface $\mathbb{Q}$ that terminate on $\mathbb{P}$; they approach $\mathbb{P}$ from the opposite directions.

A schematic illustration is shown in Fig. 3.4. The grey surface (disk) represents $\mathbb{Q}$; it is cut into two halves by a black line that represents $\mathbb{P}$. This picture shows how the surface $\mathbb{Q}$ borders on the line $\mathbb{P}$ approaching it from two sides.

Geometric description of $\mathbb{P}$ and $\mathbb{Q}$. With some degree of informality, one can say that $\mathbb{P}$ cuts right through $\mathbb{Q}$ and 'divides' it into two pieces. (This description is purely local, we do not mean to say that $\mathbb{Q}$, as a whole, consists of two pieces; in fact it is a connected manifold, as every circle can be continuously transformed into any other circle).

Actually, if one examines the equations (3.18) closely, it becomes clear that they describe not only all sets of n co-circular points in the plane, *but also* all sets of n collinear points. In other words, equations (3.18) describe both manifolds, $\mathbb{Q}$ and $\mathbb{P}$ (that is, their union $\mathbb{Q} \cup \mathbb{P}$). This is another way to convince yourself that lines must be naturally treated as particular types of circles (we can call lines, say, "degenerate circular arcs").

Now it is easy to see why the best fitting circle may not exist (if lines are not treated as circles). Given a data point $\mathscr{X} \in \mathbb{R}^{2n}$ we need to project it orthogonally onto $\mathbb{Q}$. The projection is well defined unless the point $\mathscr{X}$ happens to be right above $\mathbb{P}$; then its projection of $\mathscr{X}$ onto $\mathbb{Q}$ falls exactly into the border of $\mathbb{Q}$, i.e., into $\mathbb{P}$.

3.5   (Non)uniqueness

After we have seen multiple lines fitted to a data set in Section 2.2 it may not be surprising to find out that there can be multiple fitted circles as well. However examples are much harder to construct; we review the ones from [43].

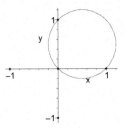

Figure 3.5: *A data set for which the objective function has four minima.*

Example of multiple circle fits. Let four data points $(\pm 1, 0)$ and $(0, \pm 1)$ make a square centered at the origin. We place another $k \geq 4$ points identically at the origin $(0,0)$ to have a total of $n = k + 4$ points.

This configuration is invariant under the rotation $\mathscr{R} = \mathscr{R}_{\pi/2}$ through the right angle about the origin. Hence, if a circle $\mathbb{S}$ minimizes $\mathscr{F}$, then by rotating that circle through $\pi/2$, π, and $3\pi/2$ we get three other circles $\mathscr{R}(\mathbb{S})$, $\mathscr{R}^2(\mathbb{S})$, and $\mathscr{R}^3(\mathbb{S})$ that minimize $\mathscr{F}$ as well. We need to check that $\mathscr{R}(\mathbb{S}) \neq \mathbb{S}$, so that we get truly distinct circles; in addition we will make sure that $\mathbb{S}$ is a circle, not a line. This involves some elementary calculations.

Note that $\mathscr{R}(\mathbb{S}) = \mathbb{S}$ if and only if the circle $\mathbb{S}$ is centered on the origin, so we need to show that such circles cannot minimize $\mathscr{F}$. If a circle has radius r and center at $(0,0)$, then $\mathscr{F} = 4(1 - r^2) + kr^2$. The minimum of this function is attained at $r = 4/(k+4)$, and it equals $\mathscr{F}_0 = 4k/(k+4)$. Assuming that $k \geq 4$ we can guarantee that the minimum is ≥ 2.

Also, the best fitting lines here pass through the origin and give $\mathscr{F}_1 = 2$, as it follows from the material of Chapter 2.

To conclude our argument it is enough to find a circle on which $\mathscr{F} < 2$. Consider the circle passing through three points $(0,0)$, $(0,1)$, and $(1,0)$. It only misses two other points, $(-1,0)$ and $(0,-1)$, and it is easy to see that for this circle $\mathscr{F} < 2$. Since $\mathscr{F}$ takes on values that are less than $\mathscr{F}_1$ and $\mathscr{F}_0$ (whenever $k \geq 4$), the best fit will be a circle (not a line), and that circle will not be centered at the origin. We do not find the best fitting circle, but our argument shows that it is not unique.

Fig. 3.5 illustrates the above example, and Fig. 3.6 plots $\mathscr{F}(a,b)$ where four distinct minima are clearly visible.

Other examples. If we change the example replacing the square with a rectangle, then we can obtain $\mathscr{F}$ that has exactly two minima. By replacing the square with a regular m-gon and increasing the number of identical points at $(0,0)$ we can construct $\mathscr{F}$ with exactly m minima for any $m \geq 3$.

In fact, for $m = 3$ the corresponding examples are quite simple; they have been described by Nievergelt [138] and Zelniker and Clarkson [196]. One can even find the centers and radii of all the best fitting circles explicitly [138].

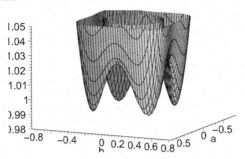

Figure 3.6: *The objective function with four minima.*

Practical remarks. Of course, if the data points are sampled randomly from a continuous probability distribution, then the chance that the objective function $\mathscr{F}$ has multiple minima is negligible (strictly speaking, it is zero). In particular, small random perturbations of the data points in our example on Fig. 3.5 will slightly change the values of $\mathscr{F}$ at its minima, so that one of them will become a global (absolute) minimum, and three others local (relative) minima.

We note, however, that while the cases of multiple minima are indeed exotic, they demonstrate that the global minimum of $\mathscr{F}$ may change abruptly if one just slightly perturbs data points, recall an example of Section 2.2. This fact was also pointed out by Nievergelt, see page 261 in [138].

3.6 Local minima

Three potential troublemakers. Recall that the minimization of the objective function (3.1) is a nonlinear problem that has no closed form solution. Hence there is no finite algorithm that computes the minimum of $\mathscr{F}$. There are plenty of iterative and approximative methods that solve this problem, which will be reviewed in the next two chapters.

Here we do a "reconnaissance:" We see what may cause trouble in practical computations. There are three major ways in which conventional iterative procedures may fail to find the (global) minimum of a given function $\mathscr{F}$:

(a) they converge to a local minimum of $\mathscr{F}$;

(b) they slow down or stall on a nearly flat plateau or in a valley;

(c) they diverge to infinity if the domain of $\mathscr{F}$ is unbounded.

Perhaps the option (a) is the most obvious and frequently discussed in the literature. In many applications iterative procedures tend to be trapped by local minima and return false solutions.

One might assume that a function defined by such a complicated expression

	5	10	15	25	50	100
0	0.879	0.843	0.883	0.935	0.967	0.979
1	0.118	0.149	0.109	0.062	0.031	0.019
≥ 2	0.003	0.008	0.008	0.003	0.002	0.002

Table 3.1 *Frequency of appearance of 0, 1, 2 or more local minima of $\mathscr{F}$ when $n = 5, \ldots, 100$ points are generated randomly with a uniform distribution.*

as (3.7) would have many local minima, and the number of local minima would grow with the sample size n. Surprisingly, detailed studies show that this is not the case, in fact quite the opposite is true.

A numerical search for local minima. An extensive investigation of local minima of the objective function $\mathscr{F}$ was undertaken in [43]. First, the authors plotted $\mathscr{F}$ (in a format similar to Fig. 3.6) for a large number of randomly generated samples and visually inspected the plots; in most cases no local minima were visible (Fig. 3.6 is exceptional; it depicts $\mathscr{F}$ for a very special, handmade configuration of points).

Next, a sweeping search for local minima was conducted [43] in a machine experiment organized as follows. First, n data points were generated randomly from a certain probability distribution. Then the most reliable iterative algorithm (Levenberg-Marquard, which is described in the next chapter) was launched starting at 1000 different, randomly selected initial points (guesses). The idea was that if there were a local minimum, then at least some of the random initial guesses would fall into its vicinity, attracting the iterations to that minimum. So every point of convergence was recorded as a minimum (local or global) of $\mathscr{F}$. If there were more than one point of convergence, then one of them was the global minimum and the others were local minima. If the algorithm converged to the same limit from all the 1000 random initial guesses, the sample was classified as having no local minima.

This search for local minima was then repeated for 10000 different random samples of n data points generated from the same probability distribution. Then the fraction of simulated samples having 0, 1, 2, etc. local minima was determined, and the results were recorded for the given n and the given probability distribution. This experiment was conducted for different values of n and different probability distributions.

Table 3.1 shows the fraction of simulated samples where $\mathscr{F}$ had 0, 1, 2 or more local minima for $n = 5, \ldots, 100$ data points; the probability distribution was uniform in the unit square $0 < x, y < 1$ (the size and location of the square do not matter, due to the invariance of the circle fit under translations, rotations, and similarities; see Section 4.11).

Rarity of local minima. These results present a remarkable picture: local minima are found in less than 15% of generated samples! The highest concentration of local minima (but still below 15%) is recorded for $n = 10$ points, and it quickly decreases as n grows; for more than 100 points, samples with local minima are virtually nonexistent. Multiple local minima turn up very rarely, if at all. The maximal number of local minima observed in that large experiment was four, and that happened only a few times in millions of random samples tested.

Generating points in a square with a uniform distribution produces completely irregular ("chaotic") samples without any predefined pattern. This is, in a sense, the worst case scenario, because different groups of random points presumably could line up along two or more different arcs thus leading to possibly distinct good fits. Still we see that this rarely happens; in > 85% of the samples the function $\mathscr{F}$ has a unique minimum.

More realistic probability distributions were tried in [43], too, in which random samples were generated along a circular arc with some noise. For example, if $n = 10$ points are sampled along a 90° circular arc of radius $R = 1$ with a Gaussian noise at level $\sigma = 0.05$, then the frequency of appearance of local minima was as low as 0.001. And this was one of the "worst" cases; in other realistically looking samples, local minima virtually never occurred.

These studies clearly demonstrate that in typical applications the objective function $\mathscr{F}$ is most likely to have a unique (global) minimum and no local minima. Thus local minima are not a real danger; it can be reasonably assumed that in $\mathscr{F}$ has a unique (global) minimum.

Does this mean that standard iterative algorithms, such as the steepest descent or the Nelder-Mead simplex or Gauss-Newton or Levenberg-Marquardt, would converge to the global minimum from any starting point? Unfortunately, this is not the case, as we demonstrate in the next section.

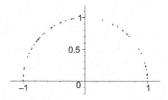

Figure 3.7: *A simulated data set of 50 points.*

3.7 Plateaus and valleys

Here we describe the shape of the objective function $\mathscr{F}(a,b)$ defined by (3.7) for typical samples in order to identify possible troubles that iterative algorithms may run into. We follow [43].

Visual inspection. Fig. 3.7 presents a generic random sample of $n = 50$ points generated along a circular arc (the upper half of the unit circle $x^2 + y^2 = 1$) with a Gaussian noise added at level $\sigma = 0.01$. Fig. 3.8 and Fig. 3.9 show the graph of $\mathscr{F}$ plotted by MAPLE in two different scales. One can clearly see that $\mathscr{F}$ has a global minimum close to $a = b = 0$ and no local minima. Fig. 3.10 presents a flat grey scale contour map, where darker colors correspond to deeper parts of the graph (smaller values of $\mathscr{F}$).

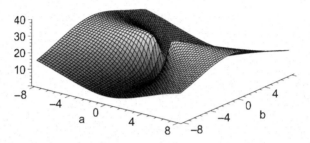

Figure 3.8: *The objective function $\mathscr{F}$ for the data set shown in Fig. 3.7 (large view).*

Plateaus. Fig. 3.8 shows that the function $\mathscr{F}$ does not grow as $a, b \to \infty$. In fact, it is bounded (see the proof of Theorem 7 below), i.e., $\mathscr{F}(a,b) \leq \mathscr{F}_{\max} < \infty$. The boundedness of $\mathscr{F}$ actually explains the appearance of large nearly flat plateaus and valleys on Fig. 3.8 that stretch out to infinity in some directions. If an iterative algorithm starts somewhere in the middle of such a plateau or a valley, or gets there by chance, it will have a hard time moving at all, since the gradient of $\mathscr{F}$ will almost vanish. We indeed observed conventional algorithms getting "stuck" on flat plateaus or valleys in our tests.

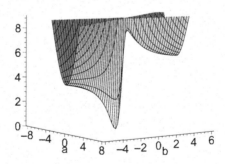

Figure 3.9 *The objective function $\mathscr{F}$ for the data set shown in Fig. 3.7 (a vicinity of the minimum).*

Two valleys. Second, there are two particularly interesting valleys that stretch roughly along the line $a = 0$ in Fig. 3.8 through Fig. 3.10. One of them, corresponding to $b < 0$, has its bottom point right at the minimum of $\mathscr{F}$. The function $\mathscr{F}$ slowly decreases along the valley as it approaches the minimum. Hence, any iterative algorithm starting in that valley or getting there by chance should, ideally, find its way downhill and arrive at the minimum of $\mathscr{F}$.

The other valley corresponds to $b > 0$, it is separated from the global minimum of $\mathscr{F}$ by a ridge. The function $\mathscr{F}$ slowly decreases along that valley as b grows. Hence, any iterative algorithm starting in *that* valley or getting there "by accident" will be forced to move up along the b axis, away from the minimum of $\mathscr{F}$, and diverge to infinity.

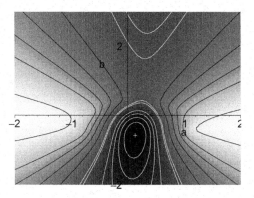

Figure 3.10 *A grey-scale contour map of the objective function* $\mathscr{F}$. *Darker colors correspond to smaller values of* $\mathscr{F}$. *The minimum is marked by a cross.*

Wrong valley perils. If an iterative algorithm starts at a randomly chosen point, it may go down either valley, and there is a chance that it will descend into the second (wrong) valley and then diverge. For the sample on Fig. 3.7, the authors of [43] applied the Levenberg-Marquardt algorithm starting at a randomly selected initial guess within the square 5×5 about the centroid $(\bar{x}, \bar{y})$ of the data. They found that the algorithm indeed escaped to infinity with probability about 50%.

Existence of the escape valley. Unfortunately, such "escape valleys" are almost inevitable: a detailed theoretical analysis shows that for *every* typical data set the graph of $\mathscr{F}$ contains an *escape valley*. This is a mathematical theorem stated below.

Theorem 7 (Two Valley Theorem) *For every typical data set, there is a pair of valleys on the graph of the objective function* $\mathscr{F}(a,b)$ *stretching out in opposite directions, so that one valley descends to the minimum of* $\mathscr{F}$, *while the other valley descends toward infinity.*

The exact meaning of the word "typical" will be clarified in the the proof. Our proof is quite long, but it reveals many useful facts beyond the existence of the two valleys, so we present it in full in the next section.

3.8   Proof of Two Valley Theorem

This section is devoted to the proof of Theorem 7. We will find the two valleys by examining the behavior of the objective function $\mathscr{F}$ for large a and b, i.e., as $a^2 + b^2 \to \infty$. In that case $R \to \infty$ as well, due to (3.6), and the fitting arc is close to a straight line, so we can use some facts established in Chapter 2.

Taylor expansion. We introduce a new variable $D^2 = a^2 + b^2$ and express $a = D\cos\varphi$ and $b = D\sin\varphi$ for some $\varphi \in [0, 2\pi)$. Then (3.3) takes form

$$d_i = \sqrt{(x_i - D\cos\varphi)^2 + (y_i - D\sin\varphi)^2} - R$$
$$= \sqrt{z_i - 2u_i D + D^2} - R,$$

where we use shorthand notation

$$z_i = x_i^2 + y_i^2 \qquad \text{and} \qquad u_i = x_i\cos\varphi + y_i\sin\varphi.$$

Expanding d_i into a Taylor series gives

$$d_i = D - u_i + \frac{z_i - u_i^2}{2D} - R + \mathscr{O}(1/D^2).$$

Note that

$$z_i - u_i^2 = v_i^2 \qquad \text{where} \qquad v_i = -x_i\sin\varphi + y_i\cos\varphi.$$

Due to (3.6) we have $\sum d_i = 0$, hence

$$D - R = \bar{u} - \frac{\overline{vv}}{2D} + \mathscr{O}(1/D^2),$$

where we use "sample mean" notation $\bar{u} = \frac{1}{n}\sum u_i$, $\overline{vv} = \frac{1}{n}\sum v_i^2$, etc. Introducing another new variable $\delta = 1/D$ we obtain

$$d_i = -(u_i - \bar{u}) + \delta(v_i^2 - \overline{vv})/2 + \mathscr{O}(\delta^2).$$

Squaring and averaging over i gives

$$\frac{1}{n}\mathscr{F} = \overline{uu} - \bar{u}^2 - \left[\overline{uvv} - \bar{u}\,\overline{vv}\right]\delta + \mathscr{O}(\delta^2). \tag{3.19}$$

This is a crucial expansion; we rewrite it, for brevity, as

$$\frac{1}{n}\mathscr{F} = f_0 + f_1\delta + \mathscr{O}(\delta^2).$$

Main term. The main term is

$$f_0 = \overline{uu} - \bar{u}^2$$
$$= s_{xx}\cos^2\varphi + 2s_{xy}\cos\varphi\sin\varphi + s_{yy}\sin^2\varphi,$$

where s_{xx}, s_{xy}, s_{yy} are the components of the scatter matrix $\mathbf{S}$ introduced in (1.5). In matrix form, we have

$$f_0 = \mathbf{A}^T\mathbf{S}\mathbf{A},$$

where $\mathbf{A} = (\cos\varphi, \sin\varphi)^T$, again as in Section 2.3, so we can use the results of that section.

Let $0 \le \lambda_1 \le \lambda_2$ denote the eigenvalues of $\mathbf{S}$ and $\mathbf{A}_1$, and $\mathbf{A}_2$ the corresponding eigenvectors. If $\lambda_1 = \lambda_2$, then f_0 (as a function of φ) is constant, but this only occurs when the data set admits infinitely many orthogonal fitting lines (Section 2.3). This constitutes an exceptional event (to which our theorem does not apply).

Setting the exceptional case aside, we assume that $\lambda_1 < \lambda_2$. Then f_0 is a nonconstant function of φ that has two properties:

(a) $f_0(\varphi)$ is a periodic function with period π;

(b) $f_0(\varphi)$ takes values in the interval $[\lambda_1, \lambda_2]$; its minimum value is λ_1 taken on $\mathbf{A} = \mathbf{A}_1$ and its maximum value is λ_2 taken on $\mathbf{A} = \mathbf{A}_2$.

See Fig. 3.11. By the way, the boundedness of f_0 demonstrates that $\mathscr{F}(a,b)$ is a bounded function, as we have mentioned earlier.

Since f_0 has period π, it takes its minimum on the interval $[0, 2\pi]$ twice: at a point φ_1 corresponding to the vector $\mathbf{A}_1$ and at $\varphi_1 + \pi$ corresponding to the opposite vector $-\mathbf{A}_1$. Thus there are two valleys on the graph of $\mathscr{F}$, they stretch in opposite directions ($\mathbf{A}_1$ and $-\mathbf{A}_1$); both valleys are parallel to the vector $\mathbf{A}_1$, i.e., they are orthogonal to the best fitting line (as φ is the direction of its normal, see Section 2.1).

Second order term. To examine the declination (direction of decrease) of those valleys, we need to analyze the behavior of the function $\mathscr{F}$ in their bottoms, which due to (3.19) is given by

$$\tfrac{1}{n}\mathscr{F} = \lambda_1 + f_1(\varphi)\delta + \mathscr{O}(\delta^2) \tag{3.20}$$

with $\varphi = \varphi_1$ for one valley and $\varphi = \varphi_1 + \pi$ for the other valley.

To simplify our formulas, we assume that the coordinate system is chosen so that its origin coincides with the centroid of the data, i.e., $\bar{x} = \bar{y} = 0$. In that case also $\bar{u} = \bar{v} = 0$, and we arrive at

$$f_1(\varphi) = -\overline{uvv}$$
$$= -\overline{xyy}\cos^3\varphi - (\overline{yyy} - 2\overline{xxy})\cos^2\varphi\sin\varphi$$
$$- \overline{xxy}\sin^3\varphi - (\overline{xxx} - 2\overline{xyy})\cos\varphi\sin^2\varphi.$$

Thus $f_1(\varphi)$ is a periodic function with period 2π, and it is *antiperiodic* with period π, i.e.

$$f_1(\varphi) = -f_1(\varphi + \pi).$$

See Fig. 3.1. In particular, $f_1(\varphi_1) = -f_1(\varphi_1 + \pi)$. Hence if f_1 is positive at the bottom of one valley, then it is negative at the bottom of the other, and vice versa.

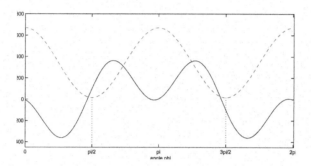

Figure 3.11 *Functions $f_0(\varphi)$ (dashed line) and $f_1(\varphi)$ (solid line) for a randomly generated data set.*

It now follows from (3.20) that if $f_1 > 0$ at the bottom of one of the two valleys, then $\mathscr{F}$ decreases as $\delta \to 0$ (i.e., as $D \to \infty$), hence the valley decreases as it stretches out to infinity. If $f_1 < 0$ along a valley, then $\mathscr{F}$ increases as $D \to \infty$, i.e., the valley increases as it stretches out to infinity.

This proves the theorem in the case $f_1(\varphi_1) \neq 0$. It remains to examine the case $f_1(\varphi_1) = 0$; this will constitute another exceptional event.

Two exceptional cases. Let us rotate the coordinate system so that its x axis coincides with the best fitting line (i.e., with the major axis of the scattering ellipse). Then $s_{xy} = 0$ and $s_{xx} > s_{yy}$, and f_0 takes its minimum at $\varphi_1 = \pi/2$, hence

$$f_1(\varphi_1) = -\overline{uvv} = -\overline{xxy}.$$

The event

$$\overline{xxy} = \frac{1}{n} \sum_{i=1}^{n} x_i^2 y_i = 0 \tag{3.21}$$

is clearly exceptional. If the data are sampled randomly from a continuous probability distribution, the event (3.21) occurs with probability zero.

This is the second exceptional even to which our theorem does not apply. Now we clarify the exact meaning of the word "typical" in the theorem's statement: it means precisely that $\lambda_1 \neq \lambda_2$ and $\overline{xxy} \neq 0$. The proof is complete. $\square$

Final remarks. We emphasize that the above theorem is very general, the "escape valley" exists whether data points are sampled along an incomplete circle (a small arc) or along an entire circle. Practically, however, in the latter case iterative algorithms are quite safe, as the escape valley lies far away from the data set, making it hard to get there.

On the other hand, for data sampled along a short arc, the escape valley gets dangerously close to the data set, so an iterative algorithm can be easily trapped in that valley.

3.9 Singular case

Here we take a closer look at the singular case arising in the circle fitting problem (described in Section 3.2), i.e., when the data are sampled along a small arc.

A word of caution. This is exactly the case where the conventional iterative algorithms are likely to be caught in the wrong ("escape") valley and diverge. This is also the case where the best fitting circle may not even exist, i.e., the best fitting line may "beat" all circles.

This case must be handled with care. One possible strategy for dealing with the singular case is described here; we use the details of the proof of Two Valley Theorem given in the previous section.

Analytic classification of valleys. First, we need to to center the coordinate system on the centroid of the data $(\bar{x}, \bar{y})$. In that coordinate system $\bar{x} = \bar{y} = 0$. Then we align the x axis with the major axis of the scattering ellipse, ensuring that $s_{xy} = 0$ and $s_{xx} > s_{yy}$. Now the best fitting line is the x axis (and on this line the function $\mathscr{F}$ takes value $\mathscr{F} = nf_0$, cf. (3.19)).

In this (adjusted) coordinate system, the value

$$\overline{xxy} - \frac{1}{n}\sum_{i=1}^{n} x_i^2 y_i$$

plays a crucial role; it is the "signature" of the data set. Precisely, we have proved the following relations:

(a) if $\overline{xxy} > 0$, then the center of the best fitting circle lies *above* the x axis; the wrong valley lies below the x axis;

(b) if $\overline{xxy} < 0$, then the center of the best fitting circle lies *below* the x axis; the wrong valley lies above the x axis.

These simple rules can be used to choose a safe initial guess for an iterative algorithm and help it avoid the wrong valley and direct it properly.

A relater issue: when lines beat circles. Another important corollary of our analysis is

Theorem 8 *In the case $\overline{xxy} \neq 0$ the best fitting circle exists, i.e., the best fitting line cannot beat the best circle.*

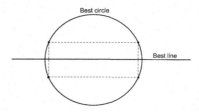

Figure 3.12: *Best circle beats the best line despite $\overline{xxy} = 0$.*

On the other hand, even if $\overline{xxy} = 0$, the best fitting circle may still exist. For instance, let $n = 4$ data points be the vertices of a rectangle, see Fig. 3.12. Then the best fitting line cuts the rectangle in half, and one can easily see that in the adjusted coordinate system $\overline{xxy} = 0$ (due to the symmetry about the x axis). However, the circumscribed circle interpolates all the four data points, so it provides a perfect fit ($\mathscr{F} = 0$), thus it beats any line.

To summarize our theoretical conclusions, we see that

(a) if $\overline{xxy} \neq 0$, then the best fitting circle always exists (and it beats any line);

(b) if $\overline{xxy} = 0$, then the best fitting circle may or may not exist, depending on the data set; i.e., it is possible that the best fitting line beats all circles.

Practically, this means that if $\overline{xxy} = 0$, then one should check the best fitting line (the x axis) as a potential solution of the circle fitting problem. If $\overline{xxy} \neq 0$, then lines can be ignored.

Practical recommendations. In fact, one should be very careful when the critical value $\overline{xxy}$ gets too close to zero. As it follows from the expansion (3.19),

$$R = \mathscr{O}(1/\overline{xxy}),$$

i.e., if $\overline{xxy}$ is too close to zero, then the radius of the best fitting circle may be extremely large; in that case an attempt to compute a, b and R may cause the catastrophic loss of accuracy described in Section 3.2.

The following simple safety checkpoint may be adopted to avoid catastrophic results. Suppose our coordinates x_i's and y_i's are of order one; in that case if an iterative algorithm reaches values $R \geq 10^8$, then further calculations would be meaningless; hence the iterations should be terminated and the algorithm should return the best fitting line (the x axis) as the solution of the circle fitting problem.

This concludes our theoretical analysis of the circle fitting problem. It will be invaluable in the assessment of practical algorithms to which we turn next.

Chapter 4

Geometric circle fits

Two types of fitting algorithms. In this and the next chapters we discuss practical solutions to the circle fitting problem. Recall that our main task is to minimize the nonlinear function $\mathscr{F}$ given by (3.1). Its minimum cannot be given in a closed form or computed by a finite algorithm. All the existing practical solutions can be divided into two large groups:

(A) iterative algorithms that are designed to converge to the minimum of

$$\mathscr{F} = \sum_{i=1}^{n} d_i^2,$$

where d_i are geometric (orthogonal) distances from the data points to the circle. The minimization of $\mathscr{F}$ is referred to as *geometric fit*;

(B) approximative algorithms that replace d_i with some other quantities, f_i, and then minimize

$$\mathscr{F}_1 = \sum_{i=1}^{n} f_i^2;$$

usually f_i are defined by simple algebraic formulas (without radicals), and the resulting solution is called *algebraic fit*.

In this chapter we describe the most popular geometric fits (type A). The next chapter is devoted to algebraic approximative methods (type B).

Comparison. The geometric fit is commonly regarded as being more accurate than algebraic fits (even though this opinion is based mostly on practical experience; our Chapter 7 provides some theoretical analysis of this issue). But every geometric fitting procedure is iterative, subject to occasional divergence, and in any case computationally expensive. On the contrary, algebraic fits are usually simple, reliable, and fast.

Algebraic fits are often used to supply an initial guess to a subsequent iterative geometric fitting routine. Some modern algebraic fits, however, are so accurate that geometric fits would not make much further improvement. In most practical applications a well designed algebraic fit (such as Taubin's or Pratt's method, see the next chapter) will do a satisfactory job.

Thus the reader who is just looking for a simple and efficient circle fit should go directly to the next chapter and check the Pratt fit or (even better) the Taubin circle fit. The reader who needs a geometric fit, regardless of its high cost, or who wants to learn this topic in general, should read this chapter.

4.1   Classical minimization schemes

We begin with a brief description of general numerical schemes used to minimize smooth functions of several variables, especially those adapted to least squares problems. First we recall two classical algorithms that are a part of any standard numerical analysis course.

Steepest descent. Suppose we need to find the minimum of a smooth function $\mathscr{G} : \mathbb{R}^k \to \mathbb{R}$, i.e.,

$$z = \mathscr{G}(\mathbf{a}), \qquad \mathbf{a} = (a_1, \ldots, a_k) \in \mathbb{R}^k \tag{4.1}$$

of k variables. Iterative procedures usually compute a sequence of points $\mathbf{a}^{(0)}$, $\mathbf{a}^{(1)}, \ldots$ that presumably converges to a point where $\mathscr{G}$ takes its minimum value. The starting point $\mathbf{a}^{(0)}$ (the *initial guess*) is assumed to be chosen somehow, and the procedure follows a certain rule to determine $\mathbf{a}^{(i+1)}$ given $\mathbf{a}^{(i)}$.

That is, to define a procedure, it suffices to describe the rule of constructing the next approximation, $\mathbf{a}'$, from the current approximation, $\mathbf{a}$.

We always assume that the derivatives of the function $\mathscr{G}$ can be evaluated; hence one can find the gradient vector $\nabla \mathscr{G}(\mathbf{a})$, and then the most logical move from $\mathbf{a}$ would be in the direction opposite to $\nabla\mathscr{G}(\mathbf{a})$, where the function $\mathscr{G}$ decreases most rapidly. This method is called the *steepest descent*. It can be described by a formula

$$\mathbf{a}' = \mathbf{a} - \eta\,\nabla\mathscr{G}(\mathbf{a}),$$

where $\eta > 0$ is a factor. The choice of η is based on the following general considerations.

Choosing the step length. If η is too large, one may "overstep" the region where $\mathscr{G}$ takes small values and land too far. If η is too small, the progress will be slow, but at least the function will decrease, i.e., one gets $\mathscr{G}(\mathbf{a}') < \mathscr{G}(\mathbf{a})$.

The simplest approach is to set $\eta = 1$, compute $\mathbf{a}'$, and then check if it is acceptable. If $\mathscr{G}(\mathbf{a}') < \mathscr{G}(\mathbf{a})$, the value $\mathbf{a}'$ is accepted, otherwise one backtracks by trying smaller values of η (for instance, $\eta = 1/2$, then $\eta = 1/4$, etc.) until $\mathbf{a}'$ is acceptable.

Other (clever but more expensive) methods of adjusting the factor η exist, such as golden section, Brent method, various line searches with derivatives; we refer the reader to standard books in numerical analysis, such as [151].

Generally, the steepest descent is a reliable method, but it usually converges slowly (at best, linearly).

Newton-Raphson method. If the second derivatives of $\mathscr{G}$ are available, one can compute both the gradient vector and the Hessian matrix of the second order partial derivatives:

$$\mathbf{D} = \nabla\mathscr{G}(\mathbf{a}) \qquad \text{and} \qquad \mathbf{H} = \nabla^2\mathscr{G}(\mathbf{a}) \tag{4.2}$$

and approximate $\mathscr{G}$ in a vicinity of $\mathbf{a}$ by the quadratic part of its Taylor polynomial;

$$\mathscr{G}(\mathbf{a}+\mathbf{h}) \approx \mathscr{G}(\mathbf{a}) + \mathbf{D}^T\mathbf{h} + \tfrac{1}{2}\mathbf{h}^T\mathbf{H}\mathbf{h}, \tag{4.3}$$

where $\mathbf{h} = \mathbf{a}' - \mathbf{a}$ denotes the step. Now one can choose $\mathbf{h}$ as the critical point of this quadratic approximation, i.e., find $\mathbf{h}$ by solving

$$\mathbf{D} + \mathbf{H}\mathbf{h} = 0. \tag{4.4}$$

This is the Newton-Raphson method.

It converges fast (quadratically) if the current iteration is already close enough to the minimum of $\mathscr{G}$. However, this method may run into various problems. First, just as the steepest descent, it may overstep the region where $\mathscr{G}$ takes small values and land too far, then one has to backtrack. Second, the matrix $\mathbf{H}$ may not be positive-definite, then the quadratic approximation in (4.3) will not even have a minimum: the solution of (4.4) will be a saddle point or a maximum. In that case the quadratic approximation in (4.3) seems to be quite useless.

Fortunately, the least squares problems allow an efficient way to get around the last trouble, see the next section.

4.2 Gauss-Newton method

Least squares problem. Consider a problem in which we are to minimize a function

$$\mathscr{G}(\mathbf{a}) = \sum_{i=1}^{n} g_i^2(\mathbf{a}) \tag{4.5}$$

of k variables $\mathbf{a} = (a_1, \ldots, a_k)$. We assume that $n > k$, and g_i have derivatives.

Newton-Raphson approach to the least squares problem. As in Newton-Raphson scheme, we start by approximating $\mathscr{G}(\mathbf{a} + \mathbf{h})$ by a quadratic part of Taylor polynomial:

$$\mathscr{G}(\mathbf{a}+\mathbf{h}) \approx \mathscr{G}(\mathbf{a}) + \mathbf{D}^T \mathbf{h} + \tfrac{1}{2}\mathbf{h}^T \mathbf{H}\mathbf{h}. \tag{4.6}$$

where

$$\mathbf{D} = \nabla \mathscr{G}(\mathbf{a}) = 2\sum_{i=1}^{n} g_i(\mathbf{a})\nabla g_i(\mathbf{a}) \tag{4.7}$$

is the gradient of $\mathscr{G}$ and

$$\mathbf{H} = \nabla^2 \mathscr{G}(\mathbf{a}) = 2\sum_{i=1}^{n}[\nabla g_i(\mathbf{a})][\nabla g_i(\mathbf{a})]^T + 2\sum_{i=1}^{n} g_i(\mathbf{a})\nabla^2 g_i(\mathbf{a}) \tag{4.8}$$

is the Hessian matrix of $\mathscr{G}$. The Newton-Raphson method (4.4) uses both $\mathbf{D}$ and $\mathbf{H}$.

Gauss-Newton for the least squares problem. The Gauss-Newton method drops the last sum of (4.8), i.e., it replaces $\mathbf{H}$ with

$$\mathbf{H}^\circ = 2\sum_{i=1}^{n}[\nabla g_i(\mathbf{a})][\nabla g_i(\mathbf{a})]^T. \tag{4.9}$$

To justify this replacement, one usually notes that in typical least squares applications $g_i(\mathbf{a})$ are small, hence the second sum in (4.8) is much smaller than the first, so its removal will not change the Hessian matrix $\mathbf{H}$ much. Also one notes that modifying $\mathbf{H}$ cannot alter the limit point of the procedure, it can only affect the path that the iterations take to approach that limit[1].

Advantages. Replacing $\mathbf{H}$ with $\mathbf{H}^\circ$ has two immediate advantages:

(a) The computation of second order derivatives of g_i is no longer necessary;

(b) Unlike $\mathbf{H}$, the new matrix $\mathbf{H}^\circ$ is always positive semi-definite.

---

[1] However it can (and does!) slow down the rate of convergence, especially if $g_i(\mathbf{a})$ are not so small, see more on that in the end of this section.

Assume for a moment that $\mathbf{H}^\circ$ is nonsingular. Then the quadratic expression

$$\mathscr{G}(\mathbf{a}) + \mathbf{D}^T\mathbf{h} + \tfrac{1}{2}\mathbf{h}^T\mathbf{H}^\circ\mathbf{h}$$

has a minimum, it is attained at $\mathbf{h}$ that satisfies equation

$$\mathbf{D} + \mathbf{H}^\circ\mathbf{h} = \mathbf{0}. \tag{4.10}$$

Hence $\mathbf{h} = -(\mathbf{H}^\circ)^{-1}\mathbf{D}$, and the next approximation is $\mathbf{a}' = \mathbf{a} + \mathbf{h}$.

Applying methods of linear algebra. Introducing matrix notation $\mathbf{g} = (g_1(\mathbf{a}), \ldots, g_n(\mathbf{a}))^T$ and

$$\mathbf{J} = \begin{bmatrix} \partial g_1/\partial a_1 & \cdots & \partial g_1/\partial a_k \\ \vdots & \ddots & \vdots \\ \partial g_n/\partial a_1 & \cdots & \partial g_n/\partial a_k \end{bmatrix} \tag{4.11}$$

we obtain $\mathbf{D} = 2\mathbf{J}^T\mathbf{g}$ and $\mathbf{H}^\circ = 2\mathbf{J}^T\mathbf{J}$. Therefore, $\mathbf{h}$ is the solution of

$$\mathbf{J}^T\mathbf{J}\mathbf{h} = -\mathbf{J}^T\mathbf{g}. \tag{4.12}$$

This equation corresponds to the overdetermined linear system

$$\mathbf{J}\mathbf{h} \approx -\mathbf{g}, \tag{4.13}$$

which is the classical least squares problem of linear algebra, cf. Section 1.7. Its (minimum-norm) solution is

$$\mathbf{h} = -\mathbf{J}^-\mathbf{g} = -(\mathbf{J}^T\mathbf{J})^-\mathbf{J}^T\mathbf{g} = -(\mathbf{H}^\circ)^-\mathbf{D}, \tag{4.14}$$

where $(\cdot)^-$ denotes the Moore-Penrose pseudoinverse. This formula works whether $\mathbf{H}^\circ$ is singular or not.

Remark. One can arrive at (4.12) differently. As our goal is to minimize $\mathscr{G} = \|\mathbf{g}\|^2$, one can replace $\mathbf{g}(\mathbf{a} + \mathbf{h})$ with its linear approximation $\mathbf{g}(\mathbf{a}) + \mathbf{J}\mathbf{h}$ and minimize $\|\mathbf{g}(\mathbf{a}) + \mathbf{J}\mathbf{h}\|^2$ with respect to $\mathbf{h}$; this is exactly the classical least squares problem (4.13), and its solution is given by (4.14).

Remark. The system of "normal equations"

$$\mathbf{J}^T\mathbf{J}\mathbf{h} = -\mathbf{J}^T\mathbf{g}$$

can be solved numerically in several ways. A simple inversion of the matrix $\mathbf{J}^T\mathbf{J}$ or the Gaussian elimination method are not recommended, since they do not take advantage of the positive semi-definiteness of $\mathbf{J}^T\mathbf{J}$ and are prone to large round-off errors. A better method is Cholesky factorization of $\mathbf{J}^T\mathbf{J}$, it is fairly fast and accurate. Its drawback is somewhat poor performance when the matrix $\mathbf{J}$ happens to be ill-conditioned. Then one can resort to numerically stable methods of linear algebra: QR decomposition or SVD; see [78].

For example, denote by $\mathbf{J} = \mathbf{U}\Sigma\mathbf{V}^T$ the (short) singular value decomposition (SVD) of the matrix $\mathbf{J}$, where $\mathbf{U}$ is a (rectangular) $n \times k$ orthogonal matrix, $\mathbf{V}$ is a (small) $k \times k$ orthogonal matrix, and Σ is a diagonal $k \times k$ matrix. Then

$$\mathbf{h} = -\mathbf{V}\Sigma^{-1}\mathbf{U}^T\mathbf{g}.$$

Speed of convergence. Many authors assert that the Gauss-Newton method, just like its Newton-Raphson prototype, converges quadratically, but this is not exactly true. The modification of $\mathbf{H}$, however small, does affect the asymptotic speed of convergence, and it becomes linear, see [23]. Precisely, if $\mathbf{a}^*$ denotes the limit point, then one can only guarantee that

$$\|\mathbf{a}' - \mathbf{a}^*\| < c\|\mathbf{a} - \mathbf{a}^*\|$$

with some $c < 1$. However, the convergence constant c here is proportional to $\mathscr{G}(\mathbf{a}^*)$, hence it is actually close to zero when $g_i(\mathbf{a}^*)$'s are small. That does not make the convergence quadratic, but with some degree of informality it can be described as *nearly* quadratic.

4.3　Levenberg-Marquardt correction

The Gauss-Newton method works well under favorable conditions, in which case its convergence is fast, but it may still overstep the region where $\mathscr{G}$ takes small values, as we noted above, and its performance may be problematic if the design matrix $\mathbf{N} = \mathbf{J}^T\mathbf{J}$ happens to be near-singular.

Augmenting the design matrix. The Levenberg-Marquardt correction aims at eliminating these drawbacks. The design matrix $\mathbf{N}$ is augmented to

$$\mathbf{N}_\lambda = \mathbf{N} + \lambda\mathbf{I}, \tag{4.15}$$

where $\lambda > 0$ is an additional control parameter and $\mathbf{I}$ is the $k \times k$ identity matrix. In other words, the diagonal entries of $\mathbf{N}$ are increased by λ. Then, instead of (4.12), one solves the new system

$$\mathbf{N}_\lambda\mathbf{h} = -\mathbf{J}^T\mathbf{g} \tag{4.16}$$

to determine $\mathbf{h}$. Note that the matrix $\mathbf{N}_\lambda$, with $\lambda > 0$, is always positive definite (while $\mathbf{N}$ is only guaranteed to be positive semi-definite), and in fact all the eigenvalues of $\mathbf{N}_\lambda$ are $\geq \lambda$.

Checkpoint. After $\mathbf{h}$ has been computed, the algorithm passes through a checkpoint. If the new approximation $\mathbf{a}' = \mathbf{a} + \mathbf{h}$ reduces the value of $\mathscr{G}$, i.e.,

---

1. Initialize $\mathbf{a}_0$ and λ_0, set $k = 0$.
2. At the current point $\mathbf{a}_k$ compute the vector $\mathbf{g}_k$ and its gradient $\mathbf{J}_k$.
3. Compute $\mathbf{h}_k$ by solving equation $(\mathbf{J}_k^T \mathbf{J}_k + \lambda_k \mathbf{I})\mathbf{h}_k = -\mathbf{J}_k^T \mathbf{g}_k$.
4. Compute the vector $\mathbf{g}'$ at the point $\mathbf{a}' = \mathbf{a}_k + \mathbf{h}_k$.
5. If $\|\mathbf{g}'\|^2 \geq \|\mathbf{g}_k\|^2$, reset $\lambda_k := \beta \lambda_k$ and return to Step 3.
6. Update $\lambda_{k+1} = \alpha \lambda_k$ and $\mathbf{a}_{k+1} = \mathbf{a}_k + \mathbf{h}_k$, increment k, return to Step 2.

Table 4.1: *Levenberg-Marquardt algorithm.*

if $\mathscr{G}(\mathbf{a}') < \mathscr{G}(\mathbf{a})$, it is accepted and λ is decreased by a certain factor α before the next iteration (suppressing the corrective term $\lambda \mathbf{I}$).

Otherwise the new value $\mathbf{a}' = \mathbf{a} + \mathbf{h}$ is rejected, λ is increased by a certain factor β and the augmented normal equations $\mathbf{N}_\lambda \mathbf{h} = -\mathbf{J}^T \mathbf{g}$ are solved again. These recursive attempts continue until the increment $\mathbf{h}$ leads to a smaller value of $\mathscr{G}$. This is bound to happen, since for large λ the method approaches the steepest descent.

Advantages. In other words, when λ increases, the recomputed vector $\mathbf{h}$ not only gets smaller but also turns (rotates) and aligns somewhat better with the negative gradient vector $-\nabla \mathscr{G}(\mathbf{a})$. As $\lambda \to \infty$, the length of $\mathbf{h}$ approaches zero and its direction converges to that of $-\nabla \mathscr{G}(\mathbf{a})$ (Fig. 4.1).

We see that the Levenberg-Marquardt correction combines two classical ideas:

(a) The quadratic approximation to the function, which works well in a vicinity of its minimum and yields a fast (nearly quadratic) convergence.

(b) The steepest descent scheme that ensures reliability in difficult cases.

As the algorithm is based on a reasonable balance between these two principles, it is sometimes referred to as Marquardt compromise.

Practical issues. In many implementations, the parameter λ is initialized to a small value, e.g., 10^{-3} or 10^{-4}. A common choice for α and β is $\alpha = 0.1$ and $\beta = 10$.

Remark. Just like equations (4.12) in the previous section, the new system $\mathbf{N}_\lambda \mathbf{h} = -\mathbf{J}^T \mathbf{g}$ can be solved numerically in many ways, including Cholesky factorization, QR decomposition, and SVD. The use of QR and SVD requires a trick, though, because the system (4.16) is not in the right form (yet). The trick is to rewrite (4.16) as

$$\mathbf{J}_\lambda^T \mathbf{J}_\lambda \mathbf{h} = -\mathbf{J}_\lambda^T \mathbf{g}_0, \tag{4.17}$$

where the matrix $\mathbf{J}_\lambda$ is obtained by appending the $k \times k$ scalar matrix $\sqrt{\lambda}\,\mathbf{I}$ to the bottom of $\mathbf{J}$, and the vector $\mathbf{g}_0$ is obtained by extending $\mathbf{g}$ with k trailing zeros:

$$\mathbf{J}_\lambda = \left[\begin{array}{c} \mathbf{J} \\ \sqrt{\lambda}\,\mathbf{I} \end{array} \right], \qquad \mathbf{g}_0 = \left[\begin{array}{c} \mathbf{g} \\ \mathbf{0} \end{array} \right].$$

Now the system (4.17) is equivalent to the least squares problem

$$\mathbf{J}_\lambda \mathbf{h} \approx -\mathbf{g}_0,$$

whose solution is $\mathbf{h} = -\mathbf{J}_\lambda^- \mathbf{g}_0$, which can be computed by QR or SVD (as we described in the end of the previous section).

History. The Levenberg-Marquardt correction was introduced in the middle of the twentieth century: it was invented by Levenberg [122] in 1944, rediscovered and enhanced by Marquardt [131] in 1963. It was popularized in the 1970s; see e.g., [71]. This method and its variant called the *trust region*, see the next section, dominate the literature on least squares applications in the past decades.

4.4 Trust region

To complete our survey of general minimization schemes, we will describe the trust region method, which currently constitutes the state of the art.

Motivation. The Levenberg-Marquardt algorithm is flexible enough to avoid obvious pitfalls of the classical minimization schemes and virtually guarantees convergence to a minimum of the objective function. Its drawback, though, is that the control parameter λ is just an abstract variable whose values have no apparent relation to the problem at hand. Therefore it is hard to properly initialize λ. Also, the Levenberg-Marquardt simplistic rules for updating λ (by arbitrarily chosen factors α and β) often cause erratic, suboptimal performance.

A variant of the Levenberg-Marquardt method was developed in the 1970s that fixed the above drawback. It was popularized by More in his well written 1978 paper [133]. Eventually this method came to be known as *trust region* and was adopted in nearly all standard software packages, such as MATLAB® Optimization Toolbox, MINPACK [134], ODRPACK [22, 23], etc. We describe its main ideas here, referring to [133] for further technical details.

Geometric description of Levenberg-Marquardt. The Levenberg-Marquardt method can be interpreted geometrically as follows. Recall that an iteration of the Gauss-Newton method consists of minimization of the quadratic function

$$\mathscr{Q}(\mathbf{a} + \mathbf{h}) = \|\mathbf{g}(\mathbf{a}) + \mathbf{J}\mathbf{h}\|^2,$$

which approximates the given function $\mathscr{G}(\mathbf{a} + \mathbf{h})$ in a vicinity of the current iteration **a**. Fig. 4.1 shows the contour map (the sets of level curves) of $\mathscr{Q}(\mathbf{a}+\mathbf{h})$ in the 2D case; the level curves are concentric ellipses.

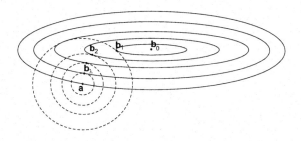

Figure 4.1 *The level curves of $\mathscr{Q}(\mathbf{a}+\mathbf{h})$ are concentric ellipses (solid ovals), the boundaries of trust regions are dashed circles around* **a**.

The "pure" Gauss-Newton step (i.e., when $\lambda = 0$) lands at the minimum of $\mathscr{Q}(\mathbf{a}+\mathbf{h})$, i.e., at the ellipses' center $\mathbf{b}_0$. When $\lambda > 0$, the Levenberg-Marquardt step (the solution of (4.16)) lands at some other point, $\mathbf{b}_\lambda = \mathbf{a}+\mathbf{h}$, closer to **a**. At that point we have, according to (4.16),

$$\operatorname{grad}\mathscr{Q}(\mathbf{a}+\mathbf{h}) = -\lambda\mathbf{h}.$$

Accordingly, the vector $\mathbf{h} = \mathbf{b}_\lambda - \mathbf{a}$ crosses the level curve of $\mathscr{Q}(\mathbf{a}+\mathbf{h})$ passing through $\mathbf{b}_\lambda$, orthogonally. This means that $\mathbf{b}_\lambda$ provides the minimum of the function $\mathscr{Q}(\mathbf{a}+\mathbf{h})$ *restricted* to the ball $\mathbb{B}(\mathbf{a},\Delta)$ whose center is **a** and whose radius is $\Delta = \|\mathbf{h}\|$. In other words, the Levenberg-Marquardt step with $\lambda > 0$ minimizes the restriction of the quadratic approximation $\mathscr{Q}(\mathbf{a}+\mathbf{h})$ to a certain ball around the current approximation **a** whose radius Δ is determined by λ.

As $\lambda > 0$ grows, the point $\mathbf{b}_\lambda$ moves closer to **a**, and Δ decreases. At the same time the vector $\mathbf{h} = \mathbf{b}_\lambda - \mathbf{a}$ rotates and makes a larger angle with the level curves. In the limit $\lambda \to \infty$, the ball $\mathbb{B}(\mathbf{a},\Delta)$ shrinks to the point **a**, i.e., $\Delta \to 0$, and the vector **h** ultimately aligns with the direction of the steepest descent.

Replacing the control parameter. Therefore, there is a one-to-one correspondence between $\lambda > 0$ and $\Delta \in (0, \|\mathbf{b}_0 - \mathbf{a}\|)$, hence one can use Δ, instead of λ, as a control parameter, i.e., adjust Δ from iteration to iteration. As Δ has a clear meaning (further explained below), its initialization and its update at each iteration can be done in a more sensible way than the way λ is treated in the Levenberg-Marquardt scheme.

One benefit of dealing with Δ is that one can directly control the region in the parameter space where the quadratic approximation $\mathscr{Q}(\mathbf{a}+\mathbf{h})$ is minimized. It is called the *trust region*, the idea behind it is that we minimize the approximation $\mathscr{Q}(\mathbf{a}+\mathbf{h})$ where it can be *trusted* and do not go too far where the approximation is not deemed reliable.

1. Initialize $\mathbf{a}_0$ and Δ_0, set $k = 0$.
2. At the current point $\mathbf{a}_k$ compute the vector $\mathbf{g}_k$ and its gradient $\mathbf{J}_k$.
3. For the current $\mathbf{a}_k$ and Δ_k, determine λ_k.
4. Compute $\mathbf{h}_k$ by solving equation $(\mathbf{J}_k^T \mathbf{J}_k + \lambda_k \mathbf{I})\mathbf{h}_k = -\mathbf{J}_k^T \mathbf{g}_k$.
5. Find the ratio $r_k = \mathrm{Ared}_k/\mathrm{Pred}_k$.
6. If $r_k < 0$, reset $\Delta_k := \frac{1}{2}\Delta_k$ and return to Step 3.
7. If $r_k < 0.25$, update $\Delta_{k+1} = \frac{1}{2}\Delta_k$; if $r > 0.75$, update $\Delta_{k+1} = 2\Delta_k$.
8. Update $\mathbf{a}_{k+1} = \mathbf{a}_k + \mathbf{h}_k$, increment k, and return to Step 2.

Table 4.2: *Trust region algorithm.*

New updating rules. This interpretation of Δ also leads to the following update strategy. After a step $\mathbf{h}$ is computed, one finds the ratio

$$r = \frac{\mathrm{Ared}}{\mathrm{Pred}} = \frac{\mathscr{G}(\mathbf{a}) - \mathscr{G}(\mathbf{a}+\mathbf{h})}{\mathscr{G}(\mathbf{a}) - \mathscr{Q}(\mathbf{a}+\mathbf{h})}$$

of the *Actual reduction*, Ared, and the *Predicted reduction*, Pred, of the objective function. One should note that $\mathscr{G}(\mathbf{a}) = \mathscr{Q}(\mathbf{a})$, hence the denominator is always positive, but the numerator is positive only if the actual reduction occurs, in which case of course we should accept the step $\mathbf{h}$.

However, we adjust Δ based on the value of r. A common strategy is as follows. If $r < 0.25$, then the quadratic approximation is not deemed quite reliable (despite the actual decrease of the objective function) and we *reduce* the size Δ before the next iteration (say, by $\Delta := \Delta/2$). Only if $r > 0.75$, then the approximation is regarded as sufficiently accurate and we *increase* Δ (by $\Delta := 2\Delta$). In the intermediate case $0.25 \leq r \leq 0.75$ we leave Δ *unchanged*. Note that these rules are quite different (in a sense, more conservative) than the simplistic rules of updating λ in the previous section!

The subproblem. We summarize the main steps of the trust region scheme in Table 4.2. An attentive reader should notice that we have yet to specify Step 3. This is known as the *trust region subproblem*: Given the size $\Delta > 0$ of the trust region around the current approximation $\mathbf{a}$, determine the corresponding parameter $\lambda \geq 0$. There is a variety of solutions of this problem in the literature, Newton-based schemes [70, 135], dog-leg procedure [148, 149], conjugate gradient [172], quadrature-type methods [26]. We found that for our particular application of fitting circles to data the early Newton-based algorithm with bracketing described by Moré [133] works best.

A MATLAB implementation of the trust region algorithm for fitting circles can be found on our web page [84].

4.5 Levenberg-Marquardt for circles: Full version

Fitting a circle to observed points involves the minimization of

$$\mathscr{F}(a,b,R) = \sum_{i=1}^{n} \left[\sqrt{(x_i - a)^2 + (y_i - b)^2} - R \right]^2. \tag{4.18}$$

This is a relatively simple expression, so that the Levenberg-Marquardt procedure goes smoothly. Let us introduce shorthand notation:

$$r_i = \sqrt{(x_i - a)^2 + (y_i - b)^2} \tag{4.19}$$

and

$$u_i = -\partial r_i / \partial a = (x_i - a)/r_i \tag{4.20}$$

and

$$v_i = -\partial r_i / \partial b = (y_i - b)/r_i. \tag{4.21}$$

Then one can compute directly

$$\mathbf{J} = \begin{bmatrix} -u_1 & -v_1 & -1 \\ \vdots & \vdots & \vdots \\ -u_n & -v_n & -1 \end{bmatrix}$$

and then

$$\mathbf{N} = n \begin{bmatrix} \overline{uu} & \overline{uv} & \bar{u} \\ \overline{uv} & \overline{vv} & \bar{v} \\ \bar{u} & \bar{v} & 1 \end{bmatrix}$$

as well as

$$\mathbf{J}^T \mathbf{g} = n \begin{bmatrix} R\bar{u} - \overline{ur} \\ R\bar{v} - \overline{vr} \\ R - \bar{r} \end{bmatrix} = n \begin{bmatrix} R\bar{u} + a - \bar{x} \\ R\bar{v} + b - \bar{y} \\ R - \bar{r} \end{bmatrix}, \tag{4.22}$$

where, according to the previous section,

$$\mathbf{g} = \left[r_1 - R, \ldots, r_n - R \right]^T.$$

Here we use our "sample mean" notation, for example $\overline{uu} = \frac{1}{n}\sum u_i^2$, etc.

1. Initialize (a_0, b_0, R_0) and λ, compute $\mathscr{F}_0 = \mathscr{F}(a_0, b_0, R_0)$.
2. Assuming that (a_k, b_k, R_k) are known, compute r_i, u_i, v_i for all i.
3. Assemble the matrix $\mathbf{N}$ and the vector $\mathbf{J}^T \mathbf{g}$.
4. Compute the matrix $\mathbf{N}_\lambda = \mathbf{N} + \lambda \mathbf{I}$.
5. Solve the system $\mathbf{N}_\lambda \mathbf{h} = -\mathbf{J}^T \mathbf{g}$ for $\mathbf{h}$ by Cholesky factorization.
6. If $\|\mathbf{h}\|/R_k < \varepsilon$ (small tolerance), then terminate the procedure.
7. Use $\mathbf{h} = (h_1, h_2, h_3)$ to update the parameters
$$a_{k+1} = a_k + h_1, \quad b_{k+1} = b_k + h_2, \quad R_{k+1} = R_k + h_3.$$
8. Compute $\mathscr{F}_{k+1} = \mathscr{F}(a_{k+1}, b_{k+1}, R_{k+1})$.
9. If $\mathscr{F}_{k+1} \geq \mathscr{F}_k$ or $R_{k+1} \leq 0$, update $\lambda \mapsto \beta\lambda$ and return to Step 4; otherwise increment k, update $\lambda \mapsto \alpha\lambda$, and return to Step 2.

Table 4.3: *Levenberg-Marquardt circle fit.*

Now we have all the necessary formulas to implement the the Levenberg-Marquardt procedure. This can be done in two ways, each has its own merit. The most efficient one is to form the matrix $\mathbf{N}_\lambda = \mathbf{N} + \lambda \mathbf{I}$ and solve the system $\mathbf{N}_\lambda \mathbf{h} = -\mathbf{J}^T \mathbf{g}$ by Cholesky factorization. For the reader's convenience we summarize this scheme in Table 4.3.

Practical remarks. We note that $u_i^2 + v_i^2 = 1$ for all i, hence $\overline{uu} + \overline{vv} = 1$, which makes 1 the dominant diagonal element of the matrix $\frac{1}{n}\mathbf{N}$; this information helps if one uses Cholesky factorization with pivoting.

The Cholesky solution is fast, but it may run into computational problems if the matrix $\mathbf{N}_\lambda$ happens to be nearly singular. We recommend a numerically stable scheme that requires a little more work but provides ultimate accuracy. It applies QR decomposition or SVD to the least squares problem $\mathbf{J}_\lambda \mathbf{h} \approx \mathbf{g}_0$ as described in Remark in Section 4.3. The corresponding MATLAB code is available from our web page [84]. Very similar implementations are described in [67, 166].

We postpone the analysis of convergence until later in this chapter and the choice of the initial guess until Section 5.13.

4.6 Levenberg-Marquardt for circles: Reduced version

One can reduce the Levenberg-Marquardt procedure for circles by eliminating one parameter, R, as in Section 3.2, and minimizing the resulting expression

$$\mathscr{F}(a,b) = \sum_{i=1}^{n} (r_i - \bar{r})^2, \tag{4.23}$$

which was constructed in (3.7). Differentiating with respect to a and b gives

$$
\mathbf{J} = \begin{bmatrix} -u_1 + \bar{u} & -v_1 + \bar{v} \\ \vdots & \vdots \\ -u_n + \bar{u} & -v_n + \bar{v} \end{bmatrix},
$$

where we use sample mean notation $\bar{u} = \frac{1}{n}\sum u_i$, etc. This yields

$$
\mathbf{N} = n \begin{bmatrix} \overline{uu} - \bar{u}^2 & \overline{uv} - \bar{u}\bar{v} \\ \overline{uv} - \bar{u}\bar{v} & \overline{vv} - \bar{v}^2 \end{bmatrix}.
$$

We also get

$$
\mathbf{J}^T \mathbf{g} = n \begin{bmatrix} -\overline{ur} + \bar{u}\bar{r} \\ -\overline{vr} + \bar{v}\bar{r} \end{bmatrix} = n \begin{bmatrix} a - \bar{x} + \bar{u}\bar{r} \\ b - \bar{y} + \bar{v}\bar{r} \end{bmatrix}.
$$

One can easily recognize elements of these matrices as sample covariances of the corresponding variables.

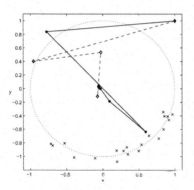

Figure 4.2 *Two versions of the Levenberg-Marquardt circle fit with initial guess at* $(1,1)$: *the full version (the solid line with star markers) and the reduced one (the dashed line with diamond markers) converge to the same limit in 7 step; here 20 simulated points are marked by crosses.*

Numerical tests. This reduced version of the Levenberg-Marquardt circle fit is slightly simpler and faster than the full version presented in the previous section. A typical example is shown in Fig. 4.2, where 20 points are randomly generated along a $130°$ arc of the unit circle $x^2 + y^2 = 1$ with a Gaussian noise at level $\sigma = 0.05$, and both versions of the Levenberg-Marquardt fit start at $a = b = 1$. They both converge to the (unique) minimum of $\mathscr{F}$ at $\hat{a} = -0.057$ and $\hat{b} = 0.041$ in 7 iterations, but take somewhat different paths to the limit.

Numerical tests conducted in the course of a computer experiment reported in [43] showed that while these two versions perform almost identically in typical cases, the reduced version appears slightly inferior in unfavorable situations: if supplied with a bad initial guess, it has an unfortunate tendency to stall in the middle of a plateau.

4.7   A modification of Levenberg-Marquardt circle fit

It is also tempting to compute the Hessian matrix $\mathbf{H}$ in (4.8) exactly, without discarding the second sum in it, thus possibly improving the convergence. Here we explore this approach.

Expanding the squares in (4.23) gives

$$\mathscr{F}(a,b) = \sum_{i=1}^{n} r_i^2 - n\bar{r}^2 = \sum_{i=1}^{n} (x_i^2 + y_i^2) + n(a^2 + b^2 - \bar{r}^2). \qquad (4.24)$$

where we assume, to shorten our formulas, that the data are centered, i.e., $\bar{x} = \bar{y} = 0$ (some authors claim [53, 176] that centering the data prior to the fit helps reduce round-off errors).

Since the first sum in (4.24) is independent of parameters, we can ignore it, along with the constant factor n in the second term, hence we arrive at the problem of minimizing a new function:

$$\mathscr{F}_1(a,b) = a^2 + b^2 - \bar{r}^2. \qquad (4.25)$$

Differentiating with respect to a and b gives

$$\frac{1}{2} \nabla \mathscr{F}_1 = \begin{bmatrix} a + \bar{u}\bar{r} \\ b + \bar{v}\bar{r} \end{bmatrix}. \qquad (4.26)$$

The second order partial derivatives are

$$\begin{aligned}
\frac{1}{2} \frac{\partial^2 \mathscr{F}_1}{\partial a^2} &= 1 - \bar{u}^2 + \bar{r} \frac{\partial \bar{u}}{\partial a} \\
\frac{1}{2} \frac{\partial^2 \mathscr{F}_1}{\partial b^2} &= 1 - \bar{v}^2 + \bar{r} \frac{\partial \bar{v}}{\partial b} \\
\frac{1}{2} \frac{\partial^2 \mathscr{F}_1}{\partial a \partial b} &= -\bar{u}\bar{v} + \bar{r} \frac{\partial \bar{u}}{\partial b}.
\end{aligned} \qquad (4.27)$$

Using again the sample mean notation we have

$$-\frac{\partial \bar{u}}{\partial a} = \frac{1}{n} \sum \frac{v_i^2}{r_i} = \overline{vv/r}$$

$$-\frac{\partial \bar{v}}{\partial b} = \frac{1}{n} \sum \frac{u_i^2}{r_i} = \overline{uu/r}$$

and

$$\frac{\partial \bar{u}}{\partial b} = \frac{1}{n}\sum \frac{u_i v_i}{r_i} = \overline{uv/r}.$$

Therefore,

$$\mathbf{N} = \frac{1}{2}\mathbf{H} = \begin{bmatrix} 1 - \bar{u}^2 - \bar{r}\,\overline{vv/r} & -\bar{u}\bar{v} + \bar{r}\,\overline{uv/r} \\ -\bar{u}\bar{v} + \bar{r}\,\overline{uv/r} & 1 - \bar{v}^2 - \bar{r}\,\overline{uu/r} \end{bmatrix}. \tag{4.28}$$

Then one augments the matrix $\mathbf{N}$ to $\mathbf{N}_\lambda$ and solves the system

$$\mathbf{N}_\lambda \mathbf{h} = -\frac{1}{2}\nabla \mathscr{F}_1.$$

We note that here, unlike the canonical Gauss-Newton algorithm, the matrix $\mathbf{N}$ is not necessarily positive semi-definite; hence $\mathbf{N}_\lambda$ is not necessarily positive definite for small $\lambda > 0$. But for large enough λ the augmented matrix $\mathbf{N}_\lambda$ computed by (4.15) will be positive definite.

Practical remarks. This version of the Levenberg-Marquardt circle fit is more complicated and expensive than the previous two: there is an additional cost of computation for the terms $\overline{uu/r}$, $\overline{vv/r}$, and $\overline{uv/r}$.

The performance of this algorithm was also tested in the course of the computer experiment reported in [43]. It was found to converge in fewer iterations than the full or reduced Levenberg-Marquardt fit, but given its higher cost per iteration the overall improvement was doubtful. On the other hand, if this algorithm diverges (along the "escape valley" described in the previous chapter), then it moves very slowly compared to the full version of the Levenberg-Marquardt circle fit. Therefore, the modified scheme presented above does not appear to be practical.

Further modification. It leads, however, to yet another reduction scheme, which is very interesting. Recall that the canonical Gauss-Newton method suppresses the second order partial derivatives of the individual terms $g_i(\mathbf{a})$ during the calculation of the Hessian, cf. (4.8) and (4.9). In that spirit, let us neglect the second order partial derivatives $\partial^2 \bar{r}/\partial a^2$, $\partial^2 \bar{r}/\partial b^2$, and $\partial^2 \bar{r}/\partial a\,\partial b$ in the computation of our Hessian (4.27). This eliminates the last term in each of the three formulas of (4.27), and now the matrix $\mathbf{N}$ reduces to

$$\mathbf{N}_0 = \begin{bmatrix} 1 - \bar{u}^2 & -\bar{u}\bar{v} \\ -\bar{u}\bar{v} & 1 - \bar{v}^2 \end{bmatrix}. \tag{4.29}$$

The good news is that the matrix $\mathbf{N}_0$, unlike $\mathbf{N}$ in (4.28), is always positive semi-definite. Indeed, using Jensen's inequality gives

$$\det \mathbf{N}_0 = 1 - \bar{u}^2 - \bar{v}^2$$
$$\geq 1 - \overline{uu} - \overline{vv} = 0.$$

In fact, the matrix N_0 is positive definite unless the data points are collinear.

Now, let us further simplify matters by setting $\lambda = 0$, i.e., by going back to the old fashioned Gauss-Newton from the modern Levenberg-Marquardt. Then the vector $\mathbf{h} = (h_1, h_2)^T$ will be the solution of

$$\mathbf{N_0 h} = -\frac{1}{2}\nabla\mathscr{F}_1.$$

In coordinate form, this equation is

$$\begin{bmatrix} 1 - \bar{u}^2 & -\bar{u}\bar{v} \\ -\bar{u}\bar{v} & 1 - \bar{v}^2 \end{bmatrix} \begin{bmatrix} h_1 \\ h_2 \end{bmatrix} = \begin{bmatrix} -a - \bar{u}\bar{r} \\ -b - \bar{v}\bar{r} \end{bmatrix}.$$

Solving this system and computing the next iteration

$$(a_{\text{new}}, b_{\text{new}}) = (a, b) + (h_1, h_2)$$

gives

$$a_{\text{new}} = -\bar{u}R, \qquad b_{\text{new}} = -\bar{v}R \tag{4.30}$$

where we use notation

$$R = \frac{a\bar{u} + b\bar{v} + \bar{r}}{1 - \bar{u}^2 - \bar{v}^2} \tag{4.31}$$

(there is a good reason to denote this fraction by R, as we will reveal in the next section). Hence we obtain a simple iterative procedure (4.30)–(4.31).

This algorithm does not seem promising since the Gauss-Newton scheme, without the Levenberg-Marquardt correction, does not guarantee convergence. Miraculously, the above method turns out to be very robust, as we will see in the next section.

4.8 Späth algorithm for circles

An original method for circle fitting was proposed by Späth in 1996, see [170, 171]. First, he enlarged the parameter space.

Use of latent parameters. For each i, he denotes by $(\hat{x}_i, \hat{y}_i)$ the closest point on the circle to the data point (x_i, y_i). The point $(\hat{x}_i, \hat{y}_i)$ can be thought of as the best estimate of the "true" point on the circle whose noisy observation is (x_i, y_i). The points $(\hat{x}_i, \hat{y}_i)$ are sometimes called the *data correction* [95]. Each point $(\hat{x}_i, \hat{y}_i)$ can be specified by an angular parameter φ_i so that

$$\hat{x}_i = a + R\cos\varphi_i, \qquad \hat{y}_i = b + R\sin\varphi_i$$

In addition to the three parameters a, b, R describing the circle, Späth proposes to estimate n angular parameters $\varphi_1, \ldots, \varphi_n$ specifying the data correction points. The objective function then takes form

$$
\begin{aligned}
\mathscr{F} &= \sum_{i=1}^{n}(x_i - \hat{x}_i)^2 + (y_i - \hat{y}_i)^2 \\
&= \sum_{i=1}^{n}(x_i - a - R\cos\varphi_i)^2 + (y_i - b - R\sin\varphi_i)^2
\end{aligned}
\tag{4.32}
$$

It now depends on $n+3$ parameters $(a,b,R,\varphi_1,\ldots,\varphi_n)$. Such an enlargement of the parameter space has been used by some other authors [67].

Alternating minimization steps. The main idea of Späth [170, 171] is to "separate" the circle parameters from the angular parameters and conduct the minimization of $\mathscr{F}$ alternatively with respect to (a,b,R) and with respect to $(\varphi_1,\ldots,\varphi_n)$. If we fix the angular parameters $(\varphi_1,\ldots,\varphi_n)$, the function $\mathscr{F}$ will be a quadratic polynomial with respect to a,b,R and so its minimum is easily found by setting its partial derivatives $\partial\mathscr{F}/\partial a$, $\partial\mathscr{F}/\partial b$, $\partial\mathscr{F}/\partial R$ to zero and solving the corresponding system of linear equations. For brevity, denote

$$
u_i = \cos\varphi_i, \qquad v_i = \sin\varphi_i
\tag{4.33}
$$

in (4.32). The differentiation of $\mathscr{F}$ and simple manipulations yield

$$
\begin{aligned}
a + \bar{u}R &= 0 \\
b + \bar{v}R &= 0 \\
\bar{u}a + \bar{v}b + R &= \overline{xu} + \overline{yv},
\end{aligned}
$$

where we again assumed that $\bar{x} = \bar{y} = 0$, i.e., the data set is centered. Solving the above system gives

$$
a = -\bar{u}R, \qquad b = -\bar{v}R
\tag{4.34}
$$

and

$$
R = \frac{\overline{xu} + \overline{yv}}{1 - \bar{u}^2 - \bar{v}^2}
\tag{4.35}
$$

We emphasize that the parameters $\varphi_1,\ldots,\varphi_n$ are fixed at this stage and the function $\mathscr{F}$ is being minimized only with respect to a,b,R. Since it is a quadratic polynomial, the above solution provides the *global* minimum of $\mathscr{F}$ with respect to a,b,R.

Let us now fix a,b,R (thus we fix the circle!) and find the minimum of $\mathscr{F}$ with respect to $\varphi_1,\ldots,\varphi_n$. This turns out to be a simple geometric exercise. In the expression (4.32), only the i-th summand depends on φ_i for each i. Hence the parameters $\varphi_1,\ldots,\varphi_n$ can be decoupled: each φ_i is determined by the minimization of

$$
d_i^2 = (x_i - a - R\cos\varphi_i)^2 + (y_i - b - R\sin\varphi_i)^2
$$

i.e., by the minimization of d_i, the distance between the data point (x_i, y_i) and the variable point $(a + R\cos\varphi_i, b + R\sin\varphi_i)$ on the circle. This minimum is reached when one takes the point on the circle closest to (x_i, y_i), and that point will determine the value of φ_i. It is easy to see that

$$\cos\varphi_i = \frac{x_i - a}{\sqrt{(x_i - a)^2 + (y_i - b)^2}}, \qquad \sin\varphi_i = \frac{y_i - b}{\sqrt{(x_i - a)^2 + (y_i - b)^2}} \qquad (4.36)$$

Again, this gives the *global* minimum of $\mathscr{F}$ with respect to φ_i (for each i), assuming that a, b, R are fixed.

The Späth algorithm then performs alternating global minimizations of $\mathscr{F}$. First it finds the global minimum of $\mathscr{F}$ with respect to $\varphi_1, \ldots, \varphi_n$ keeping a, b, R fixed, then it finds the global minimum of $\mathscr{F}$ with respect to a, b, R keeping $\varphi_1, \ldots, \varphi_n$ fixed, and so on. Since at each step a *global* minimum of $\mathscr{F}$ is found, the value of $\mathscr{F}$ *must* decrease all the time. This is a very attractive feature of the algorithm — it can never go wrong!

Combining the two steps into one. We now turn to the implementation of the Späth method. It may look like we have to keep track of $n + 3$ parameters, and in fact Späth provides [170] detailed instructions on how to compute φ_i's. However, a closer look at the procedure reveals that we do not need the angles φ_i, we only need $u_i = \cos\varphi_i$ and $v_i = \sin\varphi_i$, and these are easy to compute via (4.36), or equivalently via (4.19)–(4.21). Therefore, the first step of the algorithm—the minimization of $\mathscr{F}$ with respect to $\varphi_1, \ldots, \varphi_n$ keeping a, b, R fixed—reduces to the simple computation of u_i and v_i defined by (4.19)–(4.21).

Now each iteration of the Späth algorithm (originally, consisting of the two steps described above) can be executed as follows: one computes u_i, v_i by using the circle parameters a, b, R from the previous iteration and then updates a, b, R by the rules (4.34)–(4.35). There is no need to keep track of the additional angular parameters.

It may be noticed that the Späth algorithm now looks similar to our iterative procedure (4.30)–(4.31) developed in the end of the previous section. A close inspection reveals that they are in fact *identical*. The expression (4.31) for R (which was just an auxiliary factor in the previous algorithm) is equivalent to (4.35), as one can check directly by using algebra.

Thus, in fact we accidentally arrived at the Späth algorithm in the previous section based on the entirely different ideas. Moreover, our expressions (4.30)–(4.31) are simpler to implement than (4.34)–(4.35).

Advantages and drawbacks. The Späth algorithm thus has two attractive features: the absolute reliability (every iteration decreases the value of the objective function $\mathscr{F}$) and a low cost per iteration. One might expect that each

---

1. Initialize (a_0, b_0, R_0) and compute $\mathscr{F}_0 = \mathscr{F}(a_0, b_0, R_0)$.
2. Assuming that (a_k, b_k, R_k) are known, compute r_i, u_i, v_i for all i, then compute averages $\bar{r}, \bar{u}, \bar{v}$.
3. Update the parameters: first $R_{k+1} = (a_k \bar{u} + b_k \bar{v} + \bar{r})/(1 - \bar{u}^2 - \bar{v}^2)$, then $a_{k+1} = -\bar{u} R_{k+1}$ and $b_{k+1} = -\bar{v} R_{k+1}$.
4. If $(a_{k+1}, b_{k+1}, R_{k+1})$ are close to (a_k, b_k, R_k), then terminate the procedure, otherwise increment k and return to Step 2.

Table 4.4: *Späth algorithm.*

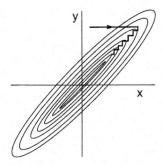

Figure 4.3: *Alternative minimization with respect to x and y, separately.*

iteration takes a big step to the minimum of $\mathscr{F}$ since it performs a *global* minimization of $\mathscr{F}$ with respect to one or the other group of parameters, alternatively. Unfortunately, such algorithms are notoriously slow: when they move along a valley that stretches diagonally in the joint parameter space, every iteration takes just a tiny step.

To illustrate this effect, consider a function $f(x, y)$ whose contour map (the set of level curves) happens to define a long, narrow valley at some angle to the coordinate axes (Fig. 4.3). If one minimizes f alternatively, with respect to x and with respect to y, then the only way "down the length of the valley" is by a series of many tiny steps. This explanation is borrowed from Section 10.5 of [151].

It was noticed in [3] that the Späth algorithm often takes 5 to 10 times more iterations to converge than the Levenberg-Marquardt method. Experiments reported in [43] reveal an even more dramatic difference. A typical example is shown in Fig. 4.4, where 20 points are randomly generated along a 130° arc

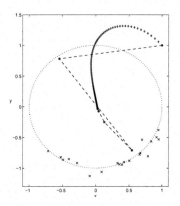

Figure 4.4 *The Späth algorithm with initial guess at* $(1,1)$ *takes 200 iterations (diamonds) to converge, while the Levenberg-Marquardt circle fit makes only 7 iterations (stars); here 20 simulated points are marked by crosses.*

of the unit circle $x^2 + y^2 = 1$ with a Gaussian noise at level $\sigma = 0.05$. The Levenberg-Marquardt fit starts at $a = b = 1$ and converges to the (unique) minimum of $\mathscr{F}$ at $\hat{a} = 0.028$ and $\hat{b} = -0.031$ in 7 iterations (marked by stars). The Späth algorithm, on the contrary, took almost 200 iterations (marked by diamonds) to converge.

4.9 Landau algorithm for circles

Yet another algorithm for circle fitting was proposed by Landau [119] and soon became quite popular; see e.g., [124]. (Berman remarks [17] that this algorithm has been known since at least 1961 when it was published by Robinson [155], but in fact Robinson's paper does not provide details of any algorithms.)

The idea of Landau is simple. The minimum of $\mathscr{F}$ corresponds to $\nabla \mathscr{F} = 0$, which is, in the notation of (4.22), equivalent to $\mathbf{J}^T \mathbf{g} = \mathbf{0}$. This can be written by using our "sample mean" notation as

$$
\begin{aligned}
a &= -\bar{u}R \\
b &= -\bar{v}R \\
R &= \bar{r},
\end{aligned}
$$

where we again assume that $\bar{x} = \bar{y} = 0$. Landau simply proposes the iterative scheme

$$\begin{aligned} R_{\text{new}} &= \bar{r} \\ a_{\text{new}} &= -\bar{u}R_{\text{new}} \\ b_{\text{new}} &= -\bar{v}R_{\text{new}} \end{aligned} \qquad (4.37)$$

where $\bar{r}, \bar{u}, \bar{v}$ are computed by using the previous iteration. This is a variant of the so called *fixed point method*, see below.

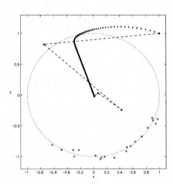

Figure 4.5 *The Landau algorithm with initial guess at* $(1,1)$ *takes 700 iterations (diamonds) to converge, while the Levenberg-Marquardt circle fit reaches the goal in 5 iterations (stars); here 20 simulated points are marked by crosses.*

Fixed point schemes. A fixed point scheme is one of the simplest ways of solving nonlinear equations. Given an equation $f(x) = 0$, one first transforms it into $x = g(x)$ with some function[2] $g(x)$, and then tries to estimate a root, call it x^*, by iterations $x_{n+1} = g(x_n)$. If this method converges at all, its limit is necessarily a root of $x = g(x)$, i.e., $f(x) = 0$.

However, the convergence depends on the function g. The method converges to x^* if $|g'(x^*)| < 1$ and diverges if $|g'(x^*)| > 1$ (similar conditions exist for multidimensional variables $x \in \mathbb{R}^n$, see [112]). Since such conditions are in practice hard to verify, the method is generally not very reliable. In other words, one has to carefully find a representation $x = g(x)$ of the given equation $f(x) = 0$ for which $|g'| < 1$ at and near the desired root.

Besides, the speed of convergence of a fixed point method is linear (unless $g'(x^*) = 0$, which rarely happens), while the convergence of the Gauss-Newton method and its Levenberg-Marquardt modification is nearly quadratic. The

---

[2]For example, the equation $x^3 - 10 = 0$ can be transformed into $x = x^3 + x - 10$ or $x = 10/x^2$. There are infinitely many ways to write it as $x = g(x)$ with different g's.

linear convergence means that if $\hat{\theta}_k$ is the parameter approximation at the k's iteration and $\hat{\theta}^* = \lim_{k \to \infty} \theta_k$, then

$$\|\theta_{k+1} - \hat{\theta}^*\| \leq c \|\theta_k - \hat{\theta}^*\|$$

with some constant $c < 1$.

Speed of convergence. Landau [119] has estimated the convergence constant c for his method and showed that $0.5 \leq c < 1$. In fact, $c \approx 0.5$ when the data are sampled along the entire circle, $c \approx 0.9$ when data are sampled along half a circle, and c quickly approaches one as the data are sampled along smaller arcs. In practice, the performance of the Landau algorithms (as is typical for fixed point schemes) is very sluggish — it happens to be even slower than the Späth method.

A typical example is shown in Fig. 4.5, where 20 points are randomly generated along a 130° arc of the unit circle $x^2 + y^2 = 1$ with a Gaussian noise at level $\sigma = 0.05$. The Levenberg-Marquardt fit starts at $a = b = 1$ and converges to the (unique) minimum of $\mathscr{F}$ at $\hat{a} = 0.014$ and $\hat{b} = -0.020$ in 5 iterations (marked by stars). The Landau algorithm, on the contrary, took almost 700 iterations (marked by diamonds) to converge.

The Späth and Landau methods sometimes take quite different routes to converge to the minimum, see another example in Fig. 4.6, where 20 points are randomly generated along a 130° arc of the unit circle $x^2 + y^2 = 1$ with a Gaussian noise at level $\sigma = 0.05$, and the initial guess is again $a = b = 1$. The Levenberg-Marquardt circle goes straight to the limit point and converges in 5 iterations. The Späth method (diamonds) approaches the limit point from the right in about 100 iterations, and the Landau method (circular markers) comes from the left (taking about 400 steps).

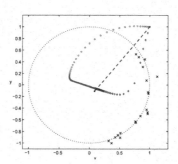

Figure 4.6: *Three algorithms converge along different routes.*

Last remarks. We only described the most popular algorithms for geometric circle fitting. Some authors use "heavy artillery," such as the Nelder-Mead simplex method [68]. Others use various heuristics, such as Fisher scoring [33]. We do not include them here.

4.10 Divergence and how to avoid it

The previous sections reviewed the most popular geometric circle fitting algorithms. They perform well when the initial guess is chosen close enough to the minimum of the objective function $\mathscr{F}$. They can also cope with the initial guesses picked far from the minimum (although sometimes iterations move slowly). But there is one condition under which all of them fail.

Divergence in the escape valley. The failure occurs if the initial guess (or one of the iterations) falls into the escape valley described in Section 3.7. Then all the iterative algorithms are driven away from the minimum of $\mathscr{F}$ and diverge.

An example is shown in Fig. 4.7, where 20 points are randomly generated along a 130° arc of the unit circle $x^2 + y^2 = 1$ with a Gaussian noise at level $\sigma = 0.05$. The Levenberg-Marquardt fit and the Späth and Landau methods start at $a = 1.1$ and $b = 0.1$. All of them move in the wrong direction (to the right) and diverge (the figure shows only the initial stage of divergence). The reason for their failure is that the initial guess happens to be in the escape valley that descends to the right, all the way to the horizon. (The divergence of the Landau algorithm may have no clear cause, as it is purely heuristic and does not use the derivatives of the objective function $\mathscr{F}$, but it follows suit anyway.)

We cannot blame iterative algorithms for their divergence here; they do what they are supposed to: detect the shape of the function at the current locality and move in the direction where the function seems to decrease. They have no way of knowing that they are in the escape valley stretching to the horizon, while the true minimum is left behind, "over the ridge."

We have established in Section 3.7 that the escape valley exists for practically every data set. However, if the data points are sampled along the full circle or a large arc, then the escape valley lies far from the true circle, and the iterative algorithms can operate safely. On the other hand, if the data points are clustered along a small arc, then the escape valley is right next to the data.

For example, the set shown in Fig. 3.1 is sampled along an arc (solid line) whose center lies above the x axis. Choosing the initial center below the x axis places the iterative procedure into the escape valley and sends it away to infinity.

One way to guard the iterative procedures against the escape valley is to use the rules for handling the singular cases described in Section 3.9. Alternatively, the iterative scheme can be initialized by any algebraic circle fit presented in the next section (which is the most common strategy in practice). Then the initial guess is almost automatically placed on the "right" side of the data set

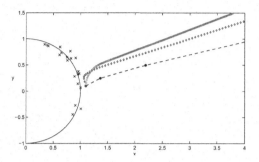

Figure 4.7 *Three algorithms move to the right, away from the true circle (along the escape valley) and diverge. The Levenberg-Marquardt circle fit dashes fast, in three steps (marked by stars) it jumps beyond the picture's boundaries. The Späth method makes 50 steps (diamonds) before escaping, and the Landau scheme crawls in 1000 steps (circular markers that quickly merge into a solid strip) before reaching the top edge; here 20 simulated points are marked by crosses.*

and the procedure avoids the escape valley (however, there are some exotic exceptions; see the W-example in Section 5.13).

On the other hand, if one has to use bad (arbitrarily chosen) initial guesses (for example, we had to do this in our experiment determining the typical number of local minima, cf. Section 3.6), then one is basically at the mercy of the initial choice.

Chernov-Lesort algorithm. In an attempt to bypass the divergence problem Chernov and Lesort [43] developed a rather unusual circle fitting algorithm, which converges to a minimum of the objective function from any (!) starting point. The trick is to adopt a different parametrization scheme in which escape valleys simply do not exist! Recall that the algebraic equation

$$A(x^2 + y^2) + Bx + Cy + D = 0$$

with four parameters A, B, C, D satisfying the constraint

$$B^2 + C^2 - 4AD = 1 \qquad (4.38)$$

describes all circles on the xy plane and, in addition, all straight lines. Circles correspond to $A \neq 0$ and lines to $A = 0$.

We have proved in Section 3.2 that if the data set lies within a bounding box $\mathbb{B}$, then the minimum of the objective function $\mathscr{F}$ is restricted to a bounded region in the parameter space

$$|A| < A_{\max}, |B| < B_{\max}, |C| < C_{\max}, |D| < D_{\max} \qquad (4.39)$$

where $A_{\max}, B_{\max}, C_{\max}, D_{\max}$ are determined by the size and the location of the box $\mathbb{B}$ and by the maximal distance between the data points, $d_{\max}$, cf. Theorem 4. Therefore, there are no "infinitely long" valleys in the A, B, C, D parameter space along which the function $\mathscr{F}$ decreases all the way to the horizon, which potentially cause divergence of iterative algorithms. The objective function $\mathscr{F}(A, B, C, D)$ actually growth as the point (A, B, C, D) moves away from the region (4.39) in any direction.

Now one can minimize $\mathscr{F}$ in the A, B, C, D parameter space by using any general algorithm for constrained or unconstrained optimization (in the latter case one has to eliminate one parameter, see below).

Chernov and Lesort [43] utilized the unconstrained Levenberg-Marquardt scheme. They introduced a new parameter, an angular coordinate θ defined by

$$B = \sqrt{1 + 4AD} \cos \theta, \qquad C = \sqrt{1 + 4AD} \sin \theta,$$

so that θ replaces B and C. We refer the reader to [43] for further details. A MATLAB code of this algorithm is available from [84].

The Chernov-Lesort algorithm is remarkable for its 100% convergence rate (verified experimentally under various conditions, see [43]). But otherwise it is rather impractical. First, it involves complicated mathematical formulas. Second, its cost per iteration is about 4 times higher than the cost of one iteration for other geometric fitting algorithms. Third, it must be safeguarded against various singularities of the objective function (points in the parameter space where $\mathscr{F}$ fails to have derivatives); see details in [43]. Last, but not the least, this algorithm heavily depends on the choice of the coordinate system on the data plane, see Section 4.11.

4.11 Invariance under translations and rotations

In Section 1.3 we determined that the geometric fitting line was invariant under translations, rotations, and similarities, but not under general scalings. This invariance is essential as it guarantees that the best fitting line is independent of the choice of the coordinates frame in the xy plane.

Invariance of the best fitting circle. Here we extend the above property to circles. We say that a circle fit is invariant under translations $T_{c,d}$, defined by (1.18), if changing the coordinates of the data points by (1.18) leaves the fitting circle unchanged, i.e., its equation in the new coordinates will be

$$(x + c - a)^2 + (y + d - b)^2 = R^2.$$

Similarly we define the invariance under rotations (1.19) and scaling (1.21).

It is quite clear that the best fitting circle, which minimizes the sum of squares of the geometric distances (3.1), is invariant under translations and rotations, but again, not under scalings (unless both coordinates x and y are scaled by the same factor). Thus the fitting circle does not depend on the choice of the coordinate frame.

Stepwise invariance of circle fitting algorithms. However, now we need more than just the independence of the best fitting circle from the choice of the coordinate system. Since in practice that circle is obtained by an iterative algorithm, we would like *every iteration* be constructed independently of the coordinate frame.

In that case the choice of the coordinate system truly would not matter. Otherwise the sequence of circles generated by the algorithm would depend on the coordinate frame; in some frames the convergence would be faster than in others; in some frames the procedure would converge to the global minimum, while in others to a local one, and yet in others it would diverge (as in the previous section). Moreover, since in practice we have to stop the iterative procedure before it reaches its limit, the exact circle returned by the procedure would really depend on the choice of the coordinate frame. This is certainly an undesirable feature.

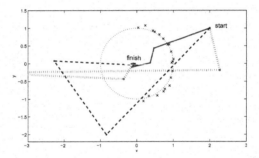

Figure 4.8: *Three different itineraries of the Chernov-Lesort procedure.*

Example. The only circle fitting algorithm discussed in this chapter which lacks the above invariance property is Chernov-Lesort (Section 4.10); Fig. 4.8 illustrates its performance. 20 points (marked by crosses) are randomly generated along a semi-circle $x^2 + y^2 = 1$, $x \geq 0$, with a Gaussian noise at level $\sigma = 0.05$. The Chernov-Lesort algorithm starts at the point $(2, 1)$, and converges to the global minimum of the objective function at $(-0.05, -0.01)$.

Note that the starting point is clearly in the escape valley, so that every other fit presented in this chapter would diverge to infinity. Only the Chernov-Lesort fit converges to the minimum of the objective function here.

But it takes a different path to the limit every time we translate or rotate the coordinate system. Once it went straight to the limit (the solid line). Another time it overshot and landed near $(-1, -2)$, then it took a detour returning to the limit (the dashed line). Yet another time it leaped far along the negative x axis and then slowly came back (the dotted line). The number of iterations varies from 5 to 20–25, depending on the choice of the coordinate frame.

Though we emphasize that the Chernov-Lesort fit converges to the minimum of the objective function from any starting point and in any coordinate system.

Next we turn to algorithms which are translation and rotation invariant.

Invariance of general minimization schemes. Since most of our algorithms are based on Gauss-Newton method and its modifications, let us examine the invariance issue in general. First, all the classical minimization algorithms (steepest descent, Newton-Raphson, Gauss-Newton, Levenberg-Marquardt) involve derivatives of the objective function, hence their iterations are invariant under translations of the coordinate frame.

To examine the invariance under rotations, or more generally under linear transformations, let us transform the coordinates $\mathbf{a} = (a_1, \ldots, a_k)$ in the underlying space $\mathbb{R}^k$, cf. (4.1), by $\mathbf{a} = \mathbf{Ab}$, where $\mathbf{A}$ denotes an arbitrary nonsingular matrix and $\mathbf{b} = (b_1, \ldots, b_k)$ new coordinates. Then by the chain rule

$$\nabla_{\mathbf{b}}\mathscr{G}(\mathbf{Ab}) = \mathbf{A}^T \nabla_{\mathbf{a}}\mathscr{G}(\mathbf{Ab}).$$

Therefore the coordinates of the direction of the steepest descent are transformed by $(\mathbf{A}^T)^{-1}$, while the data coordinates are transformed by $\mathbf{A}$. We have the desired invariance if and only if $\mathbf{A}^T = c\mathbf{A}^{-1}$, $c \neq 0$, i.e., iff $\mathbf{A}$ is a scalar multiple of an orthogonal matrix.

One can arrive at this intuitively by noticing that the direction of the steepest descent must be orthogonal to the level surface of the objective function, hence the linear transformation $\mathbf{A}$ must preserve angles.

For the Newton-Raphson method, equations (4.2) take the form

$$\mathbf{D_b} = \nabla_{\mathbf{b}}\mathscr{G}(\mathbf{Ab}) = \mathbf{A}^T \mathbf{D_a}$$

and

$$\mathbf{H_b} = \nabla_{\mathbf{b}}^2 \mathscr{G}(\mathbf{Ab}) = \mathbf{A}^T \mathbf{H_a} \mathbf{A},$$

hence (4.4) in the new coordinates $\mathbf{b}$ takes the form

$$\mathbf{A}^T \mathbf{D_a} + \mathbf{A}^T \mathbf{H_a} \mathbf{Ah_b} = 0.$$

We cancel $\mathbf{A}^T$ and conclude that $\mathbf{h_a} = \mathbf{Ah_b}$, i.e., we have the desired invariance for any nonsingular matrix $\mathbf{A}$. Intuitively, this fact is clear, because the method is based on the second order approximation (of the objective function), which survives any linear transformations.

For the Gauss-Newton method (Section 4.2) we have, in a similar way,

$$\nabla_{\mathbf{b}} g_i(\mathbf{Ab}) = \mathbf{A}^T \nabla_{\mathbf{a}} g_i(\mathbf{Ab}),$$

thus equations (4.7) and (4.9) take form

$$\mathbf{D_b} = \mathbf{A}^T \mathbf{D_a} \quad \text{and} \quad \mathbf{H_b^\circ} = \mathbf{A}^T \mathbf{H_a^\circ} \mathbf{A}.$$

Then (4.10) takes form

$$\mathbf{A}^T \mathbf{D_a} + \mathbf{A}^T \mathbf{H_a^\circ} \mathbf{A} \mathbf{h_b} = \mathbf{0}. \tag{4.40}$$

Canceling $\mathbf{A}^T$ yields $\mathbf{h_a} = \mathbf{A}\mathbf{h_b}$, i.e., we again have the invariance for any non-singular matrix $\mathbf{A}$.

The Levenberg-Marquardt correction (Section 4.2) replaces the matrix $\mathbf{H}^\circ$ with $(\mathbf{H}^\circ + \lambda \mathbf{I})$, hence equation (4.40) becomes

$$\mathbf{A}^T \mathbf{D_a} + (\mathbf{A}^T \mathbf{H_a^\circ} \mathbf{A} + \lambda \mathbf{I}) \mathbf{h_b} = \mathbf{0}.$$

If $\mathbf{A}$ is an orthogonal matrix, then $\mathbf{I} = \mathbf{A}^T \mathbf{A}$, and we again arrive at $\mathbf{h_a} = \mathbf{A}\mathbf{h_b}$, i.e., the desired invariance. Otherwise the invariance fails. This is no surprise as the Levenberg-Marquardt is a hybrid of Gauss-Newton and steepest descent, and the latter is invariant only for orthogonal $\mathbf{A}$'s.

Proof of invariance of circle fitting algorithms. As a result of our general analysis, the Levenberg-Marquardt method is always invariant under orthogonal transformations (in particular under translations and rotations). When we fit circles, we can translate or rotate the coordinate frame in the xy (data) plane; then in the 3D parameter space with coordinates (a, b, R) we translate or rotate the frame in the ab plane but keep the R axis fixed. Thus we deal with a (particular) translation or rotation of the coordinate frame in $\mathbb{R}^3$, and so the Levenberg-Marquardt algorithm will be invariant under it, as we established above.

The Späth algorithm and Landau algorithm are also invariant under translations and rotations of the coordinate frame in the xy (data) plane. In fact the invariance under translations is immediate as both algorithms require moving the origin to the centroid of the data set.

To check the invariance under rotation, one should observe that rotating the coordinate frame through angle θ results in the rotation of every (u_i, v_i) vector, cf. (4.20)–(4.21), through $-\theta$; thus their mean vector $(\bar{u}, \bar{v})$ will also rotate through $-\theta$. It is now quite clear from (4.34) and (4.37) that the vector (a, b) will rotate accordingly; we invite the reader to verify the details of this argument as an exercise.

4.12   The case of known angular differences

In Section 4.8 we used the angular coordinates $\varphi_1, \ldots, \varphi_n$ and wrote the objective function as

$$\mathscr{F} = \sum_{i=1}^{n} (x_i - a - R\cos\varphi_i)^2 + (y_i - b - R\sin\varphi_i)^2 \tag{4.41}$$

where φ_i's were treated as unknowns, in addition to the circle parameters (a,b,R). Minimization of (4.41) with respect to all its variables was the main idea of the Späth algorithm.

Berman's model. In some applications, the differences $\varphi_{i+1} - \varphi_i$ between the angular coordinates of successive data points are known. This information can be used to simplify the objective function (4.41), after which the model becomes entirely linear, leading to explicit formulas for parameter estimates.

Indeed, if we denote the known differences by

$$\tau_i = \varphi_i - \varphi_1 = \sum_{j=1}^{i-1} \varphi_{j+1} - \varphi_j$$

then we can represent $\varphi_i = \varphi + \tau_i$, where $\varphi = \varphi_1$ is the *only* unknown angular parameter. Now the objective function (4.41) takes form

$$\mathscr{F} = \sum_{i=1}^{n} (x_i - a - \alpha \cos \tau_i + \beta \sin \tau_i)^2$$
$$+ (y_i - b - \alpha \sin \tau_i - \beta \cos \tau_i)^2, \tag{4.42}$$

where

$$\alpha = R \cos \varphi \qquad \text{and} \qquad \beta = R \sin \varphi \tag{4.43}$$

can be treated as auxiliary parameters that (temporarily) replace R and φ.

Since $\mathscr{F}$ is a quadratic function in its unknowns a, b, α, β, the model becomes entirely linear. The estimates of a, b, α, β can be easily found by setting the partial derivatives of $\mathscr{F}$ to zero and solving the resulting linear system of 4 equations with 4 unknowns. We will not provide details as they are elementary.

Lastly, the radius R (and the base angle φ, if necessary) can be found from equations (4.43). There is no need for complicated iterative schemes. This model was proposed and solved in 1983 by Berman [15].

Applications of Berman's model. Berman [15] pointed out two particular applications where the angular differences are indeed known:

(a) The calibration of an impedance measuring apparatus in microwave engineering;

(b) The analysis of megalithic sites in Britain, in which archaeologists need to fit circles to stone rings.

It is interesting that so much different examples have this important feature in common.

Berman, Griffiths and Somlo [15, 19, 20] present a detailed statistical analysis of the resulting parameter estimates for the model with known angular differences.

Further extensions. Recently Yin and Wang [194] incorporated heteroscedastic errors into Berman's model; they assumed that the disturbances in the x and y directions had different (and unknown) variances. This type of data also arise in microwave engineering.

Wang and Lam [188] extend Berman's model to simultaneous fitting of several circles, where one tests the hypothesis that several round objects have identical size. Such problems arise in quality control of manufactured mechanical parts.

It seems that the situation with known angular differences is not so uncommon as it may appear, but it is still a special case of the general circle fitting problem.

Chapter 5

Algebraic circle fits

Recall that we have divided all practical solutions to the circle fitting problem into two large groups: (A) geometric fits that minimize the geometric (orthogonal) distances from the data points to the circle and (B) algebraic fits that minimize some other, mostly algebraic expressions, cf. the preamble to Chapter 4. Here we deal with type B methods.

Why are the type B methods of interest?

First, geometric fits (i.e., type A methods) are iterative, they require an initial guess be supplied prior to their work; their performance greatly depends on the accuracy of the initial guess. Algebraic fits are noniterative inexpensive procedures that provide a good initial guess.

Second, some applications are characterized by mass data processing (for instance, in high energy physics millions of circular-shaped particle tracks may come from an accelerator and need be processed). Under such conditions one can rarely afford slow iterative geometric fits, hence a noniterative algebraic fit may very well be the only option.

Last but not the least, well designed algebraic fits (such as Taubin's and Pratt's algorithms described in this chapter) appear to be nearly as accurate, statistically, as geometric fits (this will be confirmed by our detailed error analysis in Chapter 7). In many applications a good algebraic fit would do an excellent job, and subsequent geometric fits would not make noticeable improvements.

5.1 Simple algebraic fit (Kåsa method)

Perhaps the simplest and most obvious approach to the circle fitting problem is to minimize

$$\mathcal{F}_1(a,b,R) = \sum_{i=1}^{n} \left[(x_i - a)^2 + (y_i - b)^2 - R^2\right]^2. \tag{5.1}$$

In other words, we are minimizing $\mathcal{F}_1 = \sum f_i^2$, where

$$f_i = (x_i - a)^2 + (y_i - b)^2 - R^2. \tag{5.2}$$

Clearly, $f_i = 0$ if and only if the point (x_i, y_i) lies on the circle; also, f_i is small if and only if the point lies near the circle. Thus, minimizing (5.1) sounds like a reasonable idea. Some authors [62, 67, 150] call f_i the *algebraic distance* from the point (x_i, y_i) to the circle $(x - a)^2 + (y - b)^2 + R^2 = 0$.

The derivatives of the objective function (5.1) are nonlinear with respect to a, b, R, but a simple change of parameters makes them linear. Let

$$B = -2a, \qquad C = -2b, \qquad D = a^2 + b^2 - R^2. \tag{5.3}$$

Then we get

$$\mathcal{F}_1 = \sum_{i=1}^{n} (z_i + Bx_i + Cy_i + D)^2 \tag{5.4}$$

where we also denote $z_i = x_i^2 + y_i^2$ for brevity.

Now differentiating $\mathcal{F}_1$ with respect to B, C, D yields a system of *linear* equations

$$\begin{aligned}
\overline{xx}\,B + \overline{xy}\,C + \bar{x}D &= -\overline{xz} \\
\overline{xy}\,B + \overline{yy}\,C + \bar{y}D &= -\overline{yz} \\
\bar{x}B + \bar{y}C + D &= -\bar{z},
\end{aligned} \tag{5.5}$$

where we again use our "sample mean" notation

$$\overline{xx} = \frac{1}{n}\sum x_i^2, \qquad \overline{xy} = \frac{1}{n}\sum x_i y_i, \quad \text{etc.}$$

Solving this system by some methods of linear algebra (see, e.g., below) gives
B, C, D, and finally one computes a, b, R by

$$a = -\frac{B}{2}, \qquad b = -\frac{C}{2}, \qquad R = \frac{\sqrt{B^2 + C^2 - 4D}}{2}. \tag{5.6}$$

This algorithm is very fast, it can be accomplished in $13n + 35$ flops if
one solves (5.5) directly. It actually costs less than one iteration of the Gauss-
Newton or Levenberg-Marquardt methods (Chapter 4). We note, however, that
a direct solution of (5.5) is not the best option; the system (5.5) is an analogue
of normal equations that are known to cause numerical instability; a stable
solution is described in the next section.

The above algebraic fit has been proposed in the early 1970s by Delogne
[54] and Kåsa [108], and then rediscovered and published independently by
many authors [11, 24, 25, 36, 51, 53, 125, 129, 136, 170, 180, 193, 197]. Now
it is called *Delogne–Kåsa method* (see [196]) or briefly *Kåsa method* (see [52,
159, 184]), named after the first contributors. (It seems that most authors prefer
the short name "Kåsa method," so we follow this tradition, too.)

The simplicity and efficiency of the Kåsa algorithm are very attractive features. Many researchers just fall in love with it at first sight. This method is so
popular in the computer science community that we feel obliged to present its
detailed analysis here. Even though it would not be a method of our choice, we
will methodically highlight its strong and weak points.

5.2   Advantages of the Kåsa method

The use of Kåsa algorithm actually has many benefits.

First, if the points (x_i, y_i) happen to lie on a circle, the method finds that
circle right away. (In contrast, geometric fits would at best converge to that
circle iteratively.)

In statistical terms, the Kåsa method is consistent in the small-noise limit
$\sigma \to 0$ (introduced in our discussion in Section 2.5), i.e., the estimates of the
parameters (a, b, R) converge to their true values as $\sigma \to 0$. In fact, the Kåsa
fit is asymptotically efficient (optimal) in that limit, to the leading order; see
Chapter 7.

Geometric descriptions. Second, the Kåsa method can be defined geometrically in several ways. For example, observe that πf_i, where f_i is given
by (5.2), is the difference between the *areas* of two concentric circles, both
centered on (a, b): one has radius R and the other passes through the observed
point (x_i, y_i), i.e., has radius $r_i = \sqrt{(x_i - a)^2 + (y_i - b)^2}$. In other words, πf_i is
the area of the ring between the fitted circle and the observed point. Thus the
Kåsa method minimizes the combined area of such rings; this interesting fact
was noted in [180].

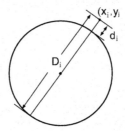

Figure 5.1 *The distance d_i from (x_i, y_i) to the nearest point and D_i to the farthest point on the circle.*

Another geometrical definition, see [51], results from the expansion

$$f_i = (r_i - R)(r_i + R), \qquad r_i = \sqrt{(x_i - a)^2 + (y_i - b)^2}. \qquad (5.7)$$

Here the first factor $d_i = r_i - R$ is the distance from (x_i, y_i) to the *nearest* point on the circle (a, b, R), while the second factor $D_i = r_i + R$ is the distance from (x_i, y_i) to the *farthest* point on the circle. Thus the Kåsa method minimizes

$$\mathscr{F}_1 = \sum_{i=1}^{n} d_i^2 D_i^2. \qquad (5.8)$$

Chord method. The Kåsa fit is also related to the *chord method* proposed in [184, 193]. While not practically efficient, this method is worth mentioning due to its relevance to the Kåsa fit.

Observe that if any two points $P_i = (x_i, y_i)$ and $P_j = (x_j, y_j)$ lie on the fitting circle, then the perpendicular bisector L_{ij} of the chord $P_i P_j$ passes through its center, see Fig. 5.2. Thus one can find the best center (a, b) by minimizing

$$\mathscr{F}_{\text{ch}} = \sum_{1 \le i < j \le n} w_{ij} D_{ij}^2,$$

where D_{ij} denotes the distance from (a, b) to the bisector L_{ij} and w_{ij} is a certain weight, see below.

By elementary geometry, $\mathscr{F}_{\text{ch}}$ is a quadratic function of a and b:

$$\mathscr{F}_{\text{ch}} = \sum_{1 \le i < j \le n} w_{ij} \frac{\left[a(x_i - x_j) + b(y_i - y_j) - \frac{1}{2}(x_i^2 - x_j^2 + y_i^2 - y_j^2) \right]^2}{(x_i - x_j)^2 + (y_i - y_j)^2}, \qquad (5.9)$$

so it can be easily minimized provided w_{ij} are specified. The choice of w_{ij} is based on the following considerations. For shorter chords $P_i P_j$, the distance D_{ij}

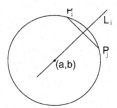

Figure 5.2 *Perpendicular bisector L_{ij} of the chord P_iP_j passes through the center (a,b).*

is less reliable, thus their influence should be suppressed. Umbach and Jones [184] propose the weight proportional to the squared chord's length:

$$w_{ij} = |P_iP_j|^2 = (x_i - x_j)^2 + (y_i - y_j)^2. \tag{5.10}$$

This choice of the w_{ij} conveniently cancels the denominator in (5.9) making the computations even simpler. Of course, generally speaking, the chord method is very impractical: as there $n(n-1)/2$ chords joining n data points, the method takes $\mathscr{O}(n^2)$ flops, while all conventional circle fits require $\mathscr{O}(n)$ flops.

With weights defined by (5.10), the chord method happens to be equivalent to the Kåsa algorithm, they produce the same circle center (a,b); this remarkable relation was noted and verified by Umbach and Jones [184].

Invariance under translations and rotations. We see now that the Kåsa method can be defined geometrically (actually, in three different ways, as described above). This fact implies that the resulting fit is invariant under translations and rotations, which is a very important feature of any curve fitting algorithm, cf. Section 4.11. Therefore, without loss of generality, we can center the data set to ensure $\bar{x} = \bar{y} = 0$ and rotate the coordinate axes to ensure $\overline{xy} = 0$ (this will make the system (5.5) diagonal).

Existence and uniqueness. Next we investigate the existence and uniqueness of the Kåsa fit (5.1). The matrix of coefficients of the system (5.5) is $\mathbf{X}^T\mathbf{X}$, where

$$\mathbf{X} = \begin{bmatrix} x_1 & y_1 & 1 \\ \vdots & \vdots & \vdots \\ x_n & y_n & 1 \end{bmatrix} \tag{5.11}$$

is the $n \times 3$ "data matrix;" thus the system (5.5) is always positive semi-definite.

It is singular if and only if there exists a nonzero vector $\mathbf{u} = (p,q,r)$ such that $\mathbf{Xu} = \mathbf{0}$, i.e., $px_i + qy_i + r = 0$ for all i. This means precisely that the data points (x_i, y_i) are collinear. Hence, unless the data are collinear, the Kåsa system (5.5) is positive definite and admits a unique solution.

In the collinear case, the Kåsa method has multiple solutions, see below.

Implementation. Practically, one can solve (5.5) by Cholesky factorization; this is perhaps the most efficient way, but it may lead to unnecessarily large round-off errors if X is nearly singular. A numerically stable (albeit somewhat more expensive) approach is to rewrite (5.5) as

$$X^T X A = -X^T Z \qquad (5.12)$$

where $A = (B, C, D)^T$ denotes the parameter vector and $Z = (z_1, \ldots, z_n)^T$. Then the solution of (5.12) is

$$A = -X^- Z,$$

where X^- denotes the Moore-Penrose pseudoinverse, and it can be computed by the QR decomposition or SVD of the matrix X (see details in the second remark in Section 4.2). The SVD solution will work smoothly even in the singular case, where $\det X^T X = 0$.

Multiple solutions. In the singular case (collinear data) the system (5.5), i.e., (5.12), has multiple solutions. Indeed, we can rotate and translate the coordinate frame to enforce $y_1 = \cdots = y_n = 0$ as well as $\bar{x} = 0$. It is then easy to see that the system (5.5) determines a uniquely, but b is left unconstrained. The SVD-based algorithm should return the so-called "minimum-norm" solution.

Nievergelt [137] points out, however, that in the collinear data case the SVD solution is *no longer* invariant under translations. He proposes some modifications to enforce invariance; see [137] for more details.

Square root issue. Lastly, note that in any case (singular or not) we have

$$B^2 + C^2 - 4D \geq 0 \qquad (5.13)$$

thus the evaluation of the radius R in (5.6) does not run into complex numbers. Indeed, by a translation of the coordinate frame we can ensure that $\bar{x} = \bar{y} = 0$, hence the third equation in (5.5) implies $D = -\bar{z} \leq 0$, so (5.13) holds.

5.3   Drawbacks of the Kåsa method

Experimental evidence of heavy bias. Now we turn to practical performance of the Kåsa method. Fig. 5.3 shows four samples of 20 random points each; they are generated along different segments of the unit circle $x^2 + y^2 = 1$: (a) full circle, (b) half circle, (c) quarter circle, and (d) 1/8 of a circle. In each case a Gaussian noise is added at level $\sigma = 0.05$.

Fig. 5.3 demonstrates that the Kåsa fit is practically identical to the geometric fit when the data are sampled along the full circle. It returns a slightly smaller circle than the geometric fit does, when the data are sampled along a semicircle. The Kåsa circle gets substantially smaller than the best one for data confined to a $90°$ arc. When the data are clustered along a $45°$ arc, the Kåsa

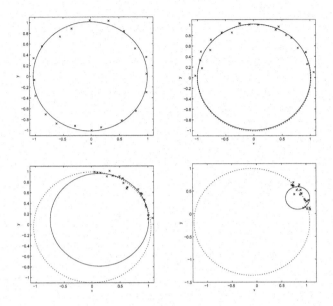

Figure 5.3 *Four samples, each of 20 points (marked by crosses), along different arcs of the unit circle* $x^2 + y^2 = 1$. *The Kåsa fit is the solid circle; the geometric fit is the dotted circle.*

method returns a badly diminished circle, while the geometric fit remains quite accurate. On smaller arcs, the Kåsa method nearly breaks down (see a rather extreme example in Fig. 5.4 and a more comprehensive numerical experiment in Section 5.7).

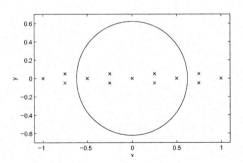

Figure 5.4 *13 data points (crosses) form a symmetric pattern stretching along a horizontal line (the x axis). The best fit is obviously the x axis itself; the Kåsa method returns a circle that is far from the best fit.*

Theoretical analysis of heavy bias. Unfortunately, it is a general fact: the Kåsa method substantially underestimates the radius and is heavily biased toward smaller circles when the data points are confined to a relatively small arc; this tendency has been noticed and documented by many authors, see [18, 45, 43, 42, 53, 67, 82, 137, 150, 159] and our Section 5.7.

The reason for this flaw of the Kåsa fit can be derived from (5.8). The method minimizes the average product of the smallest distance d_i and the largest distance D_i (squared). Reducing the size of the circle may increase d_i's but decrease D_i's, thus the minimum of $\mathscr{F}_1$ may be attained on a circle that does not necessarily minimize d_i's but rather finds a best trade-off between d_i's and D_i's.

To perform a more detailed analysis, observe that $D_i = d_i + 2R$, hence the Kåsa fit minimizes

$$\mathscr{F}_1 = \sum_{i=1}^{n} d_i^2 (d_i + 2R)^2. \tag{5.14}$$

Now if the data points are close to the circle, we have $|d_i| \ll R$ and

$$\mathscr{F}_1 \approx 4R^2 \sum_{i=1}^{n} d_i^2. \tag{5.15}$$

Suppose the data are sampled along an arc of length L (subtending a small angle $\alpha = L/R$). Let (a, b, R) denote the parameters of the true circle, then $\mathscr{F}_1(a, b, R) \approx 4R^2 n \sigma^2$. If one reduces the arc radius by a factor of 2, see Fig. 5.5, then the arc will be displaced by $\sim L^2/R$, thus distance from the data points to the new arc will be of order $\sigma + L^2/R$; hence the new value of $\mathscr{F}_1$ will be

$$\mathscr{F}_1 \approx 4(R/2)^2 n (\sigma + L^2/R)^2.$$

The algorithm will favor the new circle of the smaller radius $R/2$ as long as $\sigma > L^2/R = \alpha L$. Hence, the Kåsa method is subject to gross errors or breaking down whenever $\sigma R > L^2$.

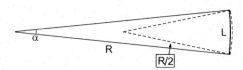

Figure 5.5 *The true arc (solid lines) and the arc after its radius is halved (dashed lines).*

Conclusions. To summarize, the Kåsa algorithm is perhaps the simplest, fastest, and most elegant of all known circle fits. Its does an excellent job whenever the data points are sampled along a full circle or at least half a circle. It also works fine if the data points lie very close to a circle, i.e., when $\sigma \approx 0$.

However, the performance of the Kåsa method may substantially deteriorate when the data points are sampled along a small arc, in which case it tends to grossly underestimate the radius.

5.4 Chernov-Ososkov modification

In the early 1980s, Chernov and Ososkov [45] proposed a modification of the Kåsa fit to improve its performance.

Modified objective function. Observe that it is the extra factor $4R^2$ in (5.15) that is responsible for the impaired performance of the Kåsa method. Chernov and Ososkov [45] suggest to get rid of that factor by minimizing a modified function $\mathscr{F}_2 = (2R)^{-2}\mathscr{F}_1$, i.e.,

$$\mathscr{F}_2(a,b,R) = (2R)^{-2}\sum_{i=1}^{n}\left[(x_i-a)^2+(y_i-b)^2-R^2\right]^2. \tag{5.16}$$

This formula has some history. It was introduced in 1984, in [45], for the purpose of fitting circular arcs to tracks of elementary particles in high energy physics. In 1991, it was employed by Karimäki [106, 107] who called it a "highly precise approximation." In 1998, it was independently derived by Kanatani [96] in the course of statistical analysis of circle fits.

Next, changing parameters a,b,R to B,C,D, as specified by (5.3) (i.e., as in the Kåsa method) gives

$$\mathscr{F}_2 = \frac{1}{B^2+C^2-4D}\sum_{i=1}^{n}(z_i+Bx_i+Cy_i+D)^2$$
$$= \frac{\overline{zz}+2B\overline{xz}+2C\overline{yz}+2D\overline{z}+B^2\overline{xx}+C^2\overline{yy}+2BC\overline{xy}+D^2}{B^2+C^2-4D}. \tag{5.17}$$

Here we assume, to shorten the above formula, that the data set has been centered so that $\bar{x}=\bar{y}=0$ (otherwise the numerator would contain two extra terms; $+2BD\bar{x}+2CD\bar{y}$).

Reduction to a quartic equation. Observe that $\mathscr{F}_2$ is a quadratically rational function of the unknown parameters. Chernov and Ososkov [45] set its derivatives to zero, eliminated B and C from the resulting equations, and arrive at a polynomial equation of degree 4 (i.e., *quartic* equation) in D:

$$D^4+a_3D^3+a_2D^2+a_1D+a_0=0, \tag{5.18}$$

where

$$a_3 = 4\bar{z}$$
$$a_2 = 3\overline{xx}^2+3\overline{yy}^2+10\overline{xx}\,\overline{yy}-4\overline{xy}^2-\overline{zz}$$
$$a_1 = 2\overline{xz}^2+2\overline{yz}^2-4\overline{zz}\,(\overline{xx}+\overline{yy}) \tag{5.19}$$
$$a_0 = 2\overline{xz}^2(\overline{xx}+3\overline{yy})+2\overline{yz}^2(3\overline{xx}+\overline{yy})-8\overline{xz}\,\overline{yz}\,\overline{xy}-\overline{zz}(a_2+\overline{zz})$$

Newton-based solution of the quartic equation. Equation (5.18) may have up to four different real roots, which are not easy to compute. However, Chernov and Ososkov [45] notice that if one sets $D = -\bar{z}$ (this is the value furnished by the Kåsa method) as the initial guess and applies a standard Newton procedure to solve (5.18), then the latter always converges to the root that corresponds to the minimum of (5.17).

Besides, the convergence is fast: it takes 3 to 5 iterations, on the average, to find the desired root (with tolerance set to 10^{-12}). It is important, though, that one uses the double-precision arithmetic to solve (5.18), as single precision round-off errors tend to ruin the accuracy of the final estimates.

However, Chernov and Ososkov [45] did not provide a theoretical analysis of their modification; in particular they did not prove that their Newton procedure described above always converges to a root of (5.18), or that the root which it finds always corresponds to the minimum of (5.16). These facts were proved much later, in 2005, by Chernov and Lesort [43] based on the Pratt circle fit discussed in the next section.

Application in high energy physics. The Chernov-Ososkov modification turned out to be more robust than the original Kåsa method in handling nearly singular cases, i.e., when the data points were sampled along small arcs (we will present experimental evidence in Section 5.7).

This modification was designed for the use in high energy physics experiments and employed is several nuclear research centers for a number of years. In those experiments, one deals with elementary particles born in an accelerator that move along circular arcs in a constant magnetic field in a detector chamber (such as a bubble chamber); physicists determine the energy of the particle by measuring the radius of its trajectory; faster particles move along arcs of larger radius (lower curvature), thus the nearly singular circle fit occurs quite often in those experiments.

Drawbacks. A disadvantage of the Chernov-Ososkov modification is the complexity of its algebraic formulas (5.17)–(5.19), which are not easy to program. Besides, the lack of theoretical analysis (until 2005) was disheartening.

These issues were resolved in the late 1980s when Pratt (independently of Chernov and Ososkov) proposed a different algebraic circle fit [150] described in the next section. The Pratt fit turns out to be mathematically equivalent to the Chernov-Ososkov method, but it admits a simpler, more elegant, and numerically stable programming implementation; it is also much easier to analyze theoretically.

5.5   Pratt circle fit

Modified objective function by Pratt. We go back to the parameters A,B,C,D introduced in Section 3.2, i.e., we describe a circle by algebraic equation

$$A(x^2+y^2)+Bx+Cy+D = 0, \qquad (5.20)$$

as it was done in (3.8). Now the function $\mathscr{F}_2$ in (5.17) can be rewritten as

$$\mathscr{F}_2 = \frac{1}{B^2+C^2-4AD} \sum_{i=1}^{n}(Az_i+Bx_i+Cy_i+D)^2 \qquad (5.21)$$

(indeed, its value does not change if we multiply the vector (A,B,C,D) by a scalar, thus it is always possible to make $A = 1$ and recover (5.17)).

Constrained minimization. Recall that $B^2+C^2-4AD > 0$ whenever the quadruple (A,B,C,D) represents a circle (Section 3.2). Thus, the minimization of (5.21) over A,B,C,D is equivalent to the minimization of the simpler function

$$\mathscr{F}_3(A,B,C,D) = \sum_{i=1}^{n}(Az_i+Bx_i+Cy_i+D)^2 \qquad (5.22)$$

subject to the constraint

$$B^2+C^2-4AD = 1, \qquad (5.23)$$

which, by the way, we have already encountered in (3.10). Under this constraint, equation (5.20) describes all circles and lines in the xy plane, see Section 3.2. Thus the Pratt algorithm conveniently incorporates lines into its selection of the best fit (unlike the Kåsa and Chernov-Ososkov methods that can only return circles).

Reduction to an eigenvalue problem. To solve the minimization problem (5.22)–(5.23), Pratt [150] employs methods of linear algebra. The function (5.22) can be written in matrix form as

$$\mathscr{F}_3 = \|\mathbf{X}\mathbf{A}\|^2 = \mathbf{A}^T(\mathbf{X}^T\mathbf{X})\mathbf{A}, \qquad (5.24)$$

where $\mathbf{A} = (A,B,C,D)^T$ is the vector of parameters and

$$\mathbf{X} = \begin{bmatrix} z_1 & x_1 & y_1 & 1 \\ \vdots & \vdots & \vdots & \vdots \\ z_n & x_n & y_n & 1 \end{bmatrix} \qquad (5.25)$$

is the $n \times 4$ "extended data matrix" (note that it is different from (5.11)). The constraint (5.23) can be written as

$$\mathbf{A}^T\mathbf{B}\mathbf{A} = 1, \qquad (5.26)$$

where

$$\mathbf{B} = \begin{bmatrix} 0 & 0 & 0 & -2 \\ 0 & 1 & 0 & 0 \\ 0 & 0 & 1 & 0 \\ -2 & 0 & 0 & 0 \end{bmatrix}. \tag{5.27}$$

Now introducing a Lagrange multiplier η we minimize the function

$$\mathscr{G}(\mathbf{A},\eta) = \mathbf{A}^T(\mathbf{X}^T\mathbf{X})\mathbf{A} - \eta(\mathbf{A}^T\mathbf{B}\mathbf{A} - 1).$$

Differentiating with respect to $\mathbf{A}$ gives the first order necessary condition

$$(\mathbf{X}^T\mathbf{X})\mathbf{A} - \eta\mathbf{B}\mathbf{A} = \mathbf{0}, \tag{5.28}$$

thus $\mathbf{A}$ must be a generalized eigenvector of the matrix pair $(\mathbf{X}^T\mathbf{X}, \mathbf{B})$. As the matrix $\mathbf{B}$ is invertible, we can rewrite (5.28) as

$$\mathbf{B}^{-1}(\mathbf{X}^T\mathbf{X})\mathbf{A} - \eta\mathbf{A} = \mathbf{0}, \tag{5.29}$$

hence η is an eigenvalue and $\mathbf{A}$ is an eigenvector of the matrix $\mathbf{B}^{-1}(\mathbf{X}^T\mathbf{X})$.

Analysis of the eigenvalue problem. Let us examine the eigensystem of $\mathbf{B}^{-1}(\mathbf{X}^T\mathbf{X})$. The matrix $\mathbf{X}^T\mathbf{X}$ is positive semi-definite, and in the generic case it is positive definite (the singular case is discussed below). The other matrix, $\mathbf{B}^{-1}$, is symmetric and has four real eigenvalues, one of them negative and three other positive (its spectrum is $\{1, 1, 0.5, -0.5\}$). Premultiplying (5.29) with $\mathbf{Y} = (\mathbf{X}^T\mathbf{X})^{1/2}$ transforms it into

$$\mathbf{Y}\mathbf{B}^{-1}\mathbf{Y}(\mathbf{Y}\mathbf{A}) - \eta(\mathbf{Y}\mathbf{A}) = \mathbf{0},$$

hence η is an eigenvalue of $\mathbf{Y}\mathbf{B}^{-1}\mathbf{Y}$; also note that $\mathbf{Y} = \mathbf{Y}^T$. Thus, by Sylvester's law of inertia [78] the matrix $\mathbf{B}^{-1}(\mathbf{X}^T\mathbf{X})$ has the same signature as $\mathbf{B}^{-1}$ does, i.e., the eigenvalues of $\mathbf{B}^{-1}(\mathbf{X}^T\mathbf{X})$ are all real, exactly three of them are positive and one is negative.

We note that the negative eigenvalue $\eta < 0$ corresponds to an eigenvector $\mathbf{A}$ satisfying $\mathbf{A}^T\mathbf{B}\mathbf{A} < 0$, according to (5.28), hence it does not represent any (real) circle or line. Three positive eigenvalues correspond to real circles (or lines).

But which eigenpair $(\eta, \mathbf{A})$ solves our problem, i.e., minimizes $\mathscr{F}_3$? Well, note that

$$\mathscr{F}_3 = \mathbf{A}^T(\mathbf{X}^T\mathbf{X})\mathbf{A} = \eta\mathbf{A}^T\mathbf{B}\mathbf{A} = \eta$$

(due to the constraint (5.26)), hence the function $\mathscr{F}_3$ is minimized when η is the *smallest positive* eigenvalue of $\mathbf{B}^{-1}(\mathbf{X}^T\mathbf{X})$.

The two larger positive eigenvalues apparently correspond to saddle points of the function $\mathscr{F}_3$, so they are of no practical interest.

Singular case of the eigenvalue problem. Lastly we turn to the singular case. The matrix $\mathbf{X}^T\mathbf{X}$ is singular if and only if there exists a nonzero vector $\mathbf{A}_0 = (A, B, C, D)$ such that $\mathbf{X}\mathbf{A}_0 = \mathbf{0}$, i.e.,

$$A(x_i^2 + y_i^2) + Bx_i + Cy_i + D = 0$$

for all $i = 1, \ldots, n$. This means precisely that the data points (x_i, y_i) either lie on a circle (this happens if $A \neq 0$) or on a line (if $A = 0$). In either case $\eta = 0$ is an eigenvalue of the matrix $\mathbf{B}^{-1}(\mathbf{X}^T\mathbf{X})$, the corresponding eigenvector is $\mathbf{A}_0$, and

$$\mathscr{F}_3(\mathbf{A}_0) = \|\mathbf{X}\mathbf{A}_0\|^2 = 0,$$

hence $\mathbf{A}_0$ minimizes $\mathscr{F}_3$. In fact it provides a perfect fit.

Conclusions. The Pratt circle fit is given by an eigenvector $\mathbf{A} = (A, B, C, D)$ of the matrix $\mathbf{B}^{-1}(\mathbf{X}^T\mathbf{X})$ that corresponds to its *smallest nonnegative eigenvalue* $\eta \geq 0$. If $\mathbf{X}$ has full rank (the generic case), then $\mathbf{A}$ can also be found by $\mathbf{A} = \mathbf{Y}^{-1}\mathbf{A}_*$, where $\mathbf{Y} = (\mathbf{X}^T\mathbf{X})^{1/2}$ and $\mathbf{A}_*$ is an eigenvector of the symmetric matrix $\mathbf{Y}\mathbf{B}^{-1}\mathbf{Y}$ that corresponds to its smallest positive eigenvalue $\eta > 0$ (as a side note: $\mathbf{B}^{-1}(\mathbf{X}^T\mathbf{X})$ has three positive and one negative eigenvalue).

5.6   Implementation of the Pratt fit

The analysis in the previous section uniquely identifies the eigenpair $(\eta, \mathbf{A})$ of the matrix $\mathbf{B}^{-1}(\mathbf{X}^T\mathbf{X})$ that provides the Pratt fit, but we also need a practical algorithm to compute it. Here we describe two such algorithms; each one has its own merit.

SVD-based Pratt fit. If one uses software with built-in matrix algebra operations (such as MATLAB), then one may be tempted to call a routine returning the eigenpairs of $\mathbf{B}^{-1}(\mathbf{X}^T\mathbf{X})$ and select the eigenvector $\mathbf{A}$ corresponding to the smallest nonnegative eigenvalue. This promises a simple, albeit not the fastest, solution.

However, this solution is not numerically stable. For one reason, the condition number of the matrix $\mathbf{X}^T\mathbf{X}$ is that of $\mathbf{X}$ squared, i.e., $\varkappa(\mathbf{X}^T\mathbf{X}) = [\varkappa(\mathbf{X})]^2$; thus assembling and using $\mathbf{X}^T\mathbf{X}$ may cause an explosion of round-off errors if $\mathbf{X}^T\mathbf{X}$ happens to be nearly singular. Second, we have observed cases where the numerically computed eigenvalues of $\mathbf{B}^{-1}(\mathbf{X}^T\mathbf{X})$ turned out complex, though theoretically they all must be real.

A more stable matrix solution is achieved by the singular value decomposition (SVD). First, we compute the (short) SVD, $\mathbf{X} = \mathbf{U}\Sigma\mathbf{V}^T$, of the matrix $\mathbf{X}$. If its smallest singular value, σ_4, is less than a predefined tolerance ε (say, $\varepsilon = 10^{-12}$), then we have a singular case at hand, and the solution $\mathbf{A}$ is simply the corresponding right singular vector, i.e., the fourth column of the $\mathbf{V}$ matrix.

1. Form the matrix $\mathbf{X}$ and compute its (short) SVD, $\mathbf{X} = \mathbf{U\Sigma V}^T$.
2. If $\sigma_4 < \varepsilon$, then $\mathbf{A}$ is the 4th column of $\mathbf{V}$. Stop.
3. If $\sigma_4 \geq \varepsilon$, form the matrix $\mathbf{Y} = \mathbf{V\Sigma V}^T$.
4. Compute the symmetric matrix $\mathbf{YB}^{-1}\mathbf{Y}$ and find its eigensystem.
5. Select the eigenpair $(\eta, \mathbf{A}_*)$ with the smallest positive eigenvalue η.
6. Compute $\mathbf{A} = \mathbf{Y}^{-1}\mathbf{A}_*$.

Table 5.1: *Pratt circle fit (SVD-based).*

In the regular case ($\sigma_4 \geq \varepsilon$), we form $\mathbf{Y} = \mathbf{V\Sigma}$ and find the eigenpairs of the symmetric matrix $\mathbf{Y}^T\mathbf{B}^{-1}\mathbf{Y}$. Selecting the eigenpair $(\eta, \mathbf{A}_*)$ with the smallest positive eigenvalue and computing $\mathbf{A} = \mathbf{V\Sigma}^{-1}\mathbf{A}_*$ completes the solution. Computing the matrix Σ^{-1} is simple — we just replace its diagonal components with their reciprocals.

We call this algorithm the *SVD-based Pratt fit*.

Newton-based Pratt fit. If matrix algebra functions are not available, or if one is hunting for speed rather than numerical stability, one can find η by solving the characteristic equation

$$\det(\mathbf{X}^T\mathbf{X} - \eta\mathbf{B}) = 0. \tag{5.30}$$

This is a polynomial equation of the 4th degree in η, which we write as $P(\eta) = 0$. In fact, this equation plays the same role as the Chernov-Ososkov quartic equation (5.18), which is not surprising as the two methods are mathematically equivalent. But (5.30) is easier to analyze and to solve than (5.18). Expanding (5.30) gives

$$P(\eta) = 4\eta^4 + c_2\eta^2 + c_1\eta + c_0 = 0, \tag{5.31}$$

where

$$\begin{aligned}
c_2 &= -\overline{zz} - 3\overline{xx}^2 - 3\overline{yy}^2 - 4\overline{xy}^2 - 2\overline{xx}\,\overline{yy} \\
c_1 &= \bar{z}(\overline{zz} - \bar{z}^2) + 4\bar{z}(\overline{xx}\,\overline{yy} - \overline{xy}^2) - \overline{xz}^2 - \overline{yz}^2 \tag{5.32} \\
c_0 &= \overline{xz}^2\overline{yy} + \overline{yz}^2\overline{xx} - 2\overline{xz}\,\overline{yz}\,\overline{xy} - (\overline{xx}\,\overline{yy} - \overline{xy}^2)(\overline{zz} - \bar{z}^2),
\end{aligned}$$

here we use the same notation system as in (5.18)–(5.19) and again assume that the data set is centered, i.e., $\bar{x} = \bar{y} = 0$. Notably, (5.31)–(5.32) look simpler and shorter than (5.18)–(5.19).

The practical solution of (5.31) is based on the following fact. Suppose the matrix $\mathbf{X}^T\mathbf{X}$ is not singular, i.e., $P(0) \neq 0$. Let

$$\eta_1 < 0 < \eta_2 \leq \eta_3 \leq \eta_4$$

1. Form the matrix $\mathbf{X}$ and compute $\mathbf{X}^T\mathbf{X}$.
2. Compute the coefficients (5.32) of the characteristic equation (5.31).
3. Initialize $\eta = 0$, and apply Newton's procedure to find a root η.
4. Compute $\mathbf{A}$ as a null vector of $\mathbf{X}^T\mathbf{X} - \eta\mathbf{B}$.

Table 5.2: *Pratt circle fit (Newton-based).*

denote the roots of (5.31).

Theorem 9 *In the nonsingular case we have $P(0) < 0$ and $P''(\eta) < 0$ for all $0 \le \eta \le \eta_2$. Thus, a simple Newton method supplied with the initial guess $\eta = 0$ always converges to the smallest positive root η_2 of $P(\eta)$.*

Proof. Denote by $\eta_1' \le \eta_2' \le \eta_3'$ the zeroes of the derivative $P'(\eta)$, and by $\eta_1'' < \eta_2''$ the zeroes of the second derivative $P''(\eta) = 48\eta^2 + c_2$. Note that the cubic term is missing in (5.31), so there is no linear term in $P''(\eta)$. Since $c_2 < 0$, it follows that $\eta_1'' < 0 < \eta_2''$. Now due to the interlacing property of the roots of a polynomial and its derivative we conclude that

$$\eta_1' \le \eta_1'' < 0 < \eta_2 \le \eta_2' \le \eta_2''.$$

This implies that $P(0) < 0$, $P'(0) > 0$, and $P(\eta)$ is a convex function (meaning that $P''(\eta) < 0$) in the interval between 0 and the smallest positive root η_2. Thus the Newton procedure is bound to converge, and its limit point will be the desired root η_2. Theorem is proved. $\square$

In practice, the Newton procedure converges in 3 to 5 iterations, on the average, with the tolerance set to 10^{-12}. Once the smallest positive root η_2 of (5.31) is located, the parameter vector $\mathbf{A}$ is chosen as a null vector of $\mathbf{X}^T\mathbf{X} - \eta\mathbf{B}$ according to (5.28).

We call this algorithm the *Newton-based Pratt fit*.

Computational issues. We have determined experimentally, working with MATLAB, that the above Newton-based implementation of the Pratt fit is just as fast as the Kåsa method (see Table 5.3). The numerically stable SVD-based implementation of the Pratt fit described earlier in this section is about 2 to 3 times slower than the Kåsa method. For comparison, all geometric fits are at least 10 times slower than the Kåsa method.

MATLAB codes of both versions of the Pratt method (as well as many other circle fits) are available from our web page [84].

	Kåsa	Pratt (SVD-based)	Pratt (Newton-based)
CPU time	1	2-3	1
Statistical bias	heavy	small	small
Numerical stability	N/A	high	low

Table 5.3: *Comparison of three algebraic circle fits.*

General remarks on numerical stability. It is common nowadays to require the use of numerically stable procedures (such as QR and SVD) in matrix computations and avoid fast but unstable solutions (e.g., forming "normal equations"). However in applied statistics the significance of numerical stability is diminished by the very nature of statistical problems, because round-off errors that occur in numerically unstable algorithms are often far smaller than uncertainties resulting from statistical errors in data.

For algebraic circle fits, the issue of numerical stability surfaces only when the relevant matrices become nearly singular, which happens in two cases: (i) the data points are almost on a circle, i.e., $\sigma \approx 0$, and (ii) the fitting circular arc is nearly flat, i.e., $R \approx \infty$. Both cases are characterized by a small ratio σ/R.

We have tested experimentally the performance of numerically stable (SVD-based) algebraic circle fits versus their unstable (Newton-based) counterparts and found that they performed almost identically under all realistic conditions. The difference between them becomes noticeable only when $\sigma/R \precsim 10^{-12}$, which is an extremely rare case.

Even in those extreme cases practical advantages of numerically stable fits are doubtful. For example, if $R \sim 1$ and $\sigma \sim 10^{-12}$, then stable fits determine the circle parameters to within $\sim 10^{-12}$, while unstable fits are accurate to within 10^{-6}, which practically looks just as good. As another extreme, let $\sigma \sim 1$ and $R \sim 10^{12}$; in that case an accurate estimation of the circle radius and center seems rather pointless anyway; one can just as well fit the data with a straight line.

In our MATLAB code bank, we supply both stable and unstable versions for algebraic circle fits.

5.7   Advantages of the Pratt algorithm

The Pratt circle fit has all the advantages of the Kåsa method we described in Section 5.2, and more!

First, if the points (x_i, y_i) happen to lie on a circle (or a line), the method finds that circle (or line) right away. (Note that the Kåsa method would fail in the case of a collinear data.)

In statistical terms, the Pratt method is consistent in the small-noise limit $\sigma \to 0$, i.e., the estimates of the parameters (a, b, R) converge to their true values as $\sigma \to 0$. Moreover, the Pratt fit is asymptotically efficient (optimal) in that limit; see Chapter 7.

Invariance under translations and rotations. Next we verify that the Pratt fit is invariant under translations and rotations. This fact is less obvious than a similar invariance of geometric fitting algorithms (established in Section 4.11) or the Kåsa method (Section 5.2), as the Pratt procedure is defined algebraically; thus we provide a detailed proof, following Taubin [176].

Let us apply a translation

$$T_{c,d} : (x, y) \mapsto (x + c, y + d)$$

by vector (c, d) to the data point coordinates. Then the data matrix $\mathbf{X}$ will be transformed to

$$\mathbf{X}_{\text{new}} = \begin{bmatrix} (x_1 + c)^2 + (y_1 + d)^2 & x_1 + c & y_1 + d & 1 \\ \vdots & \vdots & \vdots & \vdots \\ (x_n + c)^2 + (y_n + d)^2 & x_n + c & y_n + c & 1 \end{bmatrix}.$$

We can express this new data matrix as

$$\mathbf{X}_{\text{new}} = \mathbf{XF}, \tag{5.33}$$

where

$$\mathbf{F} = \begin{bmatrix} 1 & 0 & 0 & 0 \\ 2c & 1 & 0 & 0 \\ 2d & 0 & 1 & 0 \\ c^2 + d^2 & c & d & 1 \end{bmatrix}. \tag{5.34}$$

Observe also that

$$\mathbf{B} = \mathbf{F}^T \mathbf{BF} \tag{5.35}$$

Thus in the new coordinates the characteristic equation (5.28) takes form

$$\mathbf{F}^T (\mathbf{X}^T \mathbf{X}) \mathbf{FA}_{\text{new}} - \eta_{\text{new}} \mathbf{F}^T \mathbf{BFA}_{\text{new}} = \mathbf{0}.$$

Premultiplying by $\mathbf{F}^{-T}$ and comparing to (5.28) we conclude that $\eta_{\text{new}} = \eta$ and

$$\mathbf{A} = \mathbf{FA}_{\text{new}}. \tag{5.36}$$

In fact, we only need $\mathbf{A}$ up to a scalar multiple, hence we can assume that its first component is $A = 1$, hence $\mathbf{A} = (1, -2a, -2b, a^2 + b^2 - R^2)$, and the same holds for $\mathbf{A}_{\text{new}}$. Now it is easy to conclude from (5.36) that

$$a_{\text{new}} = a + c, \qquad b_{\text{new}} = b + d, \qquad R_{\text{new}} = R,$$

implying the invariance of the Pratt fit under translations.

The analysis of rotations follows the same lines. Let the coordinates be transformed by

$$R_\theta: (x,y) \mapsto (x\cos\theta + y\sin\theta, -x\sin\theta + y\cos\theta),$$

i.e., the points are rotated through angle θ. The the data matrix is transformed to $\mathbf{X}_{new} = \mathbf{XF}$ with

$$\mathbf{F} = \begin{bmatrix} 1 & 0 & 0 & 0 \\ 0 & c & -s & 0 \\ 0 & s & c & 0 \\ 0 & 0 & 0 & 1 \end{bmatrix}, \tag{5.37}$$

where we use common shorthand notation $c = \cos\theta$ and $s = \sin\theta$. Again the important relation (5.35) holds, thus we again arrive at (5.36), which easily implies

$$a_{new} = a\cos\theta + b\sin\theta, \qquad b_{new} = -a\sin\theta + b\cos\theta, \qquad R_{new} = R,$$

meaning the invariance of the Pratt fit under rotations.

A unified treatment of the Kåsa and Pratt fits. Our old Kåsa method can be described, in an equivalent manner, as minimizing the Pratt objective function

$$\mathscr{F}_3(A,B,C,D) = \sum_{i=1}^{n} (Az_i + Bx_i + Cy_i + D)^2$$

$$= \mathbf{A}^T(\mathbf{X}^T\mathbf{X})\mathbf{A} \tag{5.38}$$

subject to the constraint $|A| = 1$; the latter can be written as

$$\mathbf{AKA} = 1,$$

where the matrix $\mathbf{K}$ is given by

$$\mathbf{K} = \begin{bmatrix} 1 & 0 & 0 & 0 \\ 0 & 0 & 0 & 0 \\ 0 & 0 & 0 & 0 \\ 0 & 0 & 0 & 0 \end{bmatrix}. \tag{5.39}$$

Comparing this with (5.24)–(5.27) shows that the Pratt and Kåsa fits only differ by the constraint matrix, one uses $\mathbf{B}$ and the other $\mathbf{K}$; also both methods are invariant under translations, rotations, and similarities.

Now we may consider a general problem of minimizing the same function (5.38) subject to a constraint $\mathbf{A}^T\mathbf{NA} = 1$, where $\mathbf{N}$ is an arbitrary symmetric matrix. It is interesting to check for which $\mathbf{N}$ the resulting fit would be invariant under translations, rotations, and similarities.

General invariance under translations. We present the final results omitting tedious details. Our analysis shows that an algebraic fit with constraint matrix N is invariant under translations $T_{c,d}$ whenever $N = F^T NF$, where the matrix F is given by (5.34). This happens precisely if N is a linear combination of K and B, i.e., $N = \alpha K + \beta B$, or

$$N = \begin{bmatrix} \alpha & 0 & 0 & -2\beta \\ 0 & \beta & 0 & 0 \\ 0 & 0 & \beta & 0 \\ -2\beta & 0 & 0 & 0 \end{bmatrix}. \qquad (5.40)$$

General invariance under rotations. Next, an algebraic fit with constraint matrix N is invariant under rotations R_θ whenever $N = F^T NF$, but now the matrix F is given by (5.37). This happens precisely if N has form

$$N = \begin{bmatrix} \alpha & 0 & 0 & \gamma \\ 0 & \beta & 0 & 0 \\ 0 & 0 & \beta & 0 \\ \gamma & 0 & 0 & \delta \end{bmatrix}, \qquad (5.41)$$

where $\alpha, \beta, \gamma, \delta$ are arbitrary constants. Thus the invariance under rotations requires a little less than that under translations.

General invariance under similarities. Lastly, consider the invariance of an algebraic fit with constraint matrix N under similarities $(x, y) \mapsto (cx, cy)$. In that case the parameter vector $A = (A, B, C, D)$ changes by the rule

$$A = (A, B, C, D) \mapsto A_c = (A/c^2, B/c, C/c, D).$$

Hence we need the expression $P(c) = A_c^T NA_c$ be a homogeneous polynomial in c (i.e., all its terms must have the same degree with respect to c) so that c could be just canceled out. For the matrix N in (5.40) we have

$$P(c) = A_c^T NA_c = \alpha A^2 c^{-4} + \beta (B^2 + C^2 - 4AD)c^{-2}.$$

This expression is homogeneous in c only if $\alpha = 0$ (which gives the Pratt fit) or $\beta = 0$ (which is the Kåsa fit).

Final conclusion. The only algebraic fits with constraint matrix N that are invariant under translations, rotations, and similarities are Kåsa and Pratt fits. We should emphasize that our constraint matrix N is assumed to be constant; we will see other constraint matrices, which depend on X, later.

5.8 Experimental test

Here we present the results of an experimental comparison of the three circle fits: Kåsa, Pratt, and geometric.

	Kåsa			Pratt			Geometric		
α	q_1	q_2	q_3	q_1	q_2	q_3	q_1	q_2	q_3
360	0.995	1.002	1.010	0.997	1.005	1.012	0.994	1.001	1.009
300	0.994	1.002	1.010	0.997	1.005	1.012	0.994	1.001	1.009
240	0.991	1.001	1.011	0.994	1.004	1.015	0.991	1.001	1.011
180	0.976	0.993	1.010	0.987	1.004	1.022	0.984	1.001	1.019
120	0.908	0.943	0.981	0.962	1.003	1.047	0.960	1.001	1.046
90	0.778	0.828	0.883	0.927	1.000	1.087	0.928	1.001	1.088
60	0.449	0.502	0.561	0.833	0.990	1.224	0.840	0.999	1.237
45	0.263	0.297	0.340	0.715	0.967	1.501	0.735	0.996	1.547
30	0.166	0.176	0.190	0.421	0.694	1.433	0.482	0.804	1.701

Table 5.4 *Three quartiles of the radius estimates found by each algorithm (Kåsa, Pratt, geometric fit). The true radius is $R = 1$.*

Experiment. For each $\alpha = 360°, 350°, \ldots, 30°$ we have generated 10^6 random samples of $n = 20$ points located (equally spaced) on an arc of angle α of the unit circle $x^2 + y^2 = 1$ and corrupted by Gaussian noise at level $\sigma = 0.05$. We started with a full circle ($\alpha = 360°$) and went down to rather small arcs ($\alpha = 30°$).

For each random sample we found the best fitting circle by three methods: Kåsa, Pratt, and a geometric fit (Levenberg-Marquardt). Each method gives 10^6 estimates of the radius R, which we ordered and recorded the three quartiles: q_1, q_2 (the median), and q_3. These three characteristics roughly represent the overall performance of the method; in particular, the interval (q_1, q_3) contains 50% of all the radius estimates. The results are summarized in Table 5.4.

Results. Table 5.4 gives results in numerical format; and Fig. 5.6 presents them in graphic format. On each panel, the top curve is the third quartile q_3, the middle curve is the median, q_2, and the bottom curve is the first quartile, q_1. The grey shaded area captures the middle 50% of the radius estimates.

We see that all three methods perform nearly identically on large arcs, over the range $150° \leq \alpha \leq 360°$. A notable feature is a slight bias toward larger circles (the median always exceeds the true value $R = 1$ by a narrow margin), this phenomenon will be explained in Chapters 6 and 7.

On arcs below $150°$, the Kåsa method starts sliding down, as it consistently underestimates the radius. This tendency becomes disastrous on arcs between $90°$ and $45°$, as all the three quartiles fall sharply, and for arcs under $45°$ the Kåsa method returns absurdly small circles.

The Pratt and geometric fits fare better than Kåsa. Their errors grow as the arc decreases, but their medians stay "on mark" all the way down to $60°$ for Pratt and even to $45°$ for the geometric fit. Observe that the geometric fit

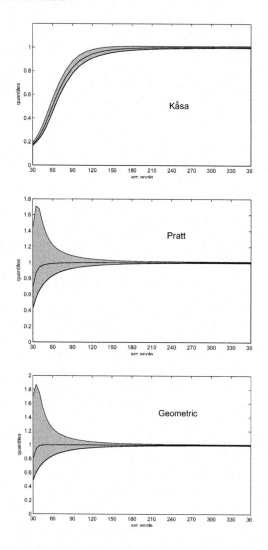

Figure 5.6 *The top curve is the third quartile, the middle curve is the median, and the bottom curve is the first quartile. The grey shaded area captures the middle 50% of the radius estimates.*

slightly outperforms the best (so far) algebraic fit (Pratt); we will return to this issue in Chapter 7.

Below 45°, all the medians go down rapidly, and on arcs smaller than 30° all the fitting procedures become quite unreliable.

Breakdown point. The reason for the breakdown of all our algorithms is

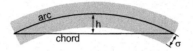

Figure 5.7: *The bow h in the arc and the noise level σ.*

illustrated in Fig. 5.7. Here

$$h = R(1 - \cos \alpha/2)$$

is a crucial parameters, the distance from the midpoint of the arc to its chord (h is called the "bow" in the arc [90]). If the noise level is smaller than the bow ($\sigma < h$), then the shape of the arc is "well defined" [90], i.e., recognizable, and circle fitting algorithms have a chance to succeed (and some do). On the other hand, if $\sigma > h$, then the noise completely blurs the arc, and circle fitting procedures consistently fail to find it. In our experiment, $R = 1$ and $\sigma = 0.05$, hence the critical arc size is

$$\alpha_{cr} = 2\cos^{-1}(1 - \sigma/R) = 36^\circ,$$

below which everybody collapses. Of course, reducing the noise level σ would allow our algorithms to perform well on smaller arcs. Generally, the fitting algorithms can handle arbitrarily small arcs, as long as σ does not exceed the bow.

Some other algebraic fits. We only describe here the most popular algebraic fitting schemes, but a few others are worth mentioning. Gander, Golub, and Strebel [67] propose to minimize the same function (5.38) that we used for the Pratt and Kåsa methods, but subject to a more straightforward constraint $\|\mathbf{A}\| = 1$, i.e.,

$$A^2 + B^2 + C^2 + D^2 = 1.$$

Our analysis in Section 5.7 shows, however, that the resulting fit would not be invariant under translations; in fact the authors [67] also admit that it is geometrically "unsatisfactory."

Nievergelt [137] proposes a fairly complicated algebraic fit, but if one inspects it closely, it turns out to be equivalent to the minimization of the same function (5.38) subject to the constraint

$$A^2 + B^2 + C^2 = 1.$$

Nievergelt argues that his fit handles certain singular cases (namely, those of collinear data) better than the Kåsa fit does. Nievergelt also recognizes that his fit would not be invariant under translations. As a remedy he uses a prior centering of the data, i.e., translating the coordinate system to enforce $\bar{x} = \bar{y} = 0$; this trick makes the final result invariant under both translations and rotations. (Prior data centering can also be applied to the Gander-Golub-Strebel fit and make it invariant.) We note, however, that neither fit is invariant under similarities, according to our Section 5.7, and their authors did not address this issue.

The Nievergelt fit was proposed independently, a year later, by Strandlie, Wroldsen, and Frühwirth [174] who explored an interesting 3D representation of the circle fitting problem. They treat $z = x^2 + y^2$ as a third coordinate, thus every observed point (x_i, y_i) has its image (x_i, y_i, z_i) in the 3D space, with $z_i = x_i^2 + y_i^2$. Note that all these spacial points lie on the paraboloid $z = x^2 + y^2$. The equation of a circle, $Az + Bx + Cy + D = 0$, now represents a plane in the xyz space; thus the problem reduces to fitting a plane to a set of 3D data points. If one fits a plane to 3D points by minimizing geometric distances, then one needs to minimize $\sum(Az_i + Bx_i + Cy_i + D)^2$ subject to constraint $A^2 + B^2 + C^2 = 1$; see [174] and our Section 8.2 for more details. This explains the choice of the constraint.

The authors of [174] call their method *Paraboloid fit*, because the data points lie on a paraboloid.

Practically, though, the above two algebraic fits perform not much better than Kåsa. Fig. 5.8 shows the corresponding plots for the Gander-Golub-Strebel (GGS) fit and the Nievergelt method (both with the prior data centering) obtained in the same experiment as the one reported in Fig. 5.6. We see that the GGS fit has only a slightly smaller bias than Kåsa (but a much greater variance when the arc is about 90°). The Nievergelt fit has smaller bias than GGS or Kåsa, but it is still more biased than the Pratt and geometric fits.

We emphasize that the above experimental results are obtained for a specific choice of the true circle center, which was placed at $(0,0)$. For different locations of the center, the bias may be larger or smaller, as these two fits are not invariant under translations.

5.9   Taubin circle fit

Another interesting variant of algebraic circle fits was proposed in 1991 by Taubin [176]. In the case of circles discussed here, the Taubin fit is very similar to the Pratt, in its design and performance, so we only describe it briefly. However, the Taubin fit (unlike Pratt) can be generalized to ellipses and other algebraic curves.

Modified objective function by Taubin. Recall that the Kåsa method minimizes the following function (5.14):

$$\mathcal{F}_1 = \sum_{i=1}^{n} d_i^2 (d_i + 2R)^2, \tag{5.42}$$

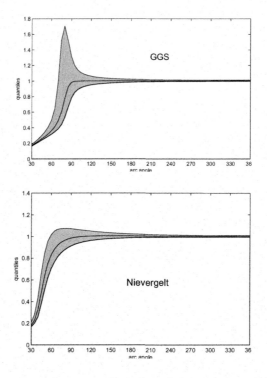

Figure 5.8 *The top curve is the third quartile, the middle curve is the median, and the bottom curve is the first quartile. The grey shaded area captures the middle 50% of the radius estimates.*

where the d_i's denote the distances from the data points to the circle (a, b, R). In Section 5.3 we used a natural assumption $|d_i| \ll R$ to derive the approximation (5.15), i.e.,

$$\mathcal{F}_1 \approx 4R^2 \sum_{i=1}^{n} d_i^2, \tag{5.43}$$

on which both Chernov-Ososkov and Pratt modifications were then built.

Now the same assumption $|d_i| \ll R$ can be used as follows:

$$R^2 \approx (d_i + R)^2 = (x_i - a)^2 + (y_i - b)^2.$$

Furthermore, we can improve this approximation by averaging

$$R^2 \approx \frac{1}{n} \sum_{i=1}^{n} (x_i - a)^2 + (y_i - b)^2,$$

in which case some positive and negative fluctuations will cancel out. Thus

(5.43) can be rewritten as

$$\mathscr{F}_1 \approx \frac{4}{n} \left[\sum_{i=1}^{n} (x_i - a)^2 + (y_i - b)^2 \right] \left[\sum_{i=1}^{n} d_i^2 \right]. \tag{5.44}$$

Since we actually want to minimize $\sum d_i^2$, we can achieve that goal (approximately) by minimizing the following function:

$$\begin{aligned} \mathscr{F}_4(a,b,R) &= \frac{\sum \left[(x_i - a)^2 + (y_i - b)^2 - R^2 \right]^2}{4n^{-1} \sum (x_i - a)^2 + (y_i - b)^2} \\ &= \frac{\sum [z_i - 2ax_i - 2by_i + a^2 + b^2 - R^2]^2}{4n^{-1} \sum [z_i - 2ax_i - 2by_i + a^2 + b^2]}. \end{aligned} \tag{5.45}$$

At this point we switch to the algebraic circle parameters A, B, C, D by the rules (3.11), i.e., we substitute

$$a = -\frac{B}{2A}, \qquad b = -\frac{C}{2A}, \qquad R^2 = \frac{B^2 + C^2 - 4AD}{4A^2}$$

and obtain

$$\mathscr{F}_4(A,B,C,D) = \frac{\sum [Az_i + Bx_i + Cy_i + D]^2}{n^{-1} \sum [4A^2 z_i + 4ABx_i + 4ACy_i + B^2 + C^2]}. \tag{5.46}$$

The minimization of (5.46) is proposed by Taubin [176]. He actually arrived at (5.46) differently, by using gradient weights; those will be discussed in Section 6.5.

Taubin fit in matrix form. The minimization of (5.46) is equivalent to the minimization of the simpler function

$$\mathscr{F}_3(A,B,C,D) = \sum_{i=1}^{n} (Az_i + Bx_i + Cy_i + D)^2, \tag{5.47}$$

which first appeared in (5.22), subject to the constraint

$$4A^2 \bar{z} + 4AB\bar{x} + 4AC\bar{y} + B^2 + C^2 = 1.$$

In matrix form, this means the minimization of

$$\mathscr{F}_3(\mathbf{A}) = \|\mathbf{XA}\|^2 = \mathbf{A}^T (\mathbf{X}^T \mathbf{X}) \mathbf{A},$$

in terms of (5.24), subject to the constraint $\mathbf{A}^T \mathbf{T} \mathbf{A} = 1$, where

$$\mathbf{T} = \begin{bmatrix} 4\bar{z} & 2\bar{x} & 2\bar{y} & 0 \\ 2\bar{x} & 1 & 0 & 0 \\ 2\bar{y} & 0 & 1 & 0 \\ 0 & 0 & 0 & 0 \end{bmatrix} \tag{5.48}$$

(here we again use our standard sample mean notation $\bar{z} = \frac{1}{n}\sum z_i$, etc.). Just like in Section 5.5, the solution $\mathbf{A}$ would be a generalized eigenvector of the matrix pair $(\mathbf{X}^T\mathbf{X}, \mathbf{T})$.

Taubin fit versus Pratt fit. Taubin's objective function (5.46) can be simplified in a different way. Observe that it is a quadratic polynomial in D, hence $\mathscr{F}_4$ has a unique global (conditional) minimum in D, when the other three parameters A, B, C are kept fixed, which gives

$$D = -A\bar{z} - B\bar{x} - C\bar{y} \tag{5.49}$$

(notably, this solves the third Kåsa equation in (5.5), as in that case $A = 1$). Substituting (5.49) into the denominator of (5.46) gives

$$\mathscr{F}_4(A,B,C,D) = \frac{\sum[Az_i + Bx_i + Cy_i + D]^2}{B^2 + C^2 - 4AD}, \tag{5.50}$$

and this function is identical to the Pratt objective function (5.21); quite a surprising observation!

So both Pratt and Taubin methods seem to minimize the same objective function $\mathscr{F}_2 = \mathscr{F}_4$, but are they really equivalent? No, because the Taubin method involves one extra relation, (5.49). We recall that Pratt's minimization of (5.21) is equivalent to the minimization of (5.47) subject to the constraint

$$B^2 + C^2 - 4AD = 1, \tag{5.51}$$

cf. (5.22)–(5.23). Now the Taubin algorithm is equivalent to the minimization of (5.47) subject to *two* constraints: (5.51) *and* (5.49).

One can think of the Pratt method as finding a minimum of the function $\mathscr{F}_3$ restricted to the 3D manifold $\mathbb{P}_1$ defined by equation (5.51), while the Taubin method finds a minimum of the same function $\mathscr{F}_3$ but restricted to a smaller domain, the 2D surface $\mathbb{P}_2$ defined by two equations, (5.51) and (5.49). Obviously, $\mathbb{P}_2 \subset \mathbb{P}_1$, hence the Pratt's minimum will be always smaller than (or equal to) the Taubin's minimum.

However, this does not mean that the Pratt fit is always better than the Taubin fit, as the real quality of each fit is measured by the sum of squared distances, $\sum d_i^2$. In our tests, we observed the Taubin fit actually beat the Pratt fit, in this sense, quite regularly. The statistical accuracy of both fits will be analyzed in Chapter 7.

5.10 Implementation of the Taubin fit

Elimination of parameter D. For simplicity, let us assume, as usual, that the data set is centered, i.e., $\bar{x} = \bar{y} = 0$. Then (5.49) takes form

$$D = -A\bar{z},$$

which allows us to easily eliminate D from the picture. Now the problem is to minimize

$$\mathscr{F}_5(A,B,C) = \sum_{i=1}^{n} \left[A(z_i - \bar{z}) + Bx_i + Cy_i \right]^2 \tag{5.52}$$

subject to the constraint

$$4\bar{z}A^2 + B^2 + C^2 = 1. \tag{5.53}$$

It is further convenient to introduce a new parameter, $A_0 = 2\bar{z}^{1/2}A$. Now we minimize

$$\mathscr{F}_5(A_0,B,C) = \sum_{i=1}^{n} \left[A_0 \frac{z_i - \bar{z}}{2\bar{z}^{1/2}} + Bx_i + Cy_i \right]^2 \tag{5.54}$$

subject to

$$A_0^2 + B^2 + C^2 = 1. \tag{5.55}$$

Reduction to eigenvalue problem. Again we employ methods of linear algebra. The function (5.54) can be written in matrix form as

$$\mathscr{F}_5 = \|\mathbf{X}_0 \mathbf{A}_0\|^2 = \mathbf{A}_0^T (\mathbf{X}_0^T \mathbf{X}_0) \mathbf{A}_0, \tag{5.56}$$

where $\mathbf{A}_0 = (A_0, B, C)^T$ is the reduced and modified vector of parameters and

$$\mathbf{X}_0 = \begin{bmatrix} (z_1 - \bar{z})/(2\bar{z}^{1/2}) & x_1 & y_1 \\ \vdots & \vdots & \vdots \\ (z_n - \bar{z})/(2\bar{z}^{1/2}) & x_n & y_n \end{bmatrix} \tag{5.57}$$

is the $n \times 3$ modified data matrix. The constraint (5.55) simply means $\|\mathbf{A}_0\| = 1$, i.e., $\mathbf{A}_0$ must be a unit vector.

The minimum of (5.56) is attained on the unit eigenvector of the matrix $\mathbf{X}_0^T \mathbf{X}_0$ corresponding to its smallest eigenvalue. This matrix is symmetric and positive-semidefinite, thus all its eigenvalues are real and nonnegative. Furthermore, this matrix is nonsingular, i.e., it is positive-definite, unless the data points lie of a circle or a line (i.e., admit a perfect fit).

SVD-based Taubin fit. Practically, one can evaluate $\mathbf{A}_0$ as follows. The simplest option is to call a matrix algebra function that returns all the eigenpairs of $\mathbf{X}_0^T \mathbf{X}_0$ and select the eigenvector corresponding to the smallest eigenvalue.

Alternatively, the eigenvectors of $\mathbf{X}_0^T \mathbf{X}_0$ can be found by singular value decomposition (SVD), as they coincide with the right singular vectors of $\mathbf{X}_0$. Accordingly, one can compute the (short) SVD of the matrix $\mathbf{X}_0$, i.e., $\mathbf{X}_0 = \mathbf{U}\Sigma\mathbf{V}^T$, and then $\mathbf{A}_0$ will be the third (last) column of $\mathbf{V}$. This procedure bypasses the evaluation of the matrix $\mathbf{X}_0^T \mathbf{X}_0$ altogether and makes computations more stable numerically.

We call this algorithm *SVD-based Taubin fit*.

Newton-based Taubin fit. If matrix algebra functions are not available, or if one is hunting for speed rather than numerical stability, one can find the smallest eigenvalue η of the matrix $\mathbf{X}_0^T \mathbf{X}_0$ by solving its characteristic equation

$$\det(\mathbf{X}_0^T \mathbf{X}_0 - \eta \mathbf{I}) = 0. \tag{5.58}$$

This is a polynomial equation of the 3rd degree in η, which can be written as

$$P(\eta) = c_3 \eta^3 + c_2 \eta^2 + c_1 \eta + c_0 = 0, \tag{5.59}$$

where

$$
\begin{aligned}
c_3 &= 4\bar{z} \\
c_2 &= -\overline{zz} - 3\bar{z}^2 \\
c_1 &= \bar{z}(\overline{zz} - \bar{z}^2) + 4\bar{z}(\overline{xx}\,\overline{yy} - \overline{xy}^2) - \overline{xz}^2 - \overline{yz}^2 \\
c_0 &= \overline{xz}^2\,\overline{yy} + \overline{yz}^2\,\overline{xx} - 2\overline{xz}\,\overline{yz}\,\overline{xy} - (\overline{xx}\,\overline{yy} - \overline{xy}^2)(\overline{zz} - \bar{z}^2)
\end{aligned}
\tag{5.60}
$$

where we use the same notation as in (5.31)–(5.32) (remember also that we again assume that the data set is centered, i.e., $\bar{x} = \bar{y} = 0$; the relation $\bar{z} = \overline{xx} + \overline{yy}$ is helpful, too). Interestingly, the formulas for c_1 and c_0 are identical in (5.32) and here.

Since the eigenvalues of $\mathbf{X}_0^T \mathbf{X}_0$ are real and nonnegative, equation (5.59) always has three nonnegative real roots. Therefore $P(0) \leq 0$. In the nonsingular case, we have $P(0) < 0$, and then $P''(\eta) < 0$ in the interval between 0 and the first (smallest) root. Thus a simple Newton method supplied with the initial guess $\eta = 0$ always converges to the desired smallest root of (5.58).

We call this algorithm *Newton-based Taubin fit*.

When choosing an algorithm for practical purposes, recall our comments on the issue of numerical stability at the end of Section 5.6.

Experimental tests. Computationally, the Taubin fit is slightly simpler, hence a bit less expensive than the Pratt fit. We also tested its accuracy experimentally, in the way described in Section 5.8. We recall that in that test, see Table 5.4, the Pratt and geometric fits showed a very similar performance, while the Kåsa method turned out to be much less reliable.

Additional tests reveal that the performance of the Taubin fit closely follows that of Pratt and geometric fits, in fact in this picture it lies "in between". For the reader's convenience, we summarize our results regarding these three algorithms in Table 5.5, where again q_1, q_2 and q_3 are the three quartiles of the radius estimates (recall that the true radius is $R = 1$).

We note that on smaller arcs Taubin's characteristics are closer to the (slightly less accurate) Pratt fit, and on larger arcs, to the (more superior) geometric fit. However, the difference between these three fits is fairly small, so in most practical application any one of them would do a good job. One should

	Pratt			Taubin			Geometric		
α	q_1	q_2	q_3	q_1	q_2	q_3	q_1	q_2	q_3
360	0.997	1.005	1.012	0.995	1.002	1.010	0.994	1.001	1.009
300	0.997	1.005	1.012	0.995	1.002	1.010	0.994	1.001	1.009
240	0.994	1.004	1.015	0.992	1.002	1.012	0.991	1.001	1.011
180	0.987	1.004	1.022	0.985	1.002	1.020	0.984	1.001	1.019
120	0.962	1.003	1.047	0.960	1.001	1.045	0.960	1.001	1.046
90	0.927	1.000	1.087	0.925	0.999	1.086	0.928	1.001	1.088
60	0.833	0.990	1.224	0.833	0.991	1.226	0.840	0.999	1.237
45	0.715	0.967	1.501	0.720	0.975	1.515	0.735	0.996	1.547
30	0.421	0.694	1.433	0.442	0.733	1.522	0.482	0.804	1.701

Table 5.5 *Three quartiles of the radius estimates found by Pratt, Taubin, and geometric fit. The true radius is $R = 1$.*

remember, though, that every version of the geometric fit (Chapter 4) is at least 5 to 10 times slower than any algebraic fit, including the numerically stable SVD-based Pratt and Taubin.

Invariance under translations and rotations. Lastly, we note that the Taubin fit, just like the Pratt and Kåsa fits, is invariant under translations, rotations, and similarities. This invariance can be proved in the same way as we did in Section 5.7 for the Pratt fit; the only difference is that now we have a new constraint matrix, T, which is data-dependent, hence it changes under coordinate transformation. We just need to check that the new matrix, T_{new}, is obtained by the rule

$$T_{new} = F^T T F$$

where F is the matrix that describes the translations, see (5.34), or rotations, see (5.37) in Section 5.7. The above identity is an analogue of (5.35); it can be easily verified directly. The invariance of under similarities follows from the fact that whenever $(x, y) \mapsto (cx, cy)$, we have $(A, B, C, D) \mapsto (A/c^2, B/c, C/c, D)$, and hence $A^T T A \mapsto c^{-2} A^T T A$. Thus the constraint $A^T T A = 1$ will be transformed into $A^T T A = c^2$, which is irrelevant as our parameter vector A only needs to be determined up to a scalar multiple.

In fact, Taubin proved that his method was invariant in a more general context of fitting ellipses to data; see his article [176].

5.11   General algebraic circle fits

We have now seen several algebraic circle fits: Kåsa, Pratt, Taubin, Gander-Golub-Strebel, and Nivergelt. Here we present a general definition of algebraic circle fits that encompasses all the known fits as particular examples.

Matrix representation. Algebraic circle fits are based on the algebraic equation of a circle

$$A(x^2 + y^2) + Bx + Cy + D = 0, \tag{5.61}$$

where $\mathbf{A} = (A, B, C, D)^T$ is the 4-parameter vector. Every algebraic fit minimizes the function

$$\mathscr{F}(A, B, C, D) = \frac{1}{n} \sum_{i=1}^{n} (Az_i + Bx_i + Cy_i + D)^2$$

$$= n^{-1} \mathbf{A}^T (\mathbf{X}^T \mathbf{X}) \mathbf{A} = \mathbf{A}^T \mathbf{M} \mathbf{A}, \tag{5.62}$$

subject to a constraint

$$\mathbf{A}^T \mathbf{N} \mathbf{A} = 1 \tag{5.63}$$

for some matrix $\mathbf{N}$. Here we use our shorthand notation $z_i = x_i^2 + y_i^2$ and

$$\mathbf{M} = \frac{1}{n} \mathbf{X}^T \mathbf{X} = \begin{bmatrix} \overline{zz} & \overline{zx} & \overline{zy} & \overline{z} \\ \overline{zx} & \overline{xx} & \overline{xy} & \overline{x} \\ \overline{zy} & \overline{xy} & \overline{yy} & \overline{y} \\ \overline{z} & \overline{x} & \overline{y} & 1 \end{bmatrix},$$

where

$$\mathbf{X} = \begin{bmatrix} z_1 & x_1 & y_1 & 1 \\ \vdots & \vdots & \vdots & \vdots \\ z_n & x_n & y_n & 1 \end{bmatrix}. \tag{5.64}$$

The constraint matrix $\mathbf{N}$ in (5.63) determines the particular algebraic fit. As we know,

$$\mathbf{N} = \mathbf{K} = \begin{bmatrix} 1 & 0 & 0 & 0 \\ 0 & 0 & 0 & 0 \\ 0 & 0 & 0 & 0 \\ 0 & 0 & 0 & 0 \end{bmatrix} \tag{5.65}$$

for the Kåsa fit, cf. (5.39),

$$\mathbf{N} = \mathbf{P} = \begin{bmatrix} 0 & 0 & 0 & -2 \\ 0 & 1 & 0 & 0 \\ 0 & 0 & 1 & 0 \\ -2 & 0 & 0 & 0 \end{bmatrix} \tag{5.66}$$

for the Pratt fit, see (5.27), and

$$\mathbf{N} = \mathbf{T} = \begin{bmatrix} 4\overline{z} & 2\overline{x} & 2\overline{y} & 0 \\ 2\overline{x} & 1 & 0 & 0 \\ 2\overline{y} & 0 & 1 & 0 \\ 0 & 0 & 0 & 0 \end{bmatrix}, \tag{5.67}$$

for the Taubin fit, see (5.48); here we again use our standard "sample mean" notation $\bar{z} = \frac{1}{n}\sum z_i$, etc.

Reduction to a generalized eigenvalue problem. To solve the constrained minimization problem (5.62)–(5.63) one uses a Lagrange multiplier η and reduces it to an unconstrained minimization of the function

$$\mathscr{G}(\mathbf{A},\eta) = \mathbf{A}^T\mathbf{M}\mathbf{A} - \eta(\mathbf{A}^T\mathbf{N}\mathbf{A} - 1).$$

Differentiating with respect to $\mathbf{A}$ and η gives

$$\mathbf{M}\mathbf{A} = \eta\mathbf{N}\mathbf{A} \qquad (5.68)$$

and

$$\mathbf{A}^T\mathbf{N}\mathbf{A} = 1, \qquad (5.69)$$

thus $\mathbf{A}$ must be a generalized eigenvector for the matrix pair $(\mathbf{M},\mathbf{N})$, which also satisfies $\mathbf{A}^T\mathbf{N}\mathbf{A} = 1$. The problem (5.68)–(5.69) may have several solutions. To choose the right one we note that for each solution $(\eta,\mathbf{A})$ we have

$$\mathbf{A}^T\mathbf{M}\mathbf{A} = \eta\mathbf{A}^T\mathbf{N}\mathbf{A} = \eta, \qquad (5.70)$$

thus for the purpose of minimizing $\mathbf{A}^T\mathbf{M}\mathbf{A}$ we should choose the solution of (5.68)–(5.69) with the smallest η. Note that η is automatically nonnegative, since $\mathbf{M} = \frac{1}{n}\mathbf{X}^T\mathbf{X}$ is a positive semi-definite matrix.

Since multiplying $\mathbf{A}$ by a scalar does not change the circle it represents, it is common in practical applications to require that $\|\mathbf{A}\| = 1$, instead of $\mathbf{A}^T\mathbf{N}\mathbf{A} = 1$. Accordingly one needs to replace the rigid constraint (5.69) with a softer one $\mathbf{A}^T\mathbf{N}\mathbf{A} > 0$. The latter can be further relaxed as follows.

For generic data sets, $\mathbf{M}$ is positive definite. Thus if $(\eta,\mathbf{A})$ is any solution of the generalized eigenvalue problem (5.68), then $\mathbf{A}^T\mathbf{M}\mathbf{A} > 0$. In that case, due to (5.70), we have $\mathbf{A}^T\mathbf{N}\mathbf{A} > 0$ if and only if $\eta > 0$. Thus it is enough to solve the problem (5.68) and choose the smallest positive η and the corresponding unit vector $\mathbf{A}$. This rule is almost universally used in practice. However, it needs to be adapted to one special case, which turns out to be quite precarious.

The singular case. The matrix $\mathbf{M}$ is singular if and only if the observed points lie on a circle (or a line); in this case the eigenvector $\mathbf{A}_0$ corresponding to $\eta = 0$ satisfies $\mathbf{X}\mathbf{A}_0 = \mathbf{0}$, i.e., it gives the interpolating circle (line), which is obviously the best possible fit. However, it may happen that for some (poorly chosen) matrices $\mathbf{N}$ we have $\mathbf{A}_0^T\mathbf{N}\mathbf{A}_0 < 0$, so that the geometrically perfect solution fails to satisfy the "soft" constraint $\mathbf{A}^T\mathbf{N}\mathbf{A} > 0$, and thus has to be rejected. Such algebraic fits are poorly designed and not worth considering.

For all the constraint matrices $\mathbf{N}$ mentioned so far we have $\mathbf{A}^T\mathbf{N}\mathbf{A} \geq 0$ whenever $\mathbf{A}^T\mathbf{M}\mathbf{A} = 0$. For example, the Kåsa matrix $\mathbf{K}$, the Taubin matrix $\mathbf{T}$, and the constraint matrices corresponding to the Gander-Golub-Strebel fit and the Nivergelt fit (Section 5.8), are all positive semi-definite, hence $\mathbf{A}^T\mathbf{N}\mathbf{A} \geq 0$ for any vector $\mathbf{A}$. The Pratt matrix $\mathbf{P}$ is not positive semi-definite, but we have seen that $\mathbf{A}^T\mathbf{P}\mathbf{A} > 0$ for any vector $\mathbf{A}$ that represents a circle or a line, so $\mathbf{P}$ is good, too.

Thus in the singular case, the eigenvector $\mathbf{A}$ corresponding to $\eta = 0$ is the solution of the constrained minimization problem (5.62)–(5.63).

Summary. We conclude that the problem (5.62)–(5.63) can be solved in two steps: first we find all solutions $(\eta, \mathbf{A})$ of the generalized eigenvalue problem (5.68), and then we pick the one with the minimal nonnegative η. If $\eta = 0$, then the method returns a perfect fit (an interpolating circle or line).

5.12 A real data example

Here we present an example of fitting circular arcs to plotted contours in archaeological field expeditions [44]. It was already mentioned in our Preface.

A common archaeological problem consists of estimating the diameter of a potsherd from a field expedition. The original diameter at specific point along the profile of a broken pot — such as the outer rim or base — is restored by fitting a circle to a sherd. Sherd profiles are often traced with a pencil on a sheet of graph paper, which is later scanned and transformed into an array of pixels (data points). A typical digitized arc tracing a circular wheelmade antefix is shown in Fig. 5.9. This image contains 7452 measured points.

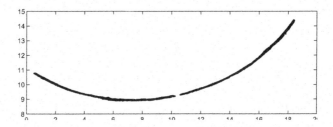

Figure 5.9 *A typical arc drawn by pencil with a profile gauge from a circular wheelmade antefix.*

The best fitting circle found by the geometric (Levenberg-Marquardt) method has parameters

$$\text{center} = (7.4487, 22.7436), \qquad \text{radius} = 13.8251, \tag{5.71}$$

which we assume to be exact (ideal).

Fig. 5.10 shows a fragment of the above image, it consists of merely $n = 22$ randomly chosen points (from the original 7452 points). Visually, they do not really appear as a circular arc, they rather look like a linear segment. Surprisingly, geometric fitting algorithms (again, we used the Levenberg-Marquardt) applied to these 22 points returned the following circle:

$$\text{center} = (7.3889, 22.6645), \qquad \text{radius} = 13.8111, \qquad (5.72)$$

which is strikingly accurate (compare this to (5.71)).

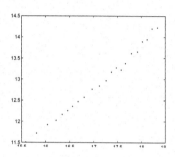

Figure 5.10: *A fragment of the arc shown in Fig. 5.9.*

Now algebraic circle fits presented in this chapter produced the following results:

$$
\begin{array}{llr}
\text{center} = (10.8452, 19.1690), & \text{radius} = 8.9208 & \textbf{(Kåsa)} \\
\text{center} = (11.2771, 18.7472), & \text{radius} = 8.3222 & \textbf{(GGS)} \\
\text{center} = (7.3584, 22.6964), & \text{radius} = 13.8552 & \textbf{(Nievergelt)} \\
\text{center} = (7.3871, 22.6674), & \text{radius} = 13.8146 & \textbf{(Pratt)} \\
\text{center} = (7.3871, 22.6674), & \text{radius} = 13.8145 & \textbf{(Taubin)}
\end{array}
$$

We see that the Kåsa and Gander-Golub-Strebel (GGS) fits greatly underestimate the radius, while the Nievergelt method overestimates it. The Pratt and Taubin fits are nearly identical and give results very close to the geometric fit. Overall, the Pratt fit found a circle closest to the ideal one (5.71).

One may naturally want to estimate errors of the returned values of the circle parameters, but this is a difficult task for most EIV regression problems, including the circle fitting problem. In particular, under the standard statistical models described in Chapter 6, the estimates of the center and radius of the circle by the Levenberg-Marquardt, Pratt, and Taubin methods happen to have infinite variances and infinite mean values!

An approximate error analysis developed in Chapter 7 can be used to assess errors in a realistic way. Then the errors of the best methods (Levenberg-Marquardt, Pratt, and Taubin) happen to be ≈ 0.1. For the Kåsa fit, there is an additional bias that needs to be taken into account, too. Since our data point span an arc of $\sim 20^\circ$, the Kåsa fit tends to return radius 2-3 times smaller than the actual radius; see Table 5.4. A detailed errors analysis is given in the next two chapters.

5.13   Initialization of iterative schemes

In the previous sections we treated each algebraic fit as a stand-alone procedure that aimed at the ultimate goal: an accurate estimation of the true circle parameters. However, in many applications a noniterative algebraic fit is used as an intermediate tool, to merely initialize a subsequent geometric fit.

In that case it is desirable that the algebraic prefit returns an estimate that lies close to the global minimum of the objective function, so that the geometric fit would quickly converge to it. This goal is different from the usual one — closeness to the true parameter value; see an illustration in Fig. 5.11.

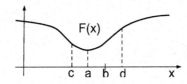

Figure 5.11 *Minimization of a function $F(x)$. Here a denotes the global minimum, b the true parameter value, and c and d are two different initial guesses. The guess d is closer to the true parameter value, thus it is better than c as a stand-alone estimate. However, c is closer to the minimum a, thus the iterative algorithm starting at c will converge to a faster than the one starting at d.*

Various initialization schemes. Recall that we have several geometric fitting procedures (Levenberg-Marquardt, Späth, Landau, all described in Chapter 4); theoretically they minimize the same objective function, (4.18), thus ideally they should return the same estimate.

However, in practice they behave quite differently (even if supplied with the same initial guess), because they choose different routes to the minimum. Hence some of them may be trapped in a local minima, some others may move too slowly and stall before reaching the global minimum, etc.

In addition, we have several algebraic fits which may supply different initial guesses to the geometric fitting procedures. We emphasize that each algebraic fit minimizes its own objective function, thus their results differ even theoretically.

Now one can combine any algebraic prefit (such as Kåsa, Pratt, Taubin) with any subsequent geometric fitting routine (such as Levenberg-Marquardt,

Späth, Landau) to obtain a full working scheme. Its performance may depend on the quality of the algebraic prefit, on the quality of the geometric fit, and on how they blend together.

Previous tests. Chernov and Lesort [43] explored various combinations of algebraic and geometric circle fits in an extensive experimental tests. Their results demonstrate that each combination is nearly 100% reliable and efficient when the data are sampled along a full circle or a large arc; this situation remains quite stable for arcs between $360°$ and about $90°$.

On smaller arcs (below $90°$), the Kåsa fit returns a heavily biased estimate that lies quite far from the global minimum of the objective function $\mathscr{F}$. Then the Späth and Landau methods would crawl to the minimum of $\mathscr{F}$ at such a slow pace that for all practical purposes they would fail to converge. The Levenberg-Marquardt fits still converge fast, within 10-20 iterations. If the initial guess is provided by the Pratt or Taubin fit, then all the geometric methods manage to converge, though again the Levenberg-Marquardt schemes are far more efficient than others.

We refer the reader to [43] for more detailed results presented by easy-to-read plots and diagrams.

Another test. We include another numerical test here similar to the one described in Section 5.8. For each $\alpha = 360°, \ldots, 30°$ we again generate 10^6 random samples of $n = 20$ points located (equally spaced) on an arc of angle α of the unit circle $x^2 + y^2 = 1$ and corrupted by Gaussian noise at level $\sigma = 0.05$. As in Section 5.8, we started with a full circle ($\alpha = 360°$) and wend down to rather small arcs ($\alpha = 30°$).

For every generated random sample we determined the distance d from the circle center estimated by the algebraic fit (Kåsa, Pratt, or Taubin) and the one found by the subsequent geometric fit (Levenberg-Marquardt). This distance shows how well the algebraic fit approximates the minimum of the objective function. We also record the number of iterations i the geometric procedure took to converge.

Table 5.6 shows the medians of d and i for each algebraic fit and for each α. We see that for large arcs all the algebraic fits provide accurate initial guesses (in one case, for the full circle, Kåsa is actually the most accurate!). On smaller arcs, the Kåsa fit falls behind the other two, due to its heavy bias, which results in a larger number of iterations the Levenberg-Marquardt has to take to recover. The Taubin fit again slightly outperforms the Pratt fit.

Initialization in nearly singular cases. To this rosy picture we must add a final, somewhat spicy remark. It concerns the singular case where the data are sampled along a *very* small arc, so that the best fitting circle would look nearly flat; see an example in Fig. 3.1. In that case it is crucial that the initial guess is selected on the right side of the data set, i.e., *inside* the best fitting circle. If the initial guess were selected *outside* the best circle, it would fall into the

α	Kåsa		Pratt		Taubin	
	d	i	d	i	d	i
360	0.00158	3	0.00159	3	0.00159	3
300	0.00190	3	0.00164	3	0.00164	3
240	0.00413	3	0.00188	3	0.00187	3
180	0.01404	4	0.00257	3	0.00256	3
120	0.06969	5	0.00478	4	0.00473	4
90	0.19039	6	0.00815	4	0.00800	4
60	0.53087	9	0.02010	6	0.01929	6
45	0.75537	12	0.04536	8	0.04265	8
30	0.58054	17	0.11519	12	0.10609	12

Table 5.6 *The median distance d from the circle center estimate provided by an algebraic fit (Kåsa, Pratt, or Taubin) and the one found by the subsequent geometric fit. In addition, i is the median number of iterations the geometric fit took to converge.*

escape valley from which geometric fitting procedures would never recover, cf. Sections 3.7 and 3.9.

According to our analysis in Section 3.9, the right side of the data set (i.e., the side on which the center of the best fitting circle lies) is determined by the sign of

$$\overline{xxy} = \frac{1}{n} \sum_{i=1}^{n} x_i^2 y_i,$$

which we called the *signature* of the data set. It is assumed here that the data set is properly aligned, i.e., it is centered so that $\bar{x} = \bar{y} = 0$ and rotated so that

$$\overline{xy} = 0 \qquad \text{and} \qquad \overline{xx} > \overline{yy}. \tag{5.73}$$

In Section 3.9 we established the following rules:

(a) if $\overline{xxy} > 0$, then the center of the best fitting circle lies *above* the x axis; the wrong valley lies below the x axis;

(b) if $\overline{xxy} < 0$, then the center of the best fitting circle lies *below* the x axis; the wrong valley lies above the x axis.

Let us see if our algebraic circle fits abide by these rules, i.e., place the center on the correct side of the data set. For the Kåsa fit, the second equation in (5.5) yields

$$b = -\frac{C}{2} = \frac{\overline{yz}}{2\overline{yy}} = \frac{\overline{xxy} + \overline{yyy}}{2\overline{yy}}. \tag{5.74}$$

Since the denominator is positive, the sign of b coincides with that of $\overline{xxy} + \overline{yyy}$. If this expression has the same sign as $\overline{xxy}$, the center is placed on the right side, otherwise it ends up on the wrong side (in the escape valley).

Usually, in accordance with (5.73), x_i's are larger than y_i's, so $\overline{xxy}$ is very likely to be greater than $\overline{yyy}$, in which case the Kåsa fit will make the right choice. However, the wrong side may be selected, too (see an example below), hence the Kåsa fit may completely distract the subsequent geometric fitting procedure.

The Pratt and Taubin fits can be analyzed similarly. In the Taubin case, we have

$$\mathbf{X}_0^T \mathbf{X}_0 \mathbf{A}_0 = \eta \mathbf{A}_0,$$

where $\eta \geq 0$ is the smallest eigenvalue of $\mathbf{X}_0^T \mathbf{X}_0$, cf. Section 5.10. Now the second equation in the above 3×3 system gives

$$b = -\frac{C}{2A} = \frac{\overline{yz}}{2(\overline{yy} - \eta)} = \frac{\overline{xxy} + \overline{yyy}}{2(\overline{yy} - \eta)}. \tag{5.75}$$

We also note that $\eta < \overline{yy}$, because the vector $\mathbf{A}_1 = (0,0,1)^T$ has Rayleigh quotient $\mathbf{A}_1^T \mathbf{X}_0^T \mathbf{X}_0 \mathbf{A}_1 = \overline{yy}$, which of course cannot be smaller than the minimal eigenvalue η. Thus again, as in the Kåsa case, the sign of b coincides with that of of $\overline{xxy} + \overline{yyy}$.

In the Pratt case, the parameter vector $\mathbf{A}$ satisfies

$$\mathbf{X}^T \mathbf{X} \mathbf{A} = \eta \mathbf{B} \mathbf{A},$$

see Section 5.5, and the third equation of this 4×4 system gives

$$b = -\frac{C}{2A} = \frac{\overline{yz}}{2(\overline{yy} - \eta)} = \frac{\overline{xxy} + \overline{yyy}}{2(\overline{yy} - \eta)},$$

which coincides with (5.75), except here η is an eigenvalue of a different matrix. But again one can easily check that $\eta < \overline{yy}$ (we omit details), hence the sign of b coincides with that of $\overline{xxy} + \overline{yyy}$.

Remark. Comparing (5.74) and (5.75) reveals why the Kåsa fit consistently returns smaller circles than the Pratt or Taubin fit in the nearly singular case: it happens because the corrective term $\eta > 0$ always decreases the denominator of (5.75).

Returning to our main issue, we see that all the three algebraic fitting algorithms follow the same rule: they place the circle center in accordance with the sign of $\overline{xxy} + \overline{yyy}$. And the correct choice should be based on the sign of $\overline{xxy}$ alone. The question is: are these two rules equivalent? The following example shows that they are not, i.e., all the algebraic algorithms can actually make the wrong choice.

W-example. Consider a set of $n = 5$ points:

$$(0,2), \quad (1, -1 - \varepsilon), \quad (-1, -1 - \varepsilon), \quad (2, \varepsilon), \quad (-2, \varepsilon)$$

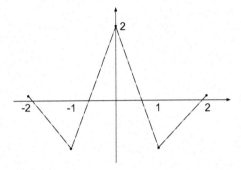

Figure 5.12 *The W-example: a five point set, on which all the algebraic fits choose the wrong side.*

where $\varepsilon > 0$ is a small number. These data resemble letter W, see Fig. 5.12, so we call it a W-example.

The conditions $\bar{x} = \bar{y} = 0$ and (5.73) hold easily. Now we see that

$$\overline{xxy} = -2 + 6\varepsilon < 0,$$

on the other hand

$$\overline{xxy} + \overline{yyy} = 4 - 6\varepsilon^2 > 0.$$

Hence the center of the best fitting circle lies below the x axis, while every algebraic circle fit returns a center on the opposite side (above the x axis).

This example demonstrates that in nearly singular cases one cannot blindly rely on algebraic circle fits to initialize a geometric fitting routine. For safety, one should check whether the "algebraic" circle abides by the rules (a) and (b), see above. If it does not, one should change the sign of b, for example, before feeding it to the geometric fitting procedure.

Chapter 6

Statistical analysis of curve fits

Thus far we discussed two sides of the the circle fitting problem: its theoretical analysis (Chapter 3) and its practical solutions (Chapters 4 and 5). In a sense, the theoretical analysis of Chapter 3 was "pre-computational," it prepared us for practical solutions. Chapters 4 and 5 were "computational." Now we turn to the third, and last ("post-computational") side of the problem: assessment of the accuracy of the circle fits. This brings us to statistical analysis of circle fits.

In this chapter we survey basic statistical properties of curve fitting algorithms when both variables are subject to errors. This is a fairly advanced topic in modern statistics, so we assume the reader is familiar with the basic concepts of probability and statistics. Our discussion need not be restricted to lines and circles anymore, it applies to general curves; though for illustrations we still use lines and circles.

We also need to make an important remark on the selection of material for this (and the next) chapter. The statistics literature devoted to the problem of fitting curves to data is vast, it encompasses all kinds of studies — from numer-

ical algorithms to highly abstract properties of related probability distributions.
It is simply impossible to present it all here.

This chapter can only serve as an introduction to the subject. When we
selected topics for it, our main criterion was practical relevance. We avoid going into the depth of theoretical analysis and only cover topics that we deem
most important for image processing applications; i.e., we describe statistical
studies which may have direct effect on practical results and which can explain
practical observations.

6.1   Statistical models

First we need to make assumptions about the underlying probability distribution of observations.

Curves. We assume that the curves are defined by an implicit equation

$$P(x,y;\Theta) = 0, \tag{6.1}$$

where $\Theta = (\theta_1, \ldots, \theta_k)^T$ denotes a vector[1] of unknown parameters to be estimated. In many cases, P is a polynomial in x and y, and its coefficients are
parameters (or some functions of parameters). For example, a circle is given
by

$$(x-a)^2 + (y-b)^2 - R^2 = 0,$$

hence $\Theta = (a,b,R)^T$ can be regarded as a 3D parameter vector.

Observed data. Each observed point (x_i, y_i) is a random (noisy) perturbation of a *true point* $(\tilde{x}_i, \tilde{y}_i)$, i.e.

$$x_i = \tilde{x}_i + \delta_i, \qquad y_i = \tilde{y}_i + \varepsilon_i, \qquad i = 1, \ldots, n, \tag{6.2}$$

where δ_i, ε_i, $i = 1, \ldots, n$ are small random errors (noise).

The true points $(\tilde{x}_i, \tilde{y}_i)$ are supposed to lie on the *true curve*, i.e., satisfy

$$P(\tilde{x}_i, \tilde{y}_i; \tilde{\Theta}) = 0, \qquad i = 1, \ldots, n, \tag{6.3}$$

where $\tilde{\Theta}$ denotes the vector of *true* (unknown) parameters. We also use vector
notation $\mathbf{x}_i = (x_i, y_i)^T$ and $\tilde{\mathbf{x}}_i = (\tilde{x}_i, \tilde{y}_i)^T$.

Disturbance vectors. The noise vectors $\mathbf{e}_i = \mathbf{x}_i - \tilde{\mathbf{x}}_i$ are assumed to be *independent* and have *zero mean*. Both assumptions are standard in the literature[2].
Two more specific assumptions on the probability distribution of the noise are
described below. We use the terminology due to Berman and Culpin [18].

---

[1] All our vectors are column vectors, thus we have to put the transposition sign T.

[2] Dependent, or correlated error vectors are occasionally treated in the literature [44, 57, 58].

Noise distribution: Cartesian model. A vast majority of statisticians assume that each $\mathbf{e}_i$ is a two-dimensional normal vector with some covariance matrix $\mathbf{V}$. In image processing studies, as we explained in Section 1.3, it is most natural to consider *isotropic noise*; then $\mathbf{V} = \sigma^2 \mathbf{I}$, where $\mathbf{I}$ denotes the identity matrix. In this case all our errors ε_i and δ_i are i.i.d. normal random variables with zero mean and a common variance σ^2. This is called Cartesian model. It will be our standard assumption.

Noise distribution: Radial model. An interesting alternative to the Cartesian model was proposed by Berman and Culpin [18]. They noted that only deviations of the observed points *from the curve* affect the fitting procedure, while their displacement *along the curve* does not matter. Thus they simplified the Cartesian model by assuming that deviations only occur in the normal direction, i.e., $\mathbf{e}_i = z_i \mathbf{n}_i$, where z_i is a Gaussian random variable with zero mean and variance σ^2, and $\mathbf{n}_i$ is a unit normal vector to the curve $P(x, y; \tilde{\Theta}) = 0$ at the point $\tilde{\mathbf{x}}_i$.

In both Cartesian and radial model, the value of σ^2 is unknown and should be regarded as a model parameter. Concerning the true points $\tilde{\mathbf{x}}_i$, $i = 1, \ldots, n$, two assumptions are possible.

True points location: Functional model. We can assume that the true points are fixed (but unobservable) and lie on the true curve. Their coordinates $(\tilde{x}_i, \tilde{y}_i)$ are treated as additional parameters of the model (called incidental or latent parameters). Since the true points are constrained to lie on the true curve (6.3), their locations can be described by n independent parameters (one per point). Now the model has the total of $k + 1 + n$ parameters: k principal parameters $\theta_1, \ldots, \theta_k$ of the unknown curve, the unknown σ^2, and the locations of the true points. This assumption is known as *functional model*.

True points location: Structural model. We can assume that the true points $\tilde{\mathbf{x}}_i$ are independent realizations of a random variable with a certain probability distribution concentrated on the true curve $P(x, y; \tilde{\Theta}) = 0$. For example, if the curve is a line, $y = \theta_0 + \theta_1 x$, then $\tilde{x}_i$ can have a normal distribution with some parameters μ_0 and σ_0^2, and $\tilde{y}_i$ is then computed by $\tilde{y}_i = \tilde{\theta}_0 + \tilde{\theta}_1 \tilde{x}_i$. In this model, the number of parameters is fixed (independent of n): these are the k principal parameters $\theta_1, \ldots, \theta_k$ of the unknown curve, the unknown σ^2, and the parameters of the underlying probability distribution on the true curve. This assumption is known as *structural model*.

True points location: Ultrastructural model. In the studies of fitting a straight line $y = \theta_0 + \theta_1 x$ to data, an interesting combination of the above two models was proposed by Dolby [56] in 1976 and quickly became popular in the statistics literature [39, 40]. Dolby assumed that every $\tilde{x}_i$ was an independent realization of a normal random variable with its own mean μ_i and a common variance $\sigma_0^2 \geq 0$. This model is a generalization of the functional and structural models: when $\sigma_0 = 0$, it reduces to the functional model (with $\mu_i = \tilde{x}_i$), and when $\mu_1 = \cdots = \mu_n (= \mu_0)$, it reduces to the structural model (with a normal distribution of $\tilde{x}_i$).

The ultrastructural model happens to be very convenient in the theoretical studies of *linear* errors-in-variables regression, but no one has yet applied it to the problem of fitting geometric curves, such as circles or ellipses. The reason for this is made clear in the next section.

6.2   Comparative analysis of statistical models

Drawbacks of the functional model. A major objection against the use of the functional model is that the number of its parameters grows with n, which makes the asymptotical analysis (as $n \to \infty$) of estimators difficult, cf. Chapters 1 and 2, where we have discussed the line fitting problem at length. However, in image processing applications, it is more natural to keep n fixed and consider the limit $\sigma \to 0$, see our discussion in Section 2.5; this asymptotic scheme causes no conflicts with the functional model.

Another possible concern with the functional model is that the statistical properties of the estimates of the principal parameters $\theta_1, \ldots, \theta_k$ (in particular, formulas for their biases, covariance matrix, and Cramer-Rao lower bounds) depend on the location of the true points, which are unobservable. This, however, is not a major issue in applications; in practice it is common to substitute either the observed points (x_i, y_i), or their projections onto the fitted curve, for the unknown $(\tilde{x}_i, \tilde{y}_i)$; and such a substitution gives a reasonable approximation.

Drawbacks of the structural model. In the studies of fitting *straight lines* to data, many statisticians prefer the more elegant structural model to the somewhat cumbersome functional model. However, when one fits nonlinear contours, especially circles or ellipses, to data, the structural model becomes more awkward, see below.

The main reason for the awkwardness of the structural model is that the true points $\tilde{x}_i$'s and $\tilde{y}_i$'s have to be chosen on the true contour (such as a circle or an ellipse); so they cannot have a normal distribution because their coordinates are restricted by the size of the contour. Thus one has to use other (nonnormal) distributions, which are concentrated on the contour (circle or ellipse). Perhaps the most natural choice would be a uniform distribution on the entire contour (with respect to the arc length on it), and indeed uniform distributions were used in early works [18, 31]. But they cannot cover many practically important cases, especially those where the data are observed along a small arc (this happens when only an occluded circle or ellipse is visible).

Von Mises distribution. Circular statistics, which deals with probability distributions concentrated on a circle, offers an analogue of the normal law: the so-called *von Mises* distribution (also called *circular normal distribution*);

it has a bell-shaped density on the circumference, and its spread over the circle is controlled by a "concentration parameter" $\varkappa \geq 0$, see [18, 16]. Its density on the circle (parameterized by an angular variable $\varphi \in [0, 2\pi]$) is

$$f(\varphi) = \frac{\exp\left[\varkappa\cos(\varphi - \mu)\right]}{2\pi I_0(\varkappa)},$$

where μ is the center of the distribution and $I_0(\varkappa)$ is the modified Bessel function of order 0. When $\varkappa = 0$, it turns to a uniform distribution on the entire circle; and when $\varkappa$ grows the density peaks at $\varphi = \mu$.

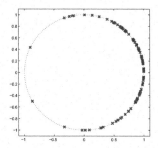

Figure 6.1 $n = 100$ *random points (crosses) on the unit circle generated by a von Mises distribution with* $\mu = 0$ *and* $\varkappa = 1$.

Simulated examples. Fig. 6.1 shows $n = 100$ points (crosses) generated on the unit circle with a von Mises distribution centered at $\mu = 0$ and having the concentration $\varkappa = 1$. We see that the points mostly cluster near the right end (where $\varphi = 0$), but occasionally appear on the left side, too. Perhaps such distributions describe some practical applications, but not such a common situation where the observed points are evenly spread along an arc, with the rest of the circle (ellipse) totally invisible, see Fig. 6.2. In the last case, a uniform distribution on the arc seems much more appropriate, but every statistician knows how inconvenient uniform densities are in theoretical analysis.

Thus it appears that there is no probability distributions concentrated on a circle or an ellipse that would adequately represent the majority of practical applications and be acceptable in theoretical studies. For these reasons practitioners prefer to rely on the functional model almost exclusively. We will restrict our further studies to the functional model (except for the end of Section 6.3).

6.3   Maximum Likelihood Estimators (MLE)

MLE in the functional model. In the previous section we described two models, which are the most appropriate for the studies of fitting geometric curves

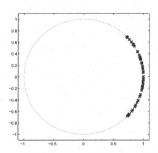

Figure 6.2 $n = 50$ *random points (crosses) on the unit circle generated by a uniform distribution on the arc* $|\varphi| \leq \pi/4$.

to data: the *Cartesian functional model* (the principal one) and the *radial functional model* (an alternative one). In both models the noise is assumed to have normal distribution.

Theorem 10 *In both functional models (Cartesian and radial), the Maximum Likelihood Estimator (MLE) of the primary parameters* $\theta_1, \ldots, \theta_k$ *is attained on the curve that minimizes the sum of squares of orthogonal distances to the data points.*

   Historical remarks. For the Cartesian functional model, the equivalence of the MLE and the geometric fit is known. In the simplest case of fitting straight lines to data, this fact is straightforward, and statisticians established it as early as the 1930s, see [113] and [126]. In the general case of fitting *nonlinear* curves to data, this fact is less apparent and its proof requires an indirect argument. To our knowledge, the first complete proof was published by Chan [31] in 1965.
   Despite this, the equivalence of the MLE and the geometric fit does not seem to be widely recognized by the applied community, it is rarely mentioned in the literature, and some authors rediscover it independently from time to time [34]. For the alternative radial functional model we could not locate a proof of this fact at all. Thus we provide a proof of the above theorem in full.

   Proof. The joint probability density function of the observations is

$$f(x_1, y_1, \ldots, x_n, y_n) = \frac{1}{(2\pi\sigma^2)^{m/2}} \exp\left[-\frac{1}{2\sigma^2} \sum_{i=1}^{n} (x_i - \tilde{x}_i)^2 + (y_i - \tilde{y}_i)^2\right]$$

$$= \frac{1}{(2\pi\sigma^2)^{m/2}} \exp\left[-\frac{1}{2\sigma^2} \sum_{i=1}^{n} \|\mathbf{x}_i - \tilde{\mathbf{x}}_i\|^2\right] \qquad (6.4)$$

In the Cartesian model, the coordinates (x_i, y_i) make $2n$ independent variables, hence $m = 2n$. In the radial model, each data point (x_i, y_i) is restricted to the line perpendicular to the curve $P(x,y;\Theta) = 0$ passing through the true point $(\tilde{x}_i, \tilde{y}_i)$, hence $m = n$.

We note that Θ is not explicitly involved in (6.4), which calls for an indirect approach to the maximum likelihood estimation. For convenience, we introduce a parameter t on the curve $P(x,y;\Theta) = 0$, then it can be expressed by parametric equations

$$x = u(t;\Theta) \qquad \text{and} \qquad y = v(t;\Theta)$$

(the functions u and v do not have to be known explicitly; it is enough that they exist). Then each $\tilde{x}_i$ can be replaced by the corresponding parameter value $\tilde{t}_i$, $i = 1,\ldots,n$. Note that $t_1,\ldots,t_n$ are unconstrained parameters.

Now the negative log-likelihood function is

$$-\log L(\Theta, \sigma^2, t_1, \ldots, t_n) = \frac{1}{2\sigma^2} \sum_{i=1}^{n} \left(x_i - u(t_i;\Theta)\right)^2 + \left(y_i - v(t_i;\Theta)\right)^2$$
$$+ \ln(2\pi\sigma^2)^{m/2}, \qquad (6.5)$$

where Θ appears explicitly, along with the nuisance parameters $t_1,\ldots,t_n$.

To minimize (6.5) we use the following simple *minimization-in-steps techniques*, as described by Chan [31]. If $h(X,Y)$ is a function on $A \times B$, where $X \in A$ and $Y \in B$, then

$$\min_{A \times B} h = \min_{A} \left(\min_{B} h\right)$$

(provided all the minima exist).

Accordingly, we can minimize (6.5) in two steps:

Step 1. Minimize (6.5) with respect to $t_1,\ldots,t_n$ when Θ and σ^2 are kept fixed.

Step 2. Minimize the resulting expression with respect to Θ and σ^2.

In Step 1 the curve $P(x,y;\Theta) = 0$ is kept fixed, and the minimum of (6.5) with respect to $t_1,\ldots,t_n$ is obviously achieved if $\left(x(t_i;\Theta), y(t_i;\Theta)\right)$ is the point on the curve *closest* to the observed point (x_i, y_i), i.e.,

$$\min_{t_1,\ldots,t_n} \sum_{i=1}^{n} \left(x_i - u(t_i;\Theta)\right)^2 + \left(y_i - v(t_i;\Theta)\right)^2 = \sum_{i=1}^{n} [d_i(\Theta)]^2, \qquad (6.6)$$

where $d_i(\Theta)$ denotes the geometric (orthogonal) distance from the point (x_i, y_i) to the curve $P(x,y;\Theta) = 0$. If the curve is smooth, then the line passing through the points (x_i, y_i) and $(u(t_i;\Theta), v(t_i;\Theta))$ crosses the curve orthogonally. The last fact is important for the radial model, as it requires that the point (x_i, y_i) lies on the line orthogonal to the curve passing through the true point; we see that this requirement will be automatically satisfied.

Lastly, in Step 2 the estimate of Θ is obtained by minimizing (6.6). Thus we get a curve that minimizes the sum of squares of the distances to the data points. The theorem is proved. $\square$

We denote by $\hat{\Theta}_{\text{MLE}}$ the maximum likelihood estimator. As we have just proved, in the functional model it is always obtained by the geometric fit, i.e., by minimizing the sum of squares of the distances from the data points to the curve.

MLE in the structural model. Though the structural model is less appropriate for image processing applications, we devote a brief discussion of the maximum likelihood estimation in its context. Quite unexpectedly, in that model the Maximum Likelihood Estimator is *different* from the geometric fit, even if the noise has a Gaussian distribution. We illustrate this fact by a rather striking example.

Example. Consider the problem of fitting a circle to data and suppose the distribution of the true points on the true circle is uniform (the least fanciful assumption). Let ρ and θ denote the polar coordinates attached to the center of the true circle. Then, in the radial structural model, the probability density function $g(\rho, \theta)$ of an observable at the point (ρ, θ) will be

$$g(\rho, \theta) = \frac{1}{2\pi} \frac{1}{(2\pi\sigma^2)^{1/2}} \exp\left[-\frac{(\rho - R)^2}{2\sigma^2}\right],$$

where R is the circle's radius. In the xy coordinates the density becomes

$$f(x,y) = \frac{1}{2\pi(d+R)} \frac{1}{(2\pi\sigma^2)^{1/2}} \exp\left[-\frac{d^2}{2\sigma^2}\right],$$

where $d = \sqrt{(x-a)^2 + (y-b)^2} - R$ denotes the signed distance from the point to the true circle. Note that the extra factor $d + R$ comes from the Jacobian.

For simplicity, assume that σ is known. Since the observations are independent, the negative log-likelihood function is

$$-\log L(a,b,R) = \sum_i \ln(d_i + R) + \sum_i \frac{d_i^2}{2\sigma^2}$$

(the constant terms are omitted).

Now suppose that the data points lie *on the true circle*. Then the geometric fit will return the true circle (and so does every algebraic fit discussed in Chapter 5). We will see next that the MLE picks a different circle.

For the true circle, $d_i = 0$, hence the negative log-likelihood is

$$-\log L(\tilde{a}, \tilde{b}, \tilde{R}) = n \ln \tilde{R}.$$

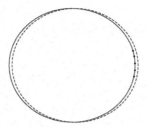

Figure 6.3 *Five data points lie on the true circle (the solid line), but the MLE returns a different circle (the dashed line).*

Suppose, in addition, that the data points are clustered along a small arc on the right hand (east) side of the circle; for example let

$$x_i = \tilde{a} + \tilde{R}\cos\big(\delta(i/n - 1/2)\big), \qquad y_i = \tilde{b} + \tilde{R}\sin\big(\delta(i/n - 1/2)\big),$$

for $i = 1,\ldots,n$, where $\delta > 0$ is small, see Fig. 6.3. Now consider a circle with parameters $(\tilde{a} + \varepsilon, \tilde{b}, \tilde{R})$. For that circle $d_i = -\varepsilon + \mathcal{O}(\varepsilon\delta^2)$, therefore

$$-\log L(\tilde{a} + \varepsilon, \tilde{b}, \tilde{R}) = n\ln(\tilde{R}) - \tilde{R}^{-1}n\varepsilon + \mathcal{O}(nR^{-1}\varepsilon\delta^2) + \mathcal{O}(n\sigma^{-2}\varepsilon^2).$$

It is clear that this function decreases when ε is a small positive number and takes its minimum at $\varepsilon \sim \sigma^2/R$. Thus the MLE fails to find the true circle, even if the true points are observed without noise.

General comments on the MLE for the structural model. Our example may give an impression that the Maximum Likelihood Estimators for the structural model are ill-behaved, but this is actually not true. General theorems in statistics guarantee that the MLE have optimal asymptotic properties, as $n \to \infty$; they are consistent, asymptotically normal, and efficient (have asymptotically minimal possible variance).

A detailed investigation of the MLE, in the context of the structural model for the circle fitting problem, was done in a series of papers by Anderson [6] and Berman and Culpin [16, 18] in the 1980s. They performed an approximative analysis, aided by numerical experiments, which confirmed the excellent behavior of the MLE in the limit $n \to \infty$. In fact the MLE were found to have smaller variance than that of the geometric circle fit, and even more so if compared to the Kåsa algebraic fit (Section 5.1). When the noise is large (and n is large), the variance of the MLE may be several times smaller than that of the geometric fit [16].

On the other hand, these studies also demonstrate difficulties in the practical computation of the MLE. Anderson [6] describes the MLE by explicit equations only in the simplest case, where the true points are uniformly distributed on the entire circle; even in that case the MLE equations involve modified Bessel functions defined by infinite power series. For other distributions of the true points on the circle, no equations describing the MLE are derived and no practical algorithms for computing the MLE are proposed.

Thus it appeared, at least back in the 1980s, that the structural model MLE, despite having superb statistical properties, were hopelessly impractical. Since then they remain abandoned. Though perhaps now, or in the near future, with more powerful computer resources and software, the structural model MLE of the circle parameters can be revisited and rehabilitated. This interesting approach promises an estimator superior to the geometric fit, at least in some situations.

6.4   Distribution and moments of the MLE

After adopting certain models for the distribution of observables (Sections 6.1 and 6.2) and deriving the corresponding MLE (Section 6.3), we can proceed to the main goals of our statistical analysis. In traditional statistics, estimators of unknown parameters are characterized by their probability distributions and especially by their moments; the moments provide such important characteristics as the *bias*, *variance*, and *mean squared error* of an estimate.

Unfortunately, this traditional approach runs into formidable difficulties arising from the nature of our problem; these difficulties are described and illustrated in this section. They prompt statisticians to develop alternative approaches based on various approximations, we will present them in the subsequent sections.

Exact distribution of the MLE. Due to the complexity of the curve fitting problems the exact distributions and moments of the maximum likelihood estimators cannot be determined in most cases. Even for the simplest task of fitting a straight line

$$y = \alpha + \beta x$$

to observed points, the distributions of the MLE $\hat{\alpha}$ and $\hat{\beta}$ are given by overly complicated expressions, which involve doubly infinite power series whose coefficients depend on incomplete Beta functions; see [7].

Exact formulas for the density functions of $\hat{\alpha}$ and $\hat{\beta}$ were derived in 1976 by Anderson [7, 10] who readily admitted that they were "not very informative" [8] and "not very useful" [7]. Instead, Anderson employed Taylor expansion to approximate the distribution function of $\hat{\beta}$, which allowed him to investigate some practical features of this estimator; see our Chapter 2 for a more detailed account.

In the case of fitting nonlinear curves, such as circles and ellipses, there are no explicit formulas of any kind for the maximum likelihood estimators themselves, let alone their probability densities. To our best knowledge, no one has ever attempted to describe the exact distributions of the corresponding MLE (although for the circle fitting problem, some related distributions were studied in [6, 18, 31]).

Now, since exact probability distributions of the MLE are unavailable, one has to follow Anderson [7, 10] and employ the Taylor expansion to construct reasonable approximations; we will do that in Section 6.6.

Moments of the MLE. Because exact distributions of the MLE's are intractable, one cannot derive explicit formulas for their moments either (with a few exceptions, though, see next). For example, if one fits a straight line

$$x \cos \varphi + y \sin \varphi + C = 0,$$

then the distribution of the MLE $\hat{\varphi}$ happens to be symmetric about its true value $\tilde{\varphi}$ (Section 2.4), hence $\mathbb{E}(\hat{\varphi}) = \tilde{\varphi}$; in this rare case the first moment can be found precisely. However, there are no explicit formulas for $\mathsf{Var}(\hat{\varphi})$ even in this case. There are no exact formulas for $\mathbb{E}(C)$ or $\mathsf{Var}(C)$ either, they can only be determined approximately, as we did in Chapter 2.

Furthermore, there is an even more disturbing fact about the moments of MLE's, which may not be easy to digest. Namely, if one describes the straight line by $y = \alpha + \beta x$, then the moments of the corresponding MLE, $\hat{\alpha}$ and $\hat{\beta}$, do not even exist, i.e.,

$$\mathbb{E}(|\hat{\alpha}|) = \mathbb{E}(|\hat{\beta}|) = \infty.$$

This means that these estimators have no mean values or variances. This fact was established by Anderson [7] in 1976, see our Section 1.6.

Anderson's discovery was rather striking; it was followed by heated discussions and a period of acute interest in the linear EIV problem. The basic question was: are the MLE $\hat{\alpha}$ and $\hat{\beta}$ statistically acceptable given that they have infinite mean squared error? (Well, not to mention infinite bias...) And what does the nonexistence of moments mean anyway, in practical terms?

To answer these questions, Anderson, Kunitomo, and Sawa [7, 10, 118] used Taylor expansions up to terms of order 3 and 4 to approximate the distribution functions of the MLE; and the resulting approximate distributions had finite moments. Those approximations remarkably agreed with numerically computed characteristics of the actual estimators (in simulated experiments), at least in all typical cases (see an exception below). Anderson and Sawa [10] noted that their approximations were "virtually exact." We will employ the approximation techniques to more general curves in Section 6.6.

An artificial example. To illustrate the idea of Anderson's approximative approach, consider a scalar estimate $\hat{\theta}$ whose density function f is a mixture,

$$f = (1 - p)f_0 + p f_1,$$

where

$$f_0(x) = \frac{1}{(2\pi\sigma_0^2)^{1/2}} \exp\left[-\frac{(x - \mu_0)^2}{2\sigma_0^2}\right]$$

is a normal density $N(\mu_0, \sigma_0^2)$ and

$$f_1(x) = \frac{1}{\pi(1 + x^2)}$$

is the standard Cauchy density. This estimate has infinite moments for any $p > 0$, but if p is very small, then in all practical terms $\hat{\theta}$ behaves as a normal random variable $N(\mu_0, \sigma_0^2)$, its Cauchy component is barely visible. For example, if $p \sim 10^{-9}$, then in every sequence of a billion observed random values of $\hat{\theta}$ typically just one (!) happens to come from the nonnormal (Cauchy) distribution, so in real data processing it would be hardly noticeable at all.

In this case the approximation $f \approx f_0$ appears to be "virtually exact," in Anderson's words. One can further speculate that the mean value $\mathbb{E}(\hat{\theta})$ can be well approximated by μ_0 and the variance $\mathrm{Var}(\hat{\theta})$ by σ_0^2; and such approximations would perfectly agree with simulated experiments and perhaps would be acceptable for all practical purposes. (On the other hand, theoretically, $\mathbb{E}(\hat{\theta})$ and $\mathrm{Var}(\hat{\theta})$ are infinite, hence they are *not* approximated by μ_0 and σ_0^2.)

Our example demonstrates that the nonexistence of moments may be an issue of a pure academic interest, whereas for practical purposes one may use the moments of properly constructed approximative distributions. Of course, if p in our example is not negligible, i.e., if the "bad" component of the distribution of $\hat{\theta}$ becomes "visible," then the above approximations will no longer be adequate, and the estimate $\hat{\theta}$ behaves poorly in practice.

In the linear fitting problem, the breakdown of Anderson's approximative approach occurs when the noise level σ becomes very large. Our tests show that this happens when σ is far beyond its level typical for most image processing applications. We recall our illustration in Fig. 1.8 demonstrating erratic behavior of the MLE $\hat{\beta}$ (thus experimentally revealing its infinite moments): there we simulated $n = 10$ data points on a stretch of length 10 along a true line, and the noise level was set to $\sigma = 2.4$. In other words, σ was 24% of the size of line segment containing the true points (while in typical computer vision applications σ does not exceed 5% of the size of the figure fitting the data set, cf. [17]). We had to set the noise level so high, i.e., to $\sigma = 2.4$, because for its smaller values, i.e., for $\sigma \leq 2.3$, the estimate $\hat{\beta}$ behaved as if it had finite moments (the solid line in Fig. 1.8 was just flat).

Moments of circle parameter estimates. It was recently discovered [41] that the MLE of the circle parameters (a, b, R) have infinite moments, too, i.e.,

$$\mathbb{E}(|\hat{a}|) = \mathbb{E}(|\hat{b}|) = \mathbb{E}(\hat{R}) = \infty, \tag{6.7}$$

thus the situation here is similar to the linear EIV regresion problem. To illustrate this fact we generated 10^6 random samples of $n = 10$ points along a semicircle of radius one with a Gaussian noise at level $\sigma = 0.58$. Fig. 6.4 plots the average maximum likelihood estimate $\hat{R}$ over k samples, as k runs from 1 to

10^6 (the solid line). It behaves erratically, like the average estimate of the slope β in Fig. 1.8. Note again that the noise level is very high, it is 29% of the circle size, because $\sigma/(2R) = 0.29$. We had to set the noise level so high because if $\sigma \leq 0.57$, the estimate $\hat{R}$ behaved as if it had finite moments (the solid line in Fig. 6.4 was just flat).

It is interesting that the circle parameter estimates $(\hat{a}_0, \hat{b}_0, \hat{R}_0)$ obtained by the Kåsa method (Section 5.1) have finite mean values whenever $n \geq 4$ and finite variances whenever $n \geq 5$; this fact was recently proved by Zelniker and Clarkson [196]. In Fig. 6.4, the average Kåsa estimate $\hat{R}_0$ is the dotted line, its value remains near a constant, 0.48, confirming that it has finite moments.

We note that both estimates (the geometric fit and the Kåsa fit) are heavily biased. The biasedness of the Kåsa fit toward smaller circles was explained in Section 5.3; the biasedness of the MLE toward larger circles will be discussed in Section 6.10 and Chapter 7.

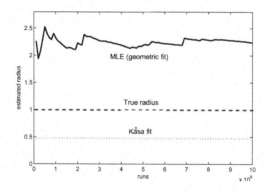

Figure 6.4: *The performance of the MLE and the Kåsa fit for circles.*

As it turns out, the circle parameter estimates obtained by the Pratt and Taubin methods (Chapter 5) also have finite or infinite moments. It appears that the nonexistence of moments of maximum likelihood estimates in geometric fitting problems is a rather general fact whenever one describes the fitted contour by naturally unbounded geometric parameters (such as the coordinates of the center, the dimensions, etc.). In such a generality, this fact remains to be proven, particularly in the context of fitting ellipses.

Of course, one can choose parameters of the fitting curve so that they are naturally bounded, and hence their estimates will have finite moments. For example, when the fitting line is described by $x\cos\varphi + y\sin\varphi + C = 0$, then $0 \leq \varphi \leq 2\pi$ is bounded. Concerning C, it represents the distance from the line to the origin (Section 2.1); though C is not truly bounded, its estimate has distribution with a tail similar to that of a normal density, hence all its moments are finite.

Similarly, a circle can be described by an algebraic equation

$$A(x^2 + y^2) + Bx + Cy + D = 0$$

with four parameters A, B, C, D subject to constraint $B^2 + C^2 - 4AD = 1$, cf. (3.8) and (3.10). We have proved in Section 3.2 that the maximum likelihood estimates of A, B, C, D are essentially restricted to a finite box, cf. (3.16); now it takes a little extra work (we omit details) to verify that the estimates of A, B, C, D have finite moments.

6.5   General algebraic fits

Here we discuss curve fitting methods that are different from maximum likelihood estimation, i.e., from geometric fitting. Such methods are used in two ways, as it is explained in the beginning of Chapter 5. First, they supply an accurate initial guess to the subsequent iterative procedures that find the MLE. Second, they provide a fast inexpensive fit in applications where the geometric fitting procedures are prohibitively slow.

Simple algebraic fit. Perhaps the simplest nongeometric fit is the one minimizing

$$\mathscr{F}_1(\Theta) = \sum_{i=1}^{n} [P(x_i, y_i; \Theta)]^2. \tag{6.8}$$

To justify this method one notes that $P(x_i, y_i; \Theta) = 0$ if and only if the point (x_i, y_i) lies on the curve; also, $[P(x_i, y_i; \Theta)]^2$ is small if and only if the point lies near the curve. If we fit a circle $P = (x-a)^2 + (y-b)^2 - R^2$, then the minimization of (6.8) brings us back to the popular Kåsa fit described in Section 5.1.

Oftentimes, P is a polynomial in Θ, then $\mathscr{F}_1(\Theta)$ is also a polynomial in Θ, so its minimization is a fairly standard algebraic problem. For this reason the method (6.8) is called a *simple algebraic fit*.

Modified algebraic fit. In many fitting problems, $P(x, y; \Theta)$ is a polynomial in x and y, too, and the parameter vector Θ is made by its coefficients. For example, a conic (ellipse, hyperbola) can be specified by

$$P(x, y; \Theta) = \theta_1 + \theta_2 x + \theta_3 y + \theta_4 x^2 + \theta_5 xy + \theta_6 y^2 = 0. \tag{6.9}$$

In such cases $\mathscr{F}_1(\Theta)$ is a quadratic form of Θ, i.e., (6.8) minimizes

$$\mathscr{F}_1(\Theta) = \Theta^T \mathbf{M} \Theta, \tag{6.10}$$

where $\mathbf{M}$ is a symmetric positive semi-definite matrix computed from the data (x_i, y_i), $1 \le i \le n$.

Note that the parameter Θ only needs to be determined up to a scalar multiple, because the desired curve $P(x, y; \Theta) = 0$ is the zero set of the function P, and it is not affected by the rescaling of Θ.

To avoid the trivial (and irrelevant) solution $\Theta = \mathbf{0}$ one can impose a constraint on Θ. For example, one can set one of the coefficients of the polynomial to one, e.g., one can put $\theta_1 = 1$ in (6.9). Alternatively, one can require $\|\Theta\| = 1$, or more generally $\Theta^T \mathbf{N}\Theta = 1$, where $\mathbf{N}$ is some symmetric *constraint matrix*. (Note that the constraint $\theta_1 = 1$ is equivalent to $\Theta^T \mathbf{N}\Theta = 1$ with $\mathbf{N} = [1,0,\ldots,0]^T \times [1,0,\ldots,0]$.) The matrix $\mathbf{N}$ may be constant (as in the Kåsa and Pratt fits) or data-dependent (as in the Taubin fit).

Now one arrives at a constrained minimization problem:

$$\hat{\Theta} = \operatorname{argmin} \Theta^T \mathbf{M}\Theta, \quad \text{subject to} \quad \Theta^T \mathbf{N}\Theta = 1. \tag{6.11}$$

Introducing a Lagrange multiplier η, as in Section 5.5, gives

$$\mathbf{M}\Theta = \eta \mathbf{N}\Theta \tag{6.12}$$

thus Θ must be a generalized eigenvector of the matrix pair $(\mathbf{M}, \mathbf{N})$. As the parameter Θ only needs to be determined up to a scalar multiple, we can solve (6.12) subject to additional constraint $\|\Theta\| = 1$.

Gradient-weighted algebraic fit (GRAF). Unfortunately, for curves other than circles, the simple algebraic fit (6.8), or a modified algebraic fit (6.10) with any constraint matrix $\mathbf{N}$, has poor statistical properties, because the values $[P(x_i, y_i; \Theta)]^2$ minimized by (6.8) have little or no relevance to the geometric distances from (x_i, y_i) to the curve, thus the resulting estimates have little or no relevance to the MLE.

To improve the accuracy of (6.8), one usually generalizes it by introducing weights w_i and minimizes

$$\mathscr{F}_2(\Theta) = \sum_{i=1}^{n} w_i \left[P(x_i, y_i; \Theta)\right]^2 \tag{6.13}$$

This modification of (6.8) is called the *weighted algebraic fit*. The weights $w_i = w(x_i, y_i; \Theta)$ may depend on the data and the parameters.

A smart way to define weights w_i results from the linear approximation

$$\frac{|P(x_i, y_i; \Theta)|}{\|\nabla_{\mathbf{x}} P(x_i, y_i; \Theta)\|} = d_i + \mathscr{O}(d_i^2)$$

where $\nabla_{\mathbf{x}} P = (\partial P/\partial x, \partial P/\partial y)$ denotes the gradient vector. Thus one can construct an "approximate geometric fit," to the leading order, by minimizing

$$\mathscr{F}_3(\Theta) = \sum_{i=1}^{n} \frac{[P(x_i, y_i; \Theta)]^2}{\|\nabla_{\mathbf{x}} P(x_i, y_i; \Theta)\|^2}. \tag{6.14}$$

This is a particular case of the weighted algebraic fit (6.13) with weights set to

$$w_i = \frac{1}{\|\nabla_{\mathbf{x}} P(x_i, y_i; \Theta)\|^2}. \tag{6.15}$$

The method (6.14) is called the *gradient weighted algebraic fit*, or GRAF for brevity.

History and implementations of GRAF. The GRAF is known since at least 1974 when it was mentioned in Turner's book [183]. It was applied specifically to quadratic curves (ellipses and hyperbolas) by Sampson [163] in 1982 and popularized by Taubin [176] in 1991. Now it has become standard in the computer vision industry; see [47, 95, 120, 198] and references therein.

If $P(x,y;\Theta)$ is a polynomial in x and y and the parameter vector Θ is made by its coefficients (such as in (6.9)), then both numerator and denominator in (6.14) are quadratic forms in Θ, i.e.,

$$\mathscr{F}_3(\Theta) = \sum_{i=1}^{n} \frac{\Theta^T \mathbf{M}_i \Theta}{\Theta^T \mathbf{N}_i \Theta}, \tag{6.16}$$

where $\mathbf{M}_i$ and $\mathbf{N}_i$ are positive semi-definite matrices computed from (x_i, y_i). Note that both numerator and denominator in (6.16) are homogeneous quadratic polynomials in the parameters, hence $\mathscr{F}_3(\Theta)$ is invariant under rescaling of Θ, so its minimization does not require any additional constraints.

In most cases, the minimization of (6.16) is a nonlinear problem that has no closed form solution. There exist several powerful iterative schemes for solving it, though, and they will be described in the context of fitting ellipses. In any case, the computational cost of GRAF if much lower than that of the geometric fit (the MLE), so currently GRAF is the method of choice for practice fitting of ellipses or more complex curves.

The statistical accuracy of GRAF is the same as that of the MLE, to the leading order, as we will see in Section 6.9.

GRAF for circles. The algebraic equation of a circle is

$$P(x,y;\Theta) = A(x^2+y^2) + Bx + Cy + D = 0,$$

hence

$$\nabla_{\mathbf{x}} P(x,y;\Theta) = (2Ax+B, 2Ay+C)$$

and

$$\begin{aligned}
\|\nabla_{\mathbf{x}} P(x_i,y_i;\Theta)\|^2 &= 4Az_i^2 + 4ABx_i + 4ACy_i + B^2 + C^2 \\
&= 4A(Az_i + Bx_i + Cy_i + D) + B^2 + C^2 - 4AD.
\end{aligned}$$

Thus the GRAF reduces to the minimization of

$$\mathscr{F}(A,B,C,D) = \sum_{i=1}^{n} \frac{[Az_i + Bx_i + Cy_i + D]^2}{4A(Az_i + Bx_i + Cy_i + D) + B^2 + C^2 - 4AD} \tag{6.17}$$

Just like (6.16), this is a nonlinear problem that can only be solved iteratively.

The Pratt and Taubin algorithms (Chapter 5) can be regarded as approximate solutions to this problem. Indeed, assuming that the data points (x_i, y_i) lie close to the circle, one gets $Az_i + Bx_i + Cy_i + D \approx 0$. Thus one can simply discard this sum from the denominator of (6.17), which gives us the Pratt fit (5.21). Alternatively, one can averages all the denominators in (6.17), which gives us the Taubin fit; see (5.45).

6.6 Error analysis: A general scheme

In Section 6.4 we have shown that the exact statistical properties of the MLE in the curve fitting problems are intractable. Unfortunately, much the same remains true for algebraic fits as well, including GRAF. Here we develop an *approximative error analysis* for curve fitting algorithms. This analysis will allow us to determine their statistical characteristics, to a reasonable extend, and draw practically valuable conclusions.

Notation and assumptions. Let $\hat{\Theta}(x_1, \ldots, x_n)$ be an estimate of the unknown parameter vector $\Theta = (\theta_1, \ldots, \theta_k)^T$ based on n independent observations $x_i = (x_i, y_i)$, $i = 1, \ldots, n$. We assume that it is a regular (at least four times differentiable) function of x_i's and y_i's.

For brevity we denote by $X = (x_1, y_1, \ldots, x_n, y_n)^T$ the vector of all our observations, so that

$$X = \tilde{X} + E,$$

where

$$\tilde{X} = (\tilde{x}_1, \tilde{y}_1, \ldots, \tilde{x}_n, \tilde{y}_n)^T$$

is the vector of the true coordinates and

$$E = (\delta_1, \varepsilon_1, \ldots, \delta_n, \varepsilon_n)^T$$

is the vector of their random perturbations. Recall that in the Cartesian functional model (Section 6.1) the components of E are i.i.d. normal random variables with mean zero and variance σ^2.

Taylor expansion. We will use Taylor expansion to the second order terms. To keep our notation simple, we work with each scalar parameter θ_m of the vector Θ separately:

$$\hat{\theta}_m(X) = \hat{\theta}_m(\tilde{X}) + G_m^T E + \tfrac{1}{2} E^T H_m E + \mathcal{O}_P(\sigma^3). \tag{6.18}$$

Here

$$G_m = \nabla \hat{\theta}_m \qquad \text{and} \qquad H_m = \nabla^2 \hat{\theta}_m$$

denote the gradient (the vector of the first order partial derivatives) and the Hessian matrix of the second order partial derivatives of $\hat{\theta}_m$, respectively, taken at the true vector $\tilde{X}$.

The remainder term $\mathcal{O}_P(\sigma^3)$ in (6.18) is a random variable $\mathcal{R}$ such that $\sigma^{-3}\mathcal{R}$ is bounded in probability; this means that for any $\varepsilon > 0$ there exists $A_\varepsilon > 0$ such that $\text{Prob}\{\sigma^{-3}\mathcal{R} > A_\varepsilon\} < \varepsilon$ for all $\sigma > 0$. Of course, the remainder term may have infinite moments, but its typical values are of order σ^3, so we will ignore it in our analysis. The first and second order terms in (6.18) have typical values of order σ and σ^2, respectively, and their moments are always finite (because $\mathbf{E}$ is a Gaussian vector).

If we do the expansion (6.18) for the whole vector $\hat{\Theta}$, then $\mathbf{G}_m$ would be replaced by an $m \times n$ matrix of the first order partial derivatives, but $\mathbf{H}_m$ would become a 3D tensor of the second partial derivatives; to avoid notational complications related to tensor products we deal with each component θ_m separately.

Geometric consistency. Expansion (6.18) shows that Expansion (6.18) shows that $\hat{\Theta}(\mathbf{X}) \to \hat{\Theta}(\tilde{\mathbf{X}})$ in probability, as $\sigma \to 0$. It is convenient to assume that

$$\hat{\Theta}(\tilde{\mathbf{X}}) = \tilde{\Theta}. \tag{6.19}$$

Precisely (6.19) means that whenever $\sigma = 0$, i.e., the true points are observed without noise, then the estimator returns the true parameter vector, i.e., finds the true curve. Geometrically, (6.19) implies that if there is a curve of type (6.1) that interpolates the data points, then the algorithm finds it.

With some degree of informality, one can assert that whenever (6.19) holds, the estimate $\hat{\Theta}$ is consistent in the limit $\sigma \to 0$. We call this property *geometric consistency*.

The assumption (6.19) is sometimes regarded a minimal requirement for any sensible fitting algorithm. For example, if the observed points lie on one circle, then every algorithm that we have discussed in Chapters 4 and 5 finds that circle uniquely. Kanatani [100] remarks that algorithms which do not enjoy this property "are not worth considering."

We recall, however, that the Maximum Likelihood Estimators of the circle parameters, in the context of the structural model, are *not* geometrically consistent (Section 6.3), thus (6.19) is not automatically ensured for every estimator of interest. But we will assume (6.19) in all that follows.

Bias and variance. Under our assumption (6.19) we rewrite (6.18) as

$$\Delta\hat{\theta}_m(\mathbf{X}) = \mathbf{G}_m^T\mathbf{E} + \tfrac{1}{2}\mathbf{E}^T\mathbf{H}_m\mathbf{E} + \mathcal{O}_P(\sigma^3), \tag{6.20}$$

where $\Delta\hat{\theta}_m(\mathbf{X}) = \hat{\theta}_m(\mathbf{X}) - \tilde{\theta}_m$ is the statistical error of the parameter estimate.

The first term in (6.20) is a linear combination of i.i.d. normal random variables (the components of $\mathbf{E}$) that have zero mean, hence it is itself a normal random variable with zero mean.

The second term is a quadratic form of i.i.d. normal variables (such forms have been studied in probability, see e.g., [64, 88, 153, 154]). Since $\mathbf{H}_m$ is a symmetric matrix, we have $\mathbf{H}_m = \mathbf{Q}_m^T \mathbf{D}_m \mathbf{Q}_m$, where $\mathbf{Q}_m$ is an orthogonal matrix and $\mathbf{D}_m = \text{diag}\{d_1, \ldots, d_{2n}\}$ is a diagonal matrix. The vector $\mathbf{E}_m = \mathbf{Q}_m \mathbf{E}$ has the same distribution as $\mathbf{E}$ does, i.e., the components of $\mathbf{E}_m$ are i.i.d. normal random variables with mean zero and variance σ^2. Thus

$$\mathbf{E}^T \mathbf{H}_m \mathbf{E} = \mathbf{E}_m^T \mathbf{D}_m \mathbf{E}_m = \sigma^2 \sum d_i Z_i^2, \tag{6.21}$$

where the Z_i's are i.i.d. standard normal random variables. One can also regard (6.21) as a generalization of the χ^2 distribution; though it is not always positive, as we may have $d_i < 0$ for some i's. In any case, the mean value of (6.21) is

$$\mathbb{E}\left(\mathbf{E}^T \mathbf{H}_m \mathbf{E}\right) = \sigma^2 \, \text{tr} \, \mathbf{D}_m = \sigma^2 \, \text{tr} \, \mathbf{H}_m.$$

Therefore, taking the mean value in (6.20) gives the bias of $\hat{\theta}_m$:

$$\mathbb{E}(\Delta \hat{\theta}_m) = \tfrac{1}{2} \sigma^2 \, \text{tr} \, \mathbf{H}_m + \mathcal{O}(\sigma^4). \tag{6.22}$$

We use the fact that the expectations of all third order terms vanish, because the components of $\mathbf{E}$ are independent and their first and third moments are zero; thus the remainder term is of order σ^4. In fact the expectations of all *odd* order terms vanish, because the distribution of the components of $\mathbf{E}$ is symmetric about zero.

Next, squaring (6.20) and again using (6.21) give the mean squared error

$$\mathbb{E}\left([\Delta \hat{\theta}_m]^2\right) = \sigma^2 \mathbf{G}_m^T \mathbf{G}_m + \tfrac{1}{4} \sigma^4 \left([\text{tr} \, \mathbf{H}_m]^2 + 2\|\mathbf{H}_m\|_F^2\right) + \mathcal{R}. \tag{6.23}$$

Here

$$\|\mathbf{H}_m\|_F^2 = \text{tr} \, \mathbf{H}_m^2 = \|\mathbf{D}_m\|_F^2 = \text{tr} \, \mathbf{D}_m^2 \tag{6.24}$$

where $\| \cdot \|_F$ stands for the Frobenius norm (it is important to note that $\mathbf{H}_m$ is symmetric). The remainder $\mathcal{R}$ includes terms of order σ^6, as well as some terms of order σ^4 that contain third order partial derivatives, such as $\partial^3 \hat{\theta}_m / \partial x_i^3$ and $\partial^3 \hat{\theta}_m / \partial x_i^2 \partial x_j$. A similar expression can be derived for $\mathbb{E}(\Delta \hat{\theta}_m \Delta \hat{\theta}_{m'})$ for $m \neq m'$, we omit it and only give the final formula below.

Comparing (6.22) and (6.23) we conclude that

$$\text{Var}(\hat{\theta}_m) = \sigma^2 \mathbf{G}_m^T \mathbf{G}_m + \tfrac{1}{2} \sigma^4 \|\mathbf{H}_m\|_F^2 + \mathcal{R}. \tag{6.25}$$

Thus, the two forth order terms in the MSE expansion (6.23) have different origins: the first one, $\tfrac{1}{4} \sigma^4 [\text{tr} \, \mathbf{H}_m]^2$, comes from the bias, and the second one, $\tfrac{1}{2} \sigma^4 \|\mathbf{H}_m\|_F^2$, comes from the variance.

Covariances. The above formulas can be easily extended to the cross product terms:

$$\begin{aligned} \mathbb{E}\left((\Delta \hat{\theta}_m)(\Delta \hat{\theta}_{m'})\right) = {} & \sigma^2 \mathbf{G}_m^T \mathbf{G}_{m'} + \tfrac{1}{4} \sigma^4 \big[(\text{tr} \, \mathbf{H}_m)(\text{tr} \, \mathbf{H}_{m'}) \\ & + 2\langle \mathbf{H}_m, \mathbf{H}_{m'} \rangle_F\big] + \mathcal{R}, \end{aligned} \tag{6.26}$$

where $\langle \mathbf{H}_m, \mathbf{H}_{m'} \rangle_F = \mathrm{tr}(\mathbf{H}_m \mathbf{H}_{m'})$ stands for the Frobenius scalar product of matrices (again we note that $\mathbf{H}_{m'}$ is symmetric). Comparing (6.22) and (6.26) we conclude that

$$\mathrm{Cov}(\hat{\theta}_m, \hat{\theta}_{m'}) = \sigma^2 \mathbf{G}_m^T \mathbf{G}_{m'} + \tfrac{1}{2}\sigma^4 \langle \mathbf{H}_m, \mathbf{H}_{m'} \rangle_F + \mathscr{R}. \qquad (6.27)$$

Just as before, the first $\mathscr{O}(\sigma^4)$ term in (6.26) comes from the biases, while the second one comes from the covariance.

6.7 A model with small noise and "moderate sample size"

Our approximate error analysis in Section 6.6 includes all the terms of order σ^2 and σ^4. In applications, one usually gets a large number of terms of order σ^4 given by complicated formulas; we will see examples in Chapter 7. Even the expression for the bias (6.22) alone may contain several terms of order σ^2. In this section we sort the higher order terms out to keep only the most significant ones. Our sorting method is motivated by Kanatani [104].

Kanatani's treatment of higher order terms. Kanatani [104] recently derived formulas for the bias of certain ellipse fitting algorithms. He kept all the terms of order σ^2, but in the end he noticed that some terms were of order σ^2 (independent of n), while the others of order σ^2/n. The magnitude of the former was clearly larger than that of the latter, and when Kanatani made his final conclusions he ignored the terms of order σ^2/n. To justify this discretion, he simply noted that "in many vision applications n is fairly large."

Formal classification of higher order terms. We formalize Kanatani's prescription for treating higher order terms in the approximate error analysis, and adopt the following rules:

(a) In the expression for the bias (6.22) we keep terms of order σ^2 (independent of n) and ignore terms of order σ^2/n.

(b) In the expression for the mean squared error (6.23) we keep terms of order σ^4 (independent of n) and ignore terms of order σ^4/n.

In the strict mathematical sense, to justify our rules we would need to assume that not only $\sigma \to 0$, but also $n \to \infty$, although n may increase rather slowly. Precisely, n need to grow slower that $1/\sigma^2$. For simplicity, however, we prefer to treat σ as small variable and n as a moderately large constant. We call this a *model with small noise and moderate sample size*.

Magnitude of various terms. The main term $\sigma^2 \mathbf{G}_m^T \mathbf{G}_m$ in our expression for the mean squared error (6.23) is always of order σ^2/n; it will never be ignored. Out of the two higher order terms $\tfrac{1}{4}\sigma^4[\mathrm{tr}\,\mathbf{H}_m]^2$ and $\tfrac{1}{2}\sigma^4\|\mathbf{H}_m\|_F^2$, the latter is always of order σ^4/n (under certain natural conditions), hence it will be discarded. The bias $\sigma^2 \mathrm{tr}\, \mathbf{H}_m$ in (6.22) is, generally, of order σ^2 (independent of n), thus its contribution to the mean squared error (6.23) is significant. However the full expression for the bias may contain terms of order σ^2 and of order σ^2/n, of which the latter will be ignored; see below.

We only explain why our terms have the above order or magnitude, omitting a formal proof, which is quite complex.

Analysis of the order of magnitude. We make a natural assumption that the estimate $\hat{\Theta}$ is balanced in the sense that every observation $\mathbf{x}_i$ contributes to it equally. More precisely, the partial derivatives of $\hat{\Theta}(\mathbf{X})$ with respect to the components of $\mathbf{X}$ must have the same order of magnitude.

For example, all our circle fits are balanced, in this sense, as they are based on averaging observations. In particular, the geometric circle fit minimizes $a^2 + b^2 - \bar{r}^2$, recall (4.25). The Kåsa fit solves the system (5.5) that involve averages, too.

Under this assumption it is easy to see that the partial derivatives of $\hat{\Theta}(\mathbf{X})$ with respect to the components of $\mathbf{X}$ are $\mathcal{O}(1/n)$. For example, if one perturbs every data point by $\varepsilon \ll \sigma$, then the center and radius of the fitting circle should change by $\mathcal{O}(\varepsilon)$. On the other hand, $\Delta\hat{\Theta} = \langle \nabla_{\mathbf{X}}\hat{\Theta}, \Delta\mathbf{X} \rangle + \mathcal{O}(\varepsilon^2)$. As one can always make the perturbation vector $\Delta\mathbf{X}$ parallel to the gradient $\nabla_{\mathbf{X}}\hat{\Theta}$, it follows that $\|\Delta\hat{\Theta}\| \geq n^{1/2}\varepsilon\|\nabla_{\mathbf{X}}\hat{\Theta}\|$, thus $\|\nabla_{\mathbf{X}}\hat{\Theta}\| \leq n^{-1/2}$, i.e., each component of $\nabla_{\mathbf{X}}\hat{\Theta}$ must be $\mathcal{O}(1/n)$.

This implies that the components of the vector $\mathbf{G}_m$ and the diagonal elements of $\mathbf{D}_m$ in (6.23)–(6.24) are of order $1/n$. As a result,

$$\mathbf{G}_m^T \mathbf{G}_m = \mathcal{O}(1/n)$$

and also

$$\mathrm{tr}\, \mathbf{H}_m = \mathrm{tr}\, \mathbf{D}_m = \mathcal{O}(1)$$

and

$$\|\mathbf{H}_m\|_F^2 = \|\mathbf{D}_m\|_F^2 = \mathcal{O}(1/n).$$

Also recall that the remainders denoted by $\mathscr{R}$ in our expansions include some terms of order σ^4 that contain third order partial derivatives; those derivatives can be shown to have order of $1/n$.

Thus the terms in (6.23) have the following orders of magnitude:

$$\mathbb{E}\big([\Delta\hat{\theta}_m]^2\big) = \mathcal{O}(\sigma^2/n) + \mathcal{O}(\sigma^4) + \mathcal{O}(\sigma^4/n) + \mathcal{O}(\sigma^6), \tag{6.28}$$

where each big-O simply indicates the order of the corresponding term in (6.23).

We see that the fourth-order term $\frac{1}{4}\sigma^4[\mathrm{tr}\, \mathbf{H}_m]^2$ coming from the bias is larger (hence more significant) than the other one, $\frac{1}{2}\sigma^4\|\mathbf{H}_m\|_F^2$, that comes from the variance.

	σ^2/n	σ^4	σ^4/n	σ^6
small samples ($n \sim 1/\sigma$)	σ^3	σ^4	σ^5	σ^6
large samples ($n \sim 1/\sigma^2$)	σ^4	σ^4	σ^6	σ^6

Table 6.1: *The order of magnitude of the four terms in* (6.23).

The cross-product expansion (6.26) has a similar structure:

$$\mathbb{E}\big((\Delta\hat{\theta}_m)(\Delta\hat{\theta}_{m'})\big) = \mathscr{O}(\sigma^2/n) + \mathscr{O}(\sigma^4) + \mathscr{O}(\sigma^4/n) + \mathscr{O}(\sigma^6), \qquad (6.29)$$

i.e., the fourth-order term coming from the biases is larger (hence more significant) than the one coming from the covariance.

Example. In typical computer vision applications, σ does not exceed 5% of the size of the image, see [17] (the value $\sigma = 0.05$ was used in most of our simulated examples). The number of data points normally varies between 10-20 (on the low end) and a few hundred (on the high end). For simplicity, we can set $n \sim 1/\sigma$ for smaller samples and $n \sim 1/\sigma^2$ for larger samples. Then Table 6.1 presents the corresponding typical magnitudes of each of the four terms in (6.23). We see that for larger samples the fourth order term coming from the bias may be just as big as the leading second-order term, hence it would be unwise to ignore it.

Some earlier studies, see e.g., [17, 42, 66, 95], focused on the leading, i.e., second-order terms only, disregarding all the fourth-order terms, and this is where our analysis is different. We make one step further — we keep all the terms of order $\mathscr{O}(\sigma^2/n)$ and $\mathscr{O}(\sigma^4)$. The less significant terms of order $\mathscr{O}(\sigma^4/n)$ and $\mathscr{O}(\sigma^6)$ would be discarded.

Complete MSE expansion. Combining (6.26) for all $1 \le m, m' \le k$ gives a matrix formula for the (total) mean squared error (MSE)

$$\mathbb{E}\Big[(\Delta\hat{\Theta})(\Delta\hat{\Theta})^T\Big] = \sigma^2 \mathbf{G}\mathbf{G}^T + \sigma^4 \mathbf{B}\mathbf{B}^T + \cdots, \qquad (6.30)$$

where $\mathbf{G}$ is the $k \times 2n$ matrix of first order partial derivatives of $\hat{\Theta}(\mathbf{X})$, its rows are $\mathbf{G}_m^T$, $1 \le m \le k$, and

$$\mathbf{B} = \tfrac{1}{2}[\operatorname{tr}\mathbf{H}_1, \dots \operatorname{tr}\mathbf{H}_k]^T$$

is the k-vector that represents the leading term of the bias of $\hat{\Theta}$, cf. (6.22). Our previous analysis shows that $\mathbf{G}\mathbf{G}^T = \mathscr{O}(1/n)$ and $\mathbf{B}\mathbf{B}^T = \mathscr{O}(1)$, hence both terms explicitly shown in (6.30) are significant, by our adopted rules; the trailing dots stand for all insignificant terms (those of order σ^4/n and σ^6).

We call the first (main) term $\sigma^2 \mathbf{GG}^T$ in (6.30) the *variance* term, as it characterizes the variance (more precisely, the covariance matrix) of the estimator $\hat{\Theta}$, to the leading order. The second term $\sigma^4 \mathbf{BB}^T$ comes from the bias $\sigma^2 \mathbf{B}$ of the estimator, again to the leading order.

For brevity we denote the variance term by $\mathbf{V} = \mathbf{GG}^T$.

Essential bias. A more detailed analysis of many particular estimators (see [104] and our Chapter 7) reveal that the bias $\sigma^2 \mathbf{B}$ is the sum of terms of two types: some of them are of order σ^2 and some others are of order σ^2/n, i.e.

$$\mathbb{E}(\Delta\hat{\Theta}) = \sigma^2 \mathbf{B} + \mathcal{O}(\sigma^4) = \sigma^2 \mathbf{B}_1 + \sigma^2 \mathbf{B}_2 + \mathcal{O}(\sigma^4),$$

where $\mathbf{B}_1 = \mathcal{O}(1)$ and $\mathbf{B}_2 = \mathcal{O}(1/n)$.

We call $\sigma^2 \mathbf{B}_1$ the *essential bias* of the estimator $\hat{\Theta}$. This is its bias to the leading order, σ^2. The other terms (i.e., $\sigma^2 \mathbf{B}_2$, and $\mathcal{O}(\sigma^4)$) constitute nonessential bias; they can be dropped according to our adopted rules. Then (6.30) can be reduced to

$$\mathbb{E}\left[(\Delta\hat{\Theta})(\Delta\hat{\Theta})^T\right] = \sigma^2 \mathbf{V} + \sigma^4 \mathbf{B}_1 \mathbf{B}_1^T + \cdots, \qquad (6.31)$$

where we only keep significant terms of order σ^2/n and σ^4 and drop the rest (remember also that $\mathbf{V} = \mathbf{GG}^T$).

6.8 Variance and essential bias of the MLE

Our approximate error analysis in Sections 6.6 and 6.7 is developed for general parameter estimators, under a very mild assumption of geometric consistency (6.19). In the circle and ellipse fitting case, our analysis applies to all geometric and algebraic fits. In this section we focus on the geometric fit only, i.e., on the Maximum Likelihood Estimators. For the MLE, explicit formulas can be derived for the asymptotic variance matrix $\mathbf{V}$ and the essential bias vector $\mathbf{B}_1$ in (6.31).

In fact, such formulas were obtained in the 1980s by Amemiya, Fuller, and Wolter [5, 192]. They studied the Maximum Likelihood Estimators for general curve parameters in the context of the Cartesian functional model. Their assumptions on n and σ were similar to ours, see below.

Amemiya-Fuller-Wolter asymptotic model. In their studies, Amemiya, Fuller, and Wolter assume that both σ and n are variable, so that $\sigma \to 0$ and $n \to \infty$. Precisely, they consider a sequence of independent experiments, indexed by $m = 1, 2, \ldots$. In each experiment $n = n_m$ points are observed along a true curve (the same for all experiments), with the noise level $\sigma = \sigma_m > 0$. They put

$$\sigma_m^{-2} n_m = m, \qquad (6.32)$$

which simply gives a precise meaning to their index m. The reason for this notation is as follows: suppose there are n_m distinct true points on the true curve, and each is observed independently k_m times, i.e., we have k_m *replicated* observations of each true point, all with unit variance $\sigma^2 = 1$. Then averaging those k_m observations of each true point gives a single observation with variance $\sigma_m^2 = 1/k_m$. Now $m = k_m n_m = \sigma_m^{-2} n_m$ represents the total number of the original (replicated) observations.

Next they assume that $n_m \to \infty$ and $\sigma_m \to 0$, as $m \to \infty$, so that

$$\sigma_m^2 n_m \to 0 \qquad \text{as } m \to \infty,$$

i.e., n grows more slowly than $1/\sigma^2$. In the strict mathematical sense, these assumptions are effectively equivalent to ours in Section 6.7, though we prefer to treat n as a (moderately large) constant, and not as a variable.

Asymptotic variance of the MLE. Amemiya, Fuller and Wolter [5, 192] investigate the Maximum Likelihood Estimator (i.e., the estimator obtained by the geometric fit); they prove asymptotic normality and explicitly compute the asymptotic covariance matrix, to the leading order (see Theorem 1 in [5]). In our notation, their main result is

$$m^{-1}\left(\hat{\Theta}_{\text{MLE}} - \tilde{\Theta}\right) \to_L N\left(0, \mathbf{V}_{\text{MLE}}^*\right), \tag{6.33}$$

where

$$\mathbf{V}_{\text{MLE}}^* = \left(\lim_{m \to \infty} \frac{1}{n_m} \sum_{i=1}^{n} \frac{P_{\Theta i} P_{\Theta i}^T}{\|P_{\mathbf{x}i}\|^2}\right)^{-1}, \tag{6.34}$$

assuming that the limit exists. Here

$$P_{\Theta i} = \left(\partial P(\tilde{\mathbf{x}}_i; \tilde{\Theta})/\partial \theta_1, \ldots, \partial P(\tilde{\mathbf{x}}_i; \tilde{\Theta})/\partial \theta_k\right)^T$$

stands for the gradient of P with respect to the model parameters $\theta_1, \ldots, \theta_k$ and

$$P_{\mathbf{x}i} = \left(\partial P(\tilde{\mathbf{x}}_i; \tilde{\Theta})/\partial x, \partial P(\tilde{\mathbf{x}}_i; \tilde{\Theta})/\partial y\right)^T$$

for the gradient with respect to the spacial variables x and y; both gradients are taken at the true point $\tilde{\mathbf{x}}_i$.

Given the equation of a curve $P(x, y; \Theta) = 0$, both gradients can be computed easily. For example in the case of fitting circles defined by

$$P = (x - a)^2 + (y - b)^2 - R^2,$$

we have

$$P_{\Theta i} = -2\left((\tilde{x}_i - \tilde{a}), (\tilde{y}_i - \tilde{b}), \tilde{R}\right)^T$$

and
$$P_{\mathbf{x}i} = 2\big((\tilde{x}_i - \tilde{a}), (\tilde{y}_i - \tilde{b})\big)^T.$$

By using (6.32) we can derive from (6.33) the asymptotic covariance matrix of the MLE:
$$\mathrm{Var}(\hat{\Theta}_{\mathrm{MLE}}) = n_m^{-1}\sigma_m^2 \mathbf{V}_{\mathrm{MLE}}^* + o(n_m^{-1}\sigma_m^2). \tag{6.35}$$

The existence of the limit in (6.34) effectively means that, as $n_m \to \infty$, there is a certain pattern in the distribution of the true points on the true curve.

Finite sample variance of the MLE. In our model, where n is a constant, we would like to describe the MLE for a fixed n (and fixed locations of the true points). Comparing (6.35) to (6.34) suggests that for finite sample size n we have
$$\mathrm{Var}(\hat{\Theta}_{\mathrm{MLE}}) = \sigma^2 \left(\sum_{i=1}^n \frac{P_{\Theta i} P_{\Theta i}^T}{\|P_{\mathbf{x}i}\|^2} \right)^{-1} + o(\sigma^2/n), \tag{6.36}$$

i.e., the matrix $\mathbf{V}$ for the MLE, in the notation of (6.31), is
$$\mathbf{V}_{\mathrm{MLE}} = \left(\sum_{i=1}^n \frac{P_{\Theta i} P_{\Theta i}^T}{\|P_{\mathbf{x}i}\|^2} \right)^{-1}. \tag{6.37}$$

This formula was indeed proved by Fuller, see Theorem 3.2.1 in [66]; it was independently derived in [42].

Bias of the MLE. Amemiya and Fuller also computed the asymptotic bias of the MLE, see (2.11) in [5], which in the above notation is
$$\mathbb{E}\big(\hat{\Theta}_{\mathrm{MLE}} - \tilde{\Theta}\big) = \frac{\sigma_m^2}{2} \sum_{i=1}^{n_m} \mathrm{tr}\left[P_{\mathbf{xx}i}\left(\mathbf{I} - \frac{P_{\mathbf{x}i} P_{\mathbf{x}i}^T}{\|P_{\mathbf{x}i}\|^2} \right) \right] \frac{\mathbf{V}_{\mathrm{MLE}} P_{\Theta i}}{\|P_{\mathbf{x}i}\|^2} + o(\sigma_m^2). \tag{6.38}$$

Here $P_{\mathbf{xx}i}$ is the 2×2 matrix of the second order partial derivatives of P with respect to x and y, taken at the point $\tilde{x}_i$. For example, in the case of fitting circles we have
$$P_{\mathbf{xx}i} = \begin{bmatrix} 2 & 0 \\ 0 & 2 \end{bmatrix}.$$

Note that the main term in (6.38) is, in our notation, of order $\mathscr{O}(\sigma^2)$, i.e., it represents the *essential bias*. The remainder term $o(\sigma_m^2)$ includes, in our notation, all the $\mathscr{O}(\sigma^2/n)$ and $\mathscr{O}(\sigma^4)$ terms, i.e., it contains the nonessential bias. Thus we can rewrite (6.38) as
$$\mathbf{B}_{1,\mathrm{MLE}} = \frac{1}{2} \sum_{i=1}^n \mathrm{tr}\left[P_{\mathbf{xx}i}\left(\mathbf{I} - \frac{P_{\mathbf{x}i} P_{\mathbf{x}i}^T}{\|P_{\mathbf{x}i}\|^2} \right) \right] \frac{\mathbf{V}_{\mathrm{MLE}} P_{\Theta i}}{\|P_{\mathbf{x}i}\|^2}. \tag{6.39}$$

6.9　Kanatani-Cramer-Rao lower bound

KCR lower bound. The matrix $\mathbf{V}$, representing the leading terms of the variance, has a natural lower bound (an analogue of the Cramer-Rao bound): there is a symmetric positive semi-definite matrix $\mathbf{V}_{min}$ such that for every geometrically consistent estimator

$$\mathbf{V} \geq \mathbf{V}_{min}, \tag{6.40}$$

in the sense that $\mathbf{V} - \mathbf{V}_{min}$ is a positive semi-definite matrix.

In fact, the matrix $\mathbf{V}_{min}$ coincides with $\mathbf{V}_{MLE}$, cf. (6.37) obtained by Fuller for the variance of the Maximum Likelihood Estimator, i.e.

$$\mathbf{V}_{min} = \mathbf{V}_{MLE} = \left(\sum_{i=1}^{n} \frac{P_{\Theta i} P_{\Theta i}^{T}}{\|P_{xi}\|^2} \right)^{-1}, \tag{6.41}$$

see the notation of (6.37).

We say that an estimator $\hat{\Theta}$ is *asymptotically efficient*, as $\sigma \to 0$, if its covariance matrix $\mathbf{V}$ satisfies $\mathbf{V} = \mathbf{V}_{min}$. Obviously, the MLE (the geometric fit) is asymptotically efficient.

History of the KCR. The inequality (6.40) with the formula (6.41) have an interesting history. First, an analogue of (6.40) was proved in the mid-1990s by Kanatani [95, 96] who showed that for any *unbiased* estimator $\hat{\Theta}$

$$\mathrm{Var}(\hat{\Theta}) \geq \sigma^2 \mathbf{V}_{min}. \tag{6.42}$$

He also derived an explicit formula (6.41) for the matrix $\mathbf{V}_{min}$.

It is important to note that the inequality (6.42) is, in a sense, stronger than (6.40), because it gives a lower bound on the *actual* covariance matrix, rather than its leading term. On the other hand, Kanatani assumed unbiasedness, hence his bound (6.42), strictly speaking, could not be applied to many practical estimators, including $\hat{\Theta}_{MLE}$, as they are biased in most cases. (Very likely it cannot be applied to *any* existing estimator of nonlinear curve parameters, as virtually all of them are biased.)

In the early 2000s Chernov and Lesort [42] realized that Kanatani's argument [95] essentially works for biased estimators, provided those are geometrically consistent. As a result, they derived a more general bound (6.40) for all geometrically consistent estimates (including the MLE). Incidentally, this proves that the MLE is asymptotically efficient (optimal to the leading order) in the much larger class of all geometrically consistent estimators.

Furthermore, Chernov and Lesort proved [42] that the gradient-weighted algebraic fit (GRAF), introduced in Section 6.5, is always asymptotically efficient, i.e., its covariance matrix $\mathbf{V}$ attains the lower bound $\mathbf{V}_{min}$. In fact, they showed in [42] that if one uses a general weighted algebraic fit (6.13), then the resulting parameter estimates will be asymptotically efficient if and only if the weights w_i are proportional to the gradient weights (6.15), i.e.,

$$\mathbf{V} = \mathbf{V}_{\min} \quad \Longleftrightarrow \quad w_i = \frac{c(\Theta)}{\|\nabla_{\mathbf{x}} P(x_i, y_i; \Theta)\|^2},$$

where $c(\Theta)$ is any function that depends on the parameters only.

In addition, Chernov and Lesort [42] derived (6.40) in the context of not only the Cartesian model, but also the radial model. Lastly, they named the inequality (6.40) Kanatani-Cramer-Rao (KCR) lower bound, recognizing the crucial contribution of Kanatani [95, 96].

Assessing the quality of estimators. Our analysis dictates the following strategy of assessing the quality of an estimator $\hat{\Theta}$: first of all, its accuracy is characterized by the matrix $\mathbf{V}$, which must be compared to the KCR lower bound $\mathbf{V}_{\min}$. We will see that for all the circle fitting algorithms and for many popular ellipse fitting algorithms the matrix $\mathbf{V}$ actually achieves its lower bound $\mathbf{V}_{\min}$, i.e., we have $\mathbf{V} = \mathbf{V}_{\min}$, hence the corresponding algorithms are optimal to the leading order.

Next, once the factor $\mathbf{V}$ is already at its natural minimum, the accuracy of an estimator should be characterized by the vector $\mathbf{B}_1$ representing the essential bias: better estimates should have smaller essential biases. It appears that there is no natural minimum for $\|\mathbf{B}_1\|$, in fact there exist estimators which have a minimum variance $\mathbf{V} = \mathbf{V}_{\min}$ and a zero essential bias, i.e., $\mathbf{B}_1 = \mathbf{0}$. We will construct them in Section 7.5.

6.10 Bias and inconsistency in the large sample limit

Thus far our analysis was restricted to the small noise asymptotic model, in which $\sigma \to 0$ and n was fixed (or assumed to be a moderately large constant); such a model is perhaps appropriate for most computer vision applications. However, many statisticians investigate curve fitting algorithms in the traditional large sample limit $n \to \infty$ keeping $\sigma > 0$ fixed. We devote the last few sections to these studies.

Bias. In many nonlinear statistical models, unbiased estimation of unknown parameters is virtually impossible, thus all practical estimators are biased. We have seen in Chapters 1 and 2 that even in the linear EIV model $y = a + bx$ all known estimators of a and b are actually biased. In nonlinear EIV regression, the bias is inevitable and much more significant than in the linear case, see e.g., [28, 79]. The biasedness of curve fitting algorithms has been pointed independently by many authors, see [17, 18, 31, 94, 189, 198], just to name a few.

The origin of the bias is geometrically obvious, see our illustration in Fig. 6.5. Under the standard assumption that the noise is centrally-symmetric, it is clear that one is more likely to observe points *outside* the true arc than *inside* it. Thus the fitted arc tends to be biased *outward*.

Figure 6.5 *The likelihood of observing a point outside the true arc is higher than that inside it.*

If one fits circles, then clearly the bias is toward larger circles, i.e., the radius estimate tends to exceed its true value. Werman and Keren [189] empirically demonstrate that the radius estimate may be as high as 150% of the true radius value, i.e., one may have $\mathbb{E}(\hat{R}) = 1.5R$. Such a heavy bias occurs, however, only if the noise is so large that $\sigma \approx R$.

A detailed analysis involving Taylor expansion [17] shows that

$$\mathbb{E}(\hat{R}) = R + \frac{\sigma^2}{2R} + \mathcal{O}(\sigma^4), \tag{6.43}$$

i.e., the ratio **bias:radius** is approximately $\frac{1}{2}(\sigma/R)^2$. As in most computer vision applications σ/R stays below 0.05, cf. [17], the **bias:radius** ratio rarely exceeds 0.001. But still it is something one has to reckon with. The worst part here is perhaps that the bias does not vanish (or even decrease) when one samples more data points, i.e., when n grows, see next.

Inconsistency. In classical statistics, an estimator $\hat{\Theta}_n$ of an unknown parameter vector Θ, based on n independent observations, is said to be *consistent* if $\hat{\Theta}_n \to \tilde{\Theta}$ in probability, as $n \to \infty$, where $\tilde{\Theta}$ denotes the true parameter vector. One upgrades this notion to *strong consistency* if the convergence occurs almost surely, i.e., with probability one.

Motivated by this notion, many authors examined the consistency of curve fitting algorithms and found that in nonlinear cases they are almost universally *inconsistent*. In fact, it was found that $\hat{\Theta}_n$ converges (both in probability and almost surely) to some limit Θ^*, which is different from the true parameter $\tilde{\Theta}$.

To determine the limit Θ^* we need to specify how the true points are chosen on the true curve, in the limit $n \to \infty$. Let $\mathbb{P}_0$ denote a probability measure on the true curve $P(x,y;\tilde{\Theta}) = 0$. If the curve is a circle or an ellipse, then $\mathbb{P}_0$ may be a uniform measure on the entire curve, or a uniform measure on an arc.

Now we can choose n true points randomly according to the measure $\mathbb{P}_0$, then we get a structural model, see Section 6.1. Alternatively, we can set n true points in some specific way but make sure that their asymptotic distribution converges to $\mathbb{P}_0$. For example, if $\mathbb{P}_0$ is a uniform measure on an arc, then the true points can be positioned equally spaced along that arc, and their distribution will converge to $\mathbb{P}_0$.

Let $\mathbb{P}_\sigma$ denote the probability distribution of an observed point $(x,y) = (\tilde{x},\tilde{y}) + (\delta,\varepsilon)$, where $(\tilde{x},\tilde{y})$ has distribution $\mathbb{P}_0$, and δ and ε are normal random variables with mean zero and variance σ^2, independent from each other and from $(\tilde{x},\tilde{y})$. In the structural model, the observed points are i.i.d. random variables with distribution $\mathbb{P}_\sigma$. (Note that if $\sigma = 0$, then $\mathbb{P}_\sigma$ actually coincides with $\mathbb{P}_0$.)

Suppose the estimate $\hat{\Theta}_n$ of the unknown parameter vector Θ is obtained by minimizing

$$\mathscr{F}(\Theta) = \sum_{i=1}^n f^2(x_i,y_i;\Theta), \tag{6.44}$$

where $f(x,y,;\Theta)$ is a certain function. The geometric fit uses $f(x,y,;\Theta) = \{$the distance from the point (x,y) to the curve $P(x,y;\Theta) = 0\}$. Algebraic fits use other functions.

Then we have the following fact:

Theorem 11 *Under certain regularity conditions, the estimator $\hat{\Theta}_n$ converges, in probability and almost surely, to the (assumed unique) value Θ^* minimizing*

$$\mathscr{F}^*(\Theta) = \int f^2(x,y;\Theta)\,d\mathbb{P}_\sigma. \tag{6.45}$$

Proof of the theorem. This theorem is a straightforward generalization of the one proved by Berman and Culpin [18] for circular fits. For the sake of completeness, we sketch its proof here (this subsection can be skipped without any harm for further reading). Our proof uses elements of measure theory.

Suppose we observe n points $(x_1,y_1), \ldots, (x_n,y_n)$. Let $\mathbb{P}_n$ denote an atomic measure on the xy plane supported by the observed points, i.e., the measure $\mathbb{P}_n$ is defined by

$$\int g(x,y)\,d\mathbb{P}_n = \frac{1}{n}\sum_{i=1}^n g(x_i,y_i)$$

for any continuous function g. In terms of $\mathbb{P}_n$, our estimate $\hat{\Theta}_n$ is obtained by minimizing the integral

$$\frac{1}{n}\mathscr{F}(\Theta) = \int f^2(x,y;\Theta)\,d\mathbb{P}_n. \tag{6.46}$$

It follows from the strong law of large numbers that $\mathbb{P}_n$ weakly converges to $\mathbb{P}_\sigma$

almost surely, i.e., with probability one. Thus the arg-minimum of the integral (6.46) converges to the (assumed unique) arg-minimum of the integral (6.45).

The mild regularity assumptions we need here are the smoothness of the curve $P(x, y; \Theta) = 0$, the compactness of the support of the measure $\mathbb{P}_0$ (which is automatically ensured for closed curves like circles or ellipses), and a certain continuity in x, y, Θ of the "cost" function f, we omit details.

Inconsistency of the MLE. In most cases, the limit value Θ^* specified by Theorem 11 is *different* from the true parameter vector $\tilde{\Theta}$. There are general mathematical theorems on the inconsistency of nonlinear EIV regression estimators. For example, Fazekas, Kukush, and Zwanzig [59] proved that under mild regularity and "genericity" conditions, the Maximum Likelihood Estimator (i.e., the geometric fit) is inconsistent; they also derived approximate formulas for the limit Θ^*.

6.11   Consistent fit and adjusted least squares

Once statisticians realized that parameter estimators in the curve fitting problems were almost universally biased, and once they derived various approximative formulas for the bias, attempts were made to eliminate the latter. Many practitioners designed various "unbiasing" schemes and tricks, and claimed that they removed the bias.

One needs to remember, however, that truly unbiased estimation in nonlinear EIV problems is practically impossible. In fact, every known "bias removal" procedure only eliminates a part of the bias; for example one can remove the leading term in an asymptotic expansion for the bias, but higher order terms will persist. We will describe several such tricks below.

Simple bias reduction. The bias of the radius estimate of the geometric circle fit is known to be $\frac{\sigma^2}{2R} + \mathcal{O}(\sigma^4)$, see (6.43). Thus one can construct an "unbiased" estimate by

$$\hat{R}_{\text{unbiased}'} = \hat{R} - \frac{\sigma^2}{2\hat{R}}. \tag{6.47}$$

Here we assume, for simplicity, that σ is known. Squaring (6.47) and ignoring higher order terms gives another version of essentially unbiased radius estimator:

$$\hat{R}^2_{\text{unbiased}'} = \hat{R}^2 - \sigma^2. \tag{6.48}$$

Similar tricks are proposed by various authors, see e.g., [189]. For other tricks, see [33, 32, 195].

Amemiya and Fuller [5] develop a more sophisticated procedure eliminating the essential bias of the maximum likelihood estimators, for a general problem of fitting arbitrary curves to data (again assuming that the noise level σ is known). One can check (we omit details) that for circles their procedure reduces to (6.48).

We note that these tricks remove most of the bias (i.e., all the terms of order σ^2), but there remain leftover terms of higher order, i.e., σ^4. They may be small, but they do not vanish as $n \to \infty$, thus the resulting estimate is not asymptotically unbiased in the large sample limit. Thus, it is not consistent either.

Several adjustments have been developed to ensure consistency as $n \to \infty$. One, designed for linear and polynomial EIV regression, is described below.

Adjusted least squares: linear case. To ensure not only a reduction of the bias, but an asymptotical unbiasedness and consistency in the large sample limit, one can make appropriate adjustments in the estimating procedure.

Suppose one fits a straight line $y = \alpha + \beta x$ to the observed points. The classical regression estimates (Section 1.1) are obtained by the least squares method minimizing

$$\mathscr{F}(\alpha, \beta) = \frac{1}{n} \sum (y_i - \alpha - \beta x_i)^2,$$

which reduces to a system of linear equations

$$\begin{align} \alpha + \beta \bar{x} &= \bar{y} \\ \alpha \bar{x} + \beta \overline{xx} &= \overline{xy}, \end{align} \tag{6.49}$$

where we use our "sample mean" notation $\bar{x} = \frac{1}{n} \sum x_i$, $\overline{xy} = \frac{1}{n} \sum x_i y_i$, etc.

Now recall that both x and y are observed with errors, i.e., $x_i = \tilde{x}_i + \delta_i$ and $y_i = \tilde{y}_i + \varepsilon_i$. In the limit $n \to \infty$, according to the central limit theorem, we have

$$\begin{align} \bar{x} &= \bar{\tilde{x}} + \mathscr{O}_P(n^{-1/2}) \\ \bar{y} &= \bar{\tilde{y}} + \mathscr{O}_P(n^{-1/2}) \\ \overline{xy} &= \overline{\tilde{x}\tilde{y}} + \mathscr{O}_P(n^{-1/2}), \end{align}$$

i.e., three random terms in (6.49) converge to their "true" values. But the fourth one does not, because

$$\begin{align} \overline{xx} &= \frac{1}{n} \sum (\tilde{x}_i + \delta_i)^2 \\ &= \frac{1}{n} \sum \tilde{x}_i^2 + \frac{1}{n} \sum \delta_i^2 + \frac{2}{n} \sum \tilde{x}_i \delta_i \\ &= \frac{1}{n} \sum \tilde{x}_i^2 + \sigma^2 + \mathscr{O}_P(n^{-1/2}), \end{align}$$

where we used the fact $\mathbb{E}(\delta_i^2) = \sigma^2$. That extra term σ^2 is the origin of inconsistency and asymptotic biasedness. To eliminate it we adjust the system (6.49) as follows:

$$\begin{align} \alpha + \beta \bar{x} &= \bar{y} \\ \alpha \bar{x} + \beta (\overline{xx} - \sigma^2) &= \overline{xy}, \end{align} \tag{6.50}$$

Now every term converges to its true value, in the limit $n \to \infty$, thus the solutions converge to the true parameter values $\tilde{\alpha}$ and $\tilde{\beta}$. The solutions of (6.50) are called the *adjusted least squares (ALS) estimators*.

The adjusted least squares estimators work well in practice, their accuracy is comparable to that of the MLE described in Chapter 1; see [37, 116]. Interestingly, both the estimators (ALS and MLE) have infinite moments, see [37]; paradoxically, this feature appears to be a "certificate of quality." We note, however, that the ALS (unlike the MLE) assumes that the noise level σ is known.

Adjusted least squares: polynomial case. The above method generalizes to polynomials as follows. Suppose one fits a polynomial of degree k,

$$y = p(x) = \beta_0 + \beta_1 x + \cdots + \beta_k x^k,$$

to observed points by the classical regression, i.e., by minimizing

$$\mathscr{F}(\beta_0, \ldots, \beta_k) = \tfrac{1}{n} \sum \left[y_i - p(x_i) \right]^2.$$

This reduces to a standard system of linear equations

$$\beta_0 \overline{x^m} + \beta_1 \overline{x^{m+1}} + \cdots + \beta_k \overline{x^{m+k}} = \overline{x^m y}, \qquad m = 0, \ldots, k \qquad (6.51)$$

We now compute the limits of the coefficients of these equations, as $n \to \infty$. According to the central limit theorem, they converge to their mean values, i.e.,

$$\overline{x^m} = \mathbb{E}\left(\overline{x^m}\right) + \mathcal{O}_P(n^{-1/2}), \qquad \overline{x^m y} = \mathbb{E}\left(\overline{x^m y}\right) + \mathcal{O}_P(n^{-1/2}).$$

To ensure the consistency it is enough to make the mean values equal to the corresponding true values, i.e., we need to replace $\overline{x^m}$ and $\overline{x^m y}$ with some other statistics, call them A_m and B_m, respectively, so that

$$\mathbb{E}\left(A_m\right) = \overline{\tilde{x}^m}, \qquad \mathbb{E}\left(B_m\right) = \overline{\tilde{x}^m \tilde{y}}.$$

Remark. Superficially, the problem appears to be about adjusting the mean values, i.e., correcting for bias. Some authors mistakenly claim that the adjustment produces unbiased estimates of the unknown parameters. In fact, the resulting estimates are only asymptotically unbiased (in the limit $n \to \infty$); they are also consistent.

We will show how to choose the desired statistics A_m and B_m for small values of m. Obviously, $A_1 = \bar{x}$ and $B_1 = \overline{xy}$. Then direct calculation gives

$$\mathbb{E}(\overline{x^2}) = \overline{\tilde{x}^2} + \sigma^2, \qquad\qquad \mathbb{E}(\overline{x^2 y}) = \overline{\tilde{x}^2 \tilde{y}} + \overline{\tilde{y}} \sigma^2,$$

$$\mathbb{E}(\overline{x^3}) = \overline{\tilde{x}^3} + 3\overline{\tilde{x}} \sigma^2, \qquad\qquad \mathbb{E}(\overline{x^3 y}) = \overline{\tilde{x}^3 \tilde{y}} + 3\overline{\tilde{x}\tilde{y}} \sigma^2,$$

$$\mathbb{E}(\overline{x^4}) = \overline{\tilde{x}^4} + 6\overline{\tilde{x}^2} \sigma^2 + 3\sigma^4, \qquad \mathbb{E}(\overline{x^4 y}) = \overline{\tilde{x}^4 \tilde{y}} + 6\overline{\tilde{x}^2 \tilde{y}} \sigma^2 + 3\overline{\tilde{y}} \sigma^4,$$

which suggest that

$$A_2 = \overline{x^2} - \sigma^2, \qquad\qquad B_2 = \overline{x^2y} - \bar{y}\sigma^2,$$
$$A_3 = \overline{x^3} - 3\bar{x}\sigma^2, \qquad\qquad B_3 = \overline{x^3y} - 3\overline{xy}\sigma^2,$$
$$A_4 = \overline{x^4} - 6\overline{x^2}\sigma^2 + 3\sigma^4, \qquad B_4 = \overline{x^4y} - 6\overline{x^2y}\sigma^2 + 3\bar{y}\sigma^4.$$

This method can be directly extended to arbitrary A_m and B_m, see [30] and Section 6.4 in [40].

Lastly, the adjusted least squares estimates are the solutions of the system

$$\beta_0 A_m + \beta_1 A_{m+1} + \cdots + \beta_k A_{m+k} = B_m, \qquad m = 0, \ldots, k \qquad (6.52)$$

Adjustments for implicit nonlinear models. This section covered explicit functional relations — linear and polynomial models. In most computer vision applications, though, one fits implicit nonlinear models to data; see Section 1.9. Those require more sophisticated approaches, discussed further in Section 7.8.

Chapter 7

Statistical analysis of circle fits

In this chapter we apply the general methods surveyed and developed in Chapter 6 to the problem of fitting circles to data.

7.1 Error analysis of geometric circle fit

First we analyze the geometric circle fit, i.e., the estimator $\hat{\Theta} = (\hat{a}, \hat{b}, \hat{R})$ of the circle parameters (the center (a, b) and the radius R) minimizing $\sum d_i^2$, where d_i is the orthogonal (geometric) distance from the data point (x_i, y_i) to the fitted circle.

Variance of the geometric circle fit. We start with the main part of our error analysis: the variance term represented by $\sigma^2 \mathbf{V}$ in (6.31).

Recall that the distance d_i is given by

$$d_i = \sqrt{(x_i - a)^2 + (y_i - b)^2} - R$$

$$= \sqrt{\left[(\tilde{x}_i - \tilde{a}) + (\delta_i - \Delta a)\right]^2 + \left[(\tilde{y}_i - \tilde{b}) + (\varepsilon_i - \Delta b)\right]^2} - \tilde{R} - \Delta R, \qquad (7.1)$$

cf. (3.3). Here $\tilde{a}, \tilde{b}, \tilde{R}$ are the parameters of the true circle. For brevity we denote

$$\tilde{u}_i = (\tilde{x}_i - \tilde{a})/\tilde{R} \qquad \text{and} \qquad \tilde{v}_i = (\tilde{y}_i - \tilde{b})/\tilde{R},$$

in fact these are the "true" values of u_i and v_i introduced earlier in (4.20)–(4.21). Note that $\tilde{u}_i^2 + \tilde{v}_i^2 = 1$ for every i.

Now, to the first order, we have

$$d_i = \sqrt{\tilde{R}^2 + 2\tilde{R}\tilde{u}_i(\delta_i - \Delta a) + 2\tilde{R}\tilde{v}_i(\varepsilon_i - \Delta b) + \mathcal{O}_P(\sigma^2)} - \tilde{R} - \Delta R$$

$$= \tilde{u}_i(\delta_i - \Delta a) + \tilde{v}_i(\varepsilon_i - \Delta b) - \Delta R + \mathcal{O}_P(\sigma^2). \qquad (7.2)$$

Minimizing $\sum d_i^2$ to the first order is equivalent to minimizing

$$\sum_{i=1}^{n} (\tilde{u}_i \Delta a + \tilde{v}_i \Delta b + \Delta R - \tilde{u}_i \delta_i - \tilde{v}_i \varepsilon_i)^2.$$

This is a classical least squares problem that can be written as

$$\mathbf{W} \Delta \Theta \approx \tilde{\mathbf{U}} \delta + \tilde{\mathbf{V}} \varepsilon, \qquad (7.3)$$

where

$$\mathbf{W} = \begin{bmatrix} \tilde{u}_1 & \tilde{v}_1 & 1 \\ \vdots & \vdots & \vdots \\ \tilde{u}_n & \tilde{v}_n & 1 \end{bmatrix}, \qquad (7.4)$$

$\Theta = (a, b, R)^T$, and

$$\delta = (\delta_1, \ldots, \delta_n)^T \qquad \text{and} \qquad \varepsilon = (\varepsilon_1, \ldots, \varepsilon_n)^T$$

and

$$\tilde{\mathbf{U}} = \text{diag}(\tilde{u}_1, \ldots, \tilde{u}_n) \qquad \text{and} \qquad \tilde{\mathbf{V}} = \text{diag}(\tilde{v}_1, \ldots, \tilde{v}_n).$$

The solution of the least squares problem (7.3) is

$$\Delta \hat{\Theta} = (\mathbf{W}^T \mathbf{W})^{-1} \mathbf{W}^T (\tilde{\mathbf{U}} \delta + \tilde{\mathbf{V}} \varepsilon), \qquad (7.5)$$

of course this does not include the $\mathcal{O}_P(\sigma^2)$ terms.

Thus the variance of our estimator, to the leading order, is

$$\mathbb{E}\left[(\Delta \hat{\Theta})(\Delta \hat{\Theta})^T\right] = (\mathbf{W}^T \mathbf{W})^{-1} \mathbf{W}^T \mathbb{E}\left[(\tilde{\mathbf{U}} \delta + \tilde{\mathbf{V}} \varepsilon)(\delta^T \tilde{\mathbf{U}} + \varepsilon^T \tilde{\mathbf{V}})\right] \mathbf{W}(\mathbf{W}^T \mathbf{W})^{-1}.$$

Now observe that

$$\mathbb{E}(\delta\varepsilon^T) = \mathbb{E}(\varepsilon\delta^T) = \mathbf{0}$$

as well as

$$\mathbb{E}(\delta\delta^T) = \mathbb{E}(\varepsilon\varepsilon^T) = \sigma^2\mathbf{I},$$

and next we have $\tilde{\mathbf{U}}^2 + \tilde{\mathbf{V}}^2 = \mathbf{I}$, thus to the leading order

$$\mathbb{E}\left[(\Delta\hat{\Theta})(\Delta\hat{\Theta})^T\right] = \sigma^2(\mathbf{W}^T\mathbf{W})^{-1}\mathbf{W}^T\mathbf{W}(\mathbf{W}^T\mathbf{W})^{-1}$$
$$= \sigma^2(\mathbf{W}^T\mathbf{W})^{-1}, \tag{7.6}$$

where the higher order (of σ^4) terms are not included. The expression (7.6) is known to the statistics community. First, it is a particular case of a more general formula (6.37), which is given in Theorem 3.2.1 of Fuller's book [66]. Second, (7.6) was explicitly derived by Kanatani [96] and others [42] by using less direct approaches.

Bias of the geometric circle fit. Now we do a second-order error analysis, which has not been previously done in the statistics literature. According to a general formula (6.18), we put

$$a = \tilde{a} + \Delta_1 a + \Delta_2 a + \mathcal{O}_P(\sigma^3),$$
$$b = \tilde{b} + \Delta_1 b + \Delta_2 b + \mathcal{O}_P(\sigma^3),$$
$$R = \tilde{R} + \Delta_1 R + \Delta_2 R + \mathcal{O}_P(\sigma^3).$$

Here $\Delta_1 a$, $\Delta_1 b$, $\Delta_1 R$ are linear combinations of ε_i's and δ_i's, which were found above, in (7.5); and $\Delta_2 a$, $\Delta_2 b$, $\Delta_2 R$ are quadratic forms of ε_i's and δ_i's to be determined next.

Expanding the distances d_i to the second order terms gives

$$d_i = \tilde{u}_i(\delta_i - \Delta_1 a) + \tilde{v}_i(\varepsilon_i - \Delta_1 b) - \Delta_1 R$$
$$- \tilde{u}_i\Delta_2 a - \tilde{v}_i\Delta_2 b - \Delta_2 R$$
$$+ \frac{\tilde{v}_i^2}{2R}(\delta_i - \Delta_1 a)^2 + \frac{\tilde{u}_i^2}{2R}(\varepsilon_i - \Delta_1 b)^2$$
$$- \frac{\tilde{u}_i\tilde{v}_i}{R}(\delta_i - \Delta_1 a)(\varepsilon_i - \Delta_1 b).$$

Since we already found $\Delta_1 a$, $\Delta_1 b$, $\Delta_1 R$, the only unknowns are $\Delta_2 a$, $\Delta_2 b$, $\Delta_2 R$. Minimizing $\sum d_i^2$ is now equivalent to minimizing

$$\sum_{i=1}^{n}(\tilde{u}_i\Delta_2 a + \tilde{v}_i\Delta_2 b + \Delta_2 R - f_i)^2,$$

where

$$f_i = \tilde{u}_i(\delta_i - \Delta_1 a) + \tilde{v}_i(\varepsilon_i - \Delta_1 b) - \Delta_1 R$$
$$+ \frac{\tilde{v}_i^2}{2R}(\delta_i - \Delta_1 a)^2 + \frac{\tilde{u}_i^2}{2R}(\varepsilon_i - \Delta_1 b)^2$$
$$- \frac{\tilde{u}_i\tilde{v}_i}{R}(\delta_i - \Delta_1 a)(\varepsilon_i - \Delta_1 b). \tag{7.7}$$

This is another classical least squares problem, and its solution is

$$\Delta_2\hat{\Theta} = (\mathbf{W}^T\mathbf{W})^{-1}\mathbf{W}^T\mathbf{F}, \tag{7.8}$$

where $\mathbf{F} = (f_1,\ldots,f_n)^T$; of course this is a quadratic approximation which does not include $\mathcal{O}_P(\sigma^3)$ terms. In fact, the contribution from the first three (linear) terms in (7.7) vanishes, quite predictably; thus only the last two (quadratic) terms will matter.

Taking the mean value gives, to the leading order,

$$\mathbb{E}(\Delta\hat{\Theta}) = \mathbb{E}(\Delta_2\hat{\Theta}) = \frac{\sigma^2}{2R}\left[(\mathbf{W}^T\mathbf{W})^{-1}\mathbf{W}^T\mathbf{1} + (\mathbf{W}^T\mathbf{W})^{-1}\mathbf{W}^T\mathbf{S}\right], \tag{7.9}$$

where $\mathbf{1} = (1,1,\ldots,1)^T$ and $\mathbf{S} = (s_1,\ldots,s_n)^T$, here s_i is a scalar given by

$$s_i = [-\tilde{v}_i, \tilde{u}_i, 0](\mathbf{W}^T\mathbf{W})^{-1}[-\tilde{v}_i, \tilde{u}_i, 0]^T.$$

The second term in (7.9) is of order $\mathcal{O}(\sigma^2/n)$, thus the essential bias is given by the first term only:

$$\mathbb{E}(\Delta\hat{\Theta}) \stackrel{\text{ess}}{=} \frac{\sigma^2}{2R}(\mathbf{W}^T\mathbf{W})^{-1}\mathbf{W}^T\mathbf{1}.$$

In fact, this expression can be simplified. Since the last column of the matrix $\mathbf{W}^T\mathbf{W}$ coincides with the vector $\mathbf{W}^T\mathbf{1}$, we have $(\mathbf{W}^T\mathbf{W})^{-1}\mathbf{W}^T\mathbf{1} = [0,0,1]^T$; hence the essential bias of the geometric circle fit is

$$\mathbb{E}(\Delta\hat{\Theta}) \stackrel{\text{ess}}{=} \frac{\sigma^2}{2R}\begin{bmatrix} 0 \\ 0 \\ 1 \end{bmatrix}. \tag{7.10}$$

The same result can also be obtained from the general formula (6.39). Thus the estimates of the circle center, $\hat{a}$ and $\hat{b}$, have *no essential bias*, while the estimate of the radius has essential bias

$$\mathbb{E}(\Delta\hat{R}) \stackrel{\text{ess}}{=} \frac{\sigma^2}{2R}, \tag{7.11}$$

which is independent of the number and location of the true points.

7.2 Cramer-Rao lower bound for the circle fit

Here we derive the classical lower bound on the covariance matrix of the circle parameter estimators. This is a fairly standard task, we just follow the lines of Section 2.8. A similar analysis was recently included in a paper by Zelniker and Clarkson [196].

We work in the context of the Cartesian functional model, in which there are $n+3$ independent parameters: three principal ones (a, b, R) and one "latent" parameter for each data point.

First we specify the coordinates of the true points by

$$\tilde{x}_i = \tilde{a} + \tilde{R}\cos\varphi_i, \qquad \tilde{y}_i = \tilde{a} + \tilde{R}\sin\varphi_i$$

where $\varphi_1, \ldots, \varphi_n$ are the latent parameters (recall that we have used this formalism in Section 4.8). Now, under our standard assumptions (Cartesian functional model described in Section 6.1), the log-likelihood function is

$$\ln L = \text{const} - \frac{1}{2\sigma^2}\sum_{i=1}^{n}(x_i - \tilde{x}_i)^2 + (y_i - \tilde{y}_i)^2. \tag{7.12}$$

According to the classical Cramer-Rao theorem, the covariance matrix of unbiased parameter estimators is bounded below by

$$\text{Cov}(\hat{a}, \hat{b}, \hat{R}, \hat{\varphi}_1, \ldots, \hat{\varphi}_n) \geq \mathbf{F}^{-1},$$

where $\mathbf{F}$ is the Fisher information matrix, $\mathbf{F} = -\mathbb{E}(\mathbf{H})$, and $\mathbf{H}$ denotes the Hessian matrix consisting of the second order partial derivatives of $\ln L$ with respect to the parameters.

Computing the second order partial derivatives of (7.12) with respect to a, b, R, and $\varphi_1, \ldots, \varphi_n$ is a routine exercise which we omit. Taking their expected values (and using the obvious relations $\mathbb{E}(x_i) = \tilde{x}_i$ and $\mathbb{E}(y_i) = \tilde{y}_i$) gives the following results (where we suppress the common factor σ^{-2} for brevity)

$$\mathbb{E}\left[\frac{\partial^2 \ln L}{\partial a^2}\right] = -n, \qquad \mathbb{E}\left[\frac{\partial^2 \ln L}{\partial b^2}\right] = -n, \qquad \mathbb{E}\left[\frac{\partial^2 \ln L}{\partial R^2}\right] = -n,$$

$$\mathbb{E}\left[\frac{\partial^2 \ln L}{\partial a\,\partial R}\right] = -\sum\cos\varphi_i, \quad \mathbb{E}\left[\frac{\partial^2 \ln L}{\partial a\,\partial b}\right] = 0, \quad \mathbb{E}\left[\frac{\partial^2 \ln L}{\partial b\,\partial R}\right] = -\sum\sin\varphi_i,$$

$$\mathbb{E}\left[\frac{\partial^2 \ln L}{\partial a\,\partial \varphi_i}\right] = R\sin\varphi_i, \quad \mathbb{E}\left[\frac{\partial^2 \ln L}{\partial b\,\partial \varphi_i}\right] = -R\cos\varphi_i, \quad \mathbb{E}\left[\frac{\partial^2 \ln L}{\partial R\,\partial \varphi_i}\right] = 0,$$

and lastly

$$\mathbb{E}\left[\frac{\partial^2 \ln L}{\partial \varphi_i\,\partial \varphi_j}\right] = -R^2\delta_{ij},$$

where δ_{ij} denotes the Kronecker delta symbol. The Fisher information matrix now can be written as

$$\mathbf{F} = \sigma^{-2}\begin{bmatrix} \mathbf{K} & \mathbf{L}^T \\ \mathbf{L} & R^2\mathbf{I}_n \end{bmatrix},$$

where

$$\mathbf{K} = \begin{bmatrix} n & 0 & \sum\cos\varphi_i \\ 0 & n & \sum\sin\varphi_i \\ \sum\cos\varphi_i & \sum\sin\varphi_i & n \end{bmatrix}$$

and

$$\mathbf{L} = \begin{bmatrix} -R\sin\varphi_1 & R\cos\varphi_1 & 0 \\ \vdots & \vdots & \vdots \\ -R\sin\varphi_n & R\cos\varphi_n & 0 \end{bmatrix}.$$

As usual, $\mathbf{I}_n$ denotes the identity matrix of order n. Next we apply a standard block matrix inversion lemma (see, e.g., page 26 of [128]) we obtain

$$\begin{bmatrix} \mathbf{K} & \mathbf{L}^T \\ \mathbf{L} & R^2\mathbf{I}_n \end{bmatrix}^{-1} = \begin{bmatrix} (\mathbf{K} - R^{-2}\mathbf{L}^T\mathbf{L})^{-1} & -(R^2\mathbf{K} - \mathbf{L}^T\mathbf{L})^{-1}\mathbf{L}^T \\ -\mathbf{L}(R^2\mathbf{K} - \mathbf{L}^T\mathbf{L})^{-1} & (R^2\mathbf{I}_n - \mathbf{L}\mathbf{K}^{-1}\mathbf{L}^T)^{-1} \end{bmatrix}.$$

One can easily verify this formula by direct multiplication. The 3×3 top left block of the last matrix is the most interesting to us as it corresponds to the principal parameters a, b, R. In terms of the matrix $\mathbf{W}$ introduced earlier in (7.4) we have $\tilde{u}_i = \cos\varphi_i$ and $\tilde{v}_i = \sin\varphi_i$, hence

$$\mathbf{K} - R^{-2}\mathbf{L}^T\mathbf{L} = \mathbf{W}^T\mathbf{W},$$

thus the Cramer-Rao bound reads

$$\mathbb{E}\left[(\Delta\hat{\Theta})(\Delta\hat{\Theta})^T\right] \geq \sigma^2(\mathbf{W}^T\mathbf{W})^{-1}. \tag{7.13}$$

This result is known to the statistics community, though it is not clear who was the first to derive it. It was apparently known to Chan in 1965, see page 53 of [31], though he did not state it explicitly. In an explicit form, it was published by Chan and Thomas [35] in 1995, but their argument was somewhat flawed. A correct argument was given in 1998 by Kanatani [96] and, independently, in 2006 by Zelniker and Clarkson [196].

The unbiasedness restriction. A few comments are in order regarding the practical relevance of the above result. The Cramer-Rao theorem only establishes the minimal variance for *unbiased* estimators. In the case of fitting non-linear curves to data, there is no such luxury as unbiased estimators; in particular *all* circle fitting algorithms are biased to some extent (sometimes heavily); see Section 6.10. Hence it is not immediately clear if the bound (7.13) has any practical value, i.e., if it applies to any circle fit.

It actually does, as we explained in Section 6.9, if one restricts the analysis to the leading terms (of order σ^2). The right-hand side of (7.13) gives the lower bound for the *leading term* of the covariance matrix, i.e.,

$$\mathbf{V} \geq (\mathbf{W}^T\mathbf{W})^{-1}. \tag{7.14}$$

This is true for any geometrically consistent circle fit, i.e., any fit satisfying (6.19). Thus (7.14) holds for the geometric circle fit and all known algebraic circle fits.

Summary. Comparing (7.6) and (7.14) shows that the geometric circle fit has a minimal possible variance, to the leading order. A similar fact for algebraic circle fits will be established shortly; then we will see that all the known circle fits have minimal possible variance $\mathbf{V}$, which is $\mathbf{V}_{\min} = (\mathbf{W}^T\mathbf{W})^{-1}$.

On the other hand, the essential bias of the geometric circle fit does not vanish, it is given by (7.9). A natural question is then: does the essential bias (7.9) also attain its natural minimum, or can it be reduced? We will see in the next sections that the essential bias *can* be reduced, and in fact *eliminated*.

7.3 Error analysis of algebraic circle fits

Here we analyze algebraic circle fits described in Chapter 5, including the Kåsa fit, the Pratt fit, and the Taubin fit.

Review of algebraic circle fits (Section 5.11). Let us again describe circles by an algebraic equation

$$A(x^2 + y^2) + Bx + Cy + D = 0, \tag{7.15}$$

where $\mathbf{A} = (A, B, C, D)^T$ is the 4-parameter vector. Recall that every algebraic fit is equivalent to the minimization of

$$\mathscr{F}(A, B, C, D) = \frac{1}{n}\sum_{i=1}^{n}(Az_i + Bx_i + Cy_i + D)^2$$

$$= n^{-1}\mathbf{A}^T(\mathbf{X}^T\mathbf{X})\mathbf{A} = \mathbf{A}^T\mathbf{MA}, \tag{7.16}$$

subject to a constraint

$$\mathbf{A}^T\mathbf{NA} = 1 \tag{7.17}$$

for some matrix $\mathbf{N}$. Here we use our shorthand notation $z_i = x_i^2 + y_i^2$ and

$$\mathbf{M} = \frac{1}{n}\mathbf{X}^T\mathbf{X} = \begin{bmatrix} \overline{zz} & \overline{zx} & \overline{zy} & \bar{z} \\ \overline{zx} & \overline{xx} & \overline{xy} & \bar{x} \\ \overline{zy} & \overline{xy} & \overline{yy} & \bar{y} \\ \bar{z} & \bar{x} & \bar{y} & 1 \end{bmatrix},$$

where

$$\mathbf{X} = \begin{bmatrix} z_1 & x_1 & y_1 & 1 \\ \vdots & \vdots & \vdots & \vdots \\ z_n & x_n & y_n & 1 \end{bmatrix}. \tag{7.18}$$

The constraint matrix $\mathbf{N}$ in (7.17) determines the particular algebraic fit. It is

$$\mathbf{N} = \mathbf{K} = \begin{bmatrix} 1 & 0 & 0 & 0 \\ 0 & 0 & 0 & 0 \\ 0 & 0 & 0 & 0 \\ 0 & 0 & 0 & 0 \end{bmatrix} \tag{7.19}$$

for the Kåsa fit, cf. (5.39),

$$\mathbf{N} = \mathbf{P} = \begin{bmatrix} 0 & 0 & 0 & -2 \\ 0 & 1 & 0 & 0 \\ 0 & 0 & 1 & 0 \\ -2 & 0 & 0 & 0 \end{bmatrix} \tag{7.20}$$

for the Pratt fit, see (5.27), and

$$\mathbf{N} = \mathbf{T} = \begin{bmatrix} 4\bar{z} & 2\bar{x} & 2\bar{y} & 0 \\ 2\bar{x} & 1 & 0 & 0 \\ 2\bar{y} & 0 & 1 & 0 \\ 0 & 0 & 0 & 0 \end{bmatrix}, \tag{7.21}$$

for the Taubin fit, see (5.48); here we again use our standard "sample mean" notation $\bar{z} = \frac{1}{n}\sum z_i$, etc.

The constrained minimization problem (7.16)–(7.17) reduces to the generalized eigenvalue problem

$$\mathbf{MA} = \eta\mathbf{NA}, \tag{7.22}$$

thus $\mathbf{A}$ must be a generalized eigenvector of the matrix pair $(\mathbf{M},\mathbf{N})$. To find the solution of the minimization problem (7.16)–(7.17) we must choose the eigenvector $\mathbf{A}$ with the smallest nonnegative eigenvalue η. This fact determines $\mathbf{A}$ up to a scalar multiple; as multiplying $\mathbf{A}$ by a scalar does not change the circle it represents, we can set $\|\mathbf{A}\| = 1$.

Matrix perturbation method. As algebraic fits can be conveniently expressed in matrix form, we perform error analysis on matrices, following Kanatani [104]. For every random variable, matrix or vector, $\mathbf{Z}$, we write

$$\mathbf{Z} = \tilde{\mathbf{Z}} + \Delta_1\mathbf{Z} + \Delta_2\mathbf{Z} + \mathcal{O}_P(\sigma^3), \tag{7.23}$$

where $\tilde{\mathbf{Z}}$ is its "true," nonrandom, value (achieved when $\sigma = 0$), $\Delta_1\mathbf{Z}$ is a linear combination of δ_i's and ε_i's, and $\Delta_2\mathbf{Z}$ is a quadratic form of δ_i's and ε_i's; all the higher order terms (cubic etc.) are represented by $\mathcal{O}_P(\sigma^3)$. For brevity, we drop the $\mathcal{O}_P(\sigma^3)$ terms in the subsequent formulas.

Also note that $\tilde{\mathbf{M}}\tilde{\mathbf{A}} = \mathbf{0}$, as well as $\tilde{\mathbf{X}}\tilde{\mathbf{A}} = \mathbf{0}$, because the true points lie on the true circle. This implies

$$\tilde{\mathbf{A}}^T\Delta_1\mathbf{M}\tilde{\mathbf{A}} = n^{-1}\tilde{\mathbf{A}}^T\left(\tilde{\mathbf{X}}^T\Delta_1\mathbf{X} + \Delta_1\mathbf{X}^T\tilde{\mathbf{X}}\right)\tilde{\mathbf{A}} = 0,$$

hence $\mathbf{A}^T\mathbf{MA} = \mathcal{O}_P(\sigma^2)$, and premultiplying (7.22) by $\mathbf{A}^T$ yields $\eta = \mathcal{O}_P(\sigma^2)$.

Next we expand (7.22) to the second order terms, according to the general rule (7.23), and omitting terms $\mathcal{O}_P(\sigma^3)$ gives

$$(\tilde{\mathbf{M}} + \Delta_1\mathbf{M} + \Delta_2\mathbf{M})(\tilde{\mathbf{A}} + \Delta_1\mathbf{A} + \Delta_2\mathbf{A}) = \eta\tilde{\mathbf{N}}\tilde{\mathbf{A}} \tag{7.24}$$

(recall that for the Taubin method $\mathbf{N}$ is data-dependent, but only its "true" value $\tilde{\mathbf{N}}$ matters, because $\eta = \mathcal{O}_P(\sigma^2)$).

The left-hand side of (7.24) consists of a linear part $(\tilde{\mathbf{M}}\Delta_1\mathbf{A} + \Delta_1\mathbf{A}\tilde{\mathbf{M}})$ and a quadratic part $(\tilde{\mathbf{M}}\Delta_2\mathbf{A} + \Delta_1\mathbf{M}\Delta_1\mathbf{A} + \Delta_2\mathbf{M}\tilde{\mathbf{A}})$. The right-hand side is quadratic. Separating linear and quadratic terms gives

$$\tilde{\mathbf{M}}\Delta_1\mathbf{A} + n^{-1}\tilde{\mathbf{X}}^T\Delta_1\mathbf{X}\tilde{\mathbf{A}} = \mathbf{0} \tag{7.25}$$

and, again omitting terms $\mathcal{O}_P(\sigma^3)$,

$$\tilde{\mathbf{M}}\Delta_2\mathbf{A} + \Delta_1\mathbf{M}\Delta_1\mathbf{A} + \Delta_2\mathbf{M}\tilde{\mathbf{A}} = \eta\tilde{\mathbf{N}}\tilde{\mathbf{A}}. \tag{7.26}$$

Note that $\tilde{\mathbf{M}}$ is a singular matrix (because $\tilde{\mathbf{M}}\tilde{\mathbf{A}} = \mathbf{0}$), but whenever there are at least three distinct true points, they determine a unique true circle, thus the kernel of $\tilde{\mathbf{M}}$ is one-dimensional, and it coincides with the one-dimensional vector space $\text{span}(\tilde{\mathbf{A}})$. Also, since we set $\|\mathbf{A}\| = 1$, we see that

$$1 = \|\tilde{\mathbf{A}} + \Delta_1\mathbf{A}\|^2 = \|\tilde{\mathbf{A}}\|^2 + 2\tilde{\mathbf{A}}^T\Delta_1\mathbf{A} + \mathcal{O}_P(\sigma^2),$$

hence $\tilde{\mathbf{A}}^T\Delta_1\mathbf{A} = 0$, i.e., $\Delta_1\mathbf{A}$ is orthogonal to $\tilde{\mathbf{A}}$. Thus we can write

$$\Delta_1\mathbf{A} = -n^{-1}\tilde{\mathbf{M}}^-\tilde{\mathbf{X}}^T\Delta_1\mathbf{X}\tilde{\mathbf{A}}, \tag{7.27}$$

where $\tilde{\mathbf{M}}^-$ denotes the Moore-Penrose pseudoinverse. Note that $\mathbb{E}(\Delta_1\mathbf{A}) = 0$.

7.4 Variance and bias of algebraic circle fits

Here we use our previous error analysis to derive explicit formulas for the variance and bias of the algebraic circle fits in the parameters A, B, C, D. The natural geometric parameters a, b, R will be treated in Section 7.6.

Variance of algebraic circle fits. The variance, to the leading order, is

$$\mathbb{E}\left[(\Delta_1\mathbf{A})(\Delta_1\mathbf{A})^T\right] = n^{-2}\tilde{\mathbf{M}}^-\mathbb{E}(\tilde{\mathbf{X}}^T\Delta_1\mathbf{X}\tilde{\mathbf{A}}\tilde{\mathbf{A}}^T\Delta_1\mathbf{X}^T\tilde{\mathbf{X}})\tilde{\mathbf{M}}^- \tag{7.28}$$

$$= n^{-2}\tilde{\mathbf{M}}^-\mathbb{E}\left[\left(\sum_i\tilde{\mathbf{X}}_i\Delta_1\mathbf{X}_i^T\right)\tilde{\mathbf{A}}\tilde{\mathbf{A}}^T\left(\sum_j\Delta_1\mathbf{X}_j^T\tilde{\mathbf{X}}_j^T\right)\right]\tilde{\mathbf{M}}^-,$$

where

$$\tilde{\mathbf{X}}_i = \begin{bmatrix} \tilde{z}_i \\ \tilde{x}_i \\ \tilde{y}_i \\ 1 \end{bmatrix} \quad \text{and} \quad \Delta_1\mathbf{X}_i = \begin{bmatrix} 2\tilde{x}_i\delta_i + 2\tilde{y}_i\varepsilon_i \\ \delta_i \\ \varepsilon_i \\ 0 \end{bmatrix}$$

denote the columns of the matrices $\tilde{\mathbf{X}}^T$ and $\Delta_1\mathbf{X}^T$, respectively. Observe that

$$\mathbb{E}\left[(\Delta_1\mathbf{X}_i)(\Delta_1\mathbf{X}_j)^T\right] = \begin{cases} 0 & \text{whenever} \quad i \neq j \\ \sigma^2\tilde{\mathbf{T}}_i & \text{whenever} \quad i = j \end{cases}$$

where

$$\tilde{\mathbf{T}}_i = \begin{bmatrix} 4\tilde{z}_i & 2\tilde{x}_i & 2\tilde{y}_i & 0 \\ 2\tilde{x}_i & 1 & 0 & 0 \\ 2\tilde{y}_i & 0 & 1 & 0 \\ 0 & 0 & 0 & 0 \end{bmatrix}.$$

Now we rewrite (7.28) as

$$\mathbb{E}\left[(\Delta_1 \mathbf{A})(\Delta_1 \mathbf{A})^T\right] = n^{-2}\tilde{\mathbf{M}}^-\left[\sum_{i,j}\tilde{\mathbf{X}}_i\tilde{\mathbf{A}}^T\mathbb{E}\left(\Delta_1 \mathbf{X}_i\Delta_1 \mathbf{X}_j^T\right)\tilde{\mathbf{A}}\tilde{\mathbf{X}}_j^T\right]\tilde{\mathbf{M}}^-$$

$$= n^{-2}\sigma^2\tilde{\mathbf{M}}^-\left[\sum_i\tilde{\mathbf{X}}_i\tilde{\mathbf{A}}^T\tilde{\mathbf{T}}_i\tilde{\mathbf{A}}\tilde{\mathbf{X}}_i^T\right]\tilde{\mathbf{M}}^-$$

Note that

$$\tilde{\mathbf{A}}^T\tilde{\mathbf{T}}_i\tilde{\mathbf{A}} = \tilde{\mathbf{A}}^T\mathbf{P}\tilde{\mathbf{A}} = \tilde{B}^2 + \tilde{C}^2 - 4\tilde{A}\tilde{D}$$

for each i, hence

$$\sum_i\tilde{\mathbf{X}}_i\tilde{\mathbf{A}}^T\tilde{\mathbf{T}}_i\tilde{\mathbf{A}}\tilde{\mathbf{X}}_i^T = \sum_i(\tilde{\mathbf{A}}^T\mathbf{P}\tilde{\mathbf{A}})\tilde{\mathbf{X}}_i\tilde{\mathbf{X}}_i^T = n(\tilde{\mathbf{A}}^T\mathbf{P}\tilde{\mathbf{A}})\tilde{\mathbf{M}}.$$

This gives

$$\mathbb{E}\left[(\Delta_1 \mathbf{A})(\Delta_1 \mathbf{A})^T\right] = n^{-1}\sigma^2\tilde{\mathbf{M}}^-(\tilde{\mathbf{A}}^T\mathbf{P}\tilde{\mathbf{A}}). \tag{7.29}$$

Remarkably, the variance of algebraic fits does not depend on the constraint matrix $\mathbf{N}$, hence all algebraic fits have the same variance (to the leading order). In Section 7.6 we will derive the variance of algebraic fits in the natural circle parameters (a, b, R) and see that it coincides with the variance of the geometric fit (7.6).

Bias of algebraic circle fits. Since $\mathbb{E}(\Delta_1 \mathbf{A}) = 0$, it will be enough to find $\mathbb{E}(\Delta_2 \mathbf{A})$, and this will be done via (7.26). First we can write, following (7.23)

$$\eta = \tilde{\eta} + \Delta_1 \eta + \Delta_2 \eta + \mathscr{O}_P(\sigma^3).$$

Premultiplying (7.22) by $\tilde{\mathbf{A}}^T$ yields

$$\eta = \frac{\tilde{\mathbf{A}}^T\mathbf{M}\mathbf{A}}{\tilde{\mathbf{A}}^T\mathbf{N}\mathbf{A}} = \frac{\tilde{\mathbf{A}}^T\Delta_2\mathbf{M}\tilde{\mathbf{A}} + \tilde{\mathbf{A}}^T\Delta_1\mathbf{M}\Delta_1\mathbf{A}}{\tilde{\mathbf{A}}^T\tilde{\mathbf{N}}\tilde{\mathbf{A}}} + \mathscr{O}_P(\sigma^3),$$

in particular we see that $\tilde{\eta} = \Delta_1 \eta = 0$ and $\eta = \mathscr{O}_P(\sigma^2)$. Thus by (7.26)

$$\Delta_2 \mathbf{A} = \tilde{\mathbf{M}}^-\left(\frac{\tilde{\mathbf{A}}^T\mathbf{L}}{\tilde{\mathbf{A}}^T\tilde{\mathbf{N}}\tilde{\mathbf{A}}}\tilde{\mathbf{N}}\tilde{\mathbf{A}} - \mathbf{L}\right) + \mathscr{O}_P(\sigma^3), \tag{7.30}$$

where

$$\mathbf{L} = \Delta_2\mathbf{M}\tilde{\mathbf{A}} + \Delta_1\mathbf{M}\Delta_1\mathbf{A}. \tag{7.31}$$

We note that (7.30) actually gives the component of $\Delta_2 \mathbf{A}$ orthogonal to $\tilde{\mathbf{A}}$, but it is exactly what we need, because we set $\|\mathbf{A}\| = 1$. We can rewrite (7.31) as

$$\mathbf{L} = \mathbf{R}\tilde{\mathbf{A}}$$

where, according to (7.27),

$$\mathbf{R} = \tfrac{1}{n} \left[\Delta_1 \mathbf{X}^T \Delta_1 \mathbf{X} + \tilde{\mathbf{X}}^T \Delta_2 \mathbf{X} + \Delta_2 \mathbf{X}^T \tilde{\mathbf{X}} \right.$$
$$\left. - \tfrac{1}{n} (\tilde{\mathbf{X}}^T \Delta_1 \mathbf{X} + \Delta_1 \mathbf{X}^T \tilde{\mathbf{X}}) \tilde{\mathbf{M}}^- \tilde{\mathbf{X}}^T \Delta_1 \mathbf{X} \right]. \tag{7.32}$$

Thus the formula for the bias becomes

$$\mathbb{E}(\Delta_2 \mathbf{A}) = \tilde{\mathbf{M}}^- \left[\frac{\tilde{\mathbf{A}}^T \mathbb{E}(\mathbf{R}) \tilde{\mathbf{A}}}{\tilde{\mathbf{A}}^T \tilde{\mathbf{N}} \tilde{\mathbf{A}}} \tilde{\mathbf{N}} \tilde{\mathbf{A}} - \mathbb{E}(\mathbf{R}) \tilde{\mathbf{A}} \right] + \mathcal{O}(\sigma^4). \tag{7.33}$$

Next we find the mean value of (7.32), term by term. First,

$$\tfrac{1}{n} \mathbb{E}(\Delta_1 \mathbf{X}^T \Delta_1 \mathbf{X}) = \tfrac{1}{n} \mathbb{E}\left(\sum_i (\Delta_1 \mathbf{X}_i)(\Delta_1 \mathbf{X}_i)^T \right) = \tfrac{\sigma^2}{n} \sum_i \tilde{\mathbf{T}}_i = \sigma^2 \tilde{\mathbf{T}}. \tag{7.34}$$

Second, since $\mathbb{E}(\Delta_2 \mathbf{X}_i) = (2\sigma^2, 0, 0, 0)^T$, we obtain

$$\tfrac{1}{n} \mathbb{E}(\tilde{\mathbf{X}}^T \Delta_2 \mathbf{X} + \Delta_2 \mathbf{X}^T \tilde{\mathbf{X}}) = \frac{2\sigma^2}{n} \sum_i \begin{bmatrix} 2\tilde{z}_i & \tilde{x}_i & \tilde{y}_i & 1 \\ \tilde{x}_i & 0 & 0 & 0 \\ \tilde{y}_i & 0 & 0 & 0 \\ 1 & 0 & 0 & 0 \end{bmatrix} = \sigma^2(\tilde{\mathbf{T}} - \mathbf{P}). \tag{7.35}$$

The above two terms combined constitute the essential bias, i.e.,

$$\mathbb{E}(\Delta_2 \mathbf{A}) \overset{\text{ess}}{=} \sigma^2 \tilde{\mathbf{M}}^- \left[\frac{\tilde{\mathbf{A}}^T (2\tilde{\mathbf{T}} - \mathbf{P}) \tilde{\mathbf{A}}}{\tilde{\mathbf{A}}^T \tilde{\mathbf{N}} \tilde{\mathbf{A}}} \tilde{\mathbf{N}} \tilde{\mathbf{A}} - (2\tilde{\mathbf{T}} - \mathbf{P}) \tilde{\mathbf{A}} \right]. \tag{7.36}$$

The entire $\mathcal{O}(\sigma^2)$ bias. We continue computing the mean value of (7.32):

$$\mathbb{E}(\tilde{\mathbf{X}}^T \Delta_1 \mathbf{X} \tilde{\mathbf{M}}^- \tilde{\mathbf{X}}^T \Delta_1 \mathbf{X}) = \sum_i \mathbb{E}(\tilde{\mathbf{X}}_i \Delta_1 \mathbf{X}_i^T \tilde{\mathbf{M}}^- \tilde{\mathbf{X}}_i \Delta_1 \mathbf{X}_i^T)$$
$$= \sum_i \mathbb{E}(\tilde{\mathbf{X}}_i \tilde{\mathbf{X}}_i^T \tilde{\mathbf{M}}^- \Delta_1 \mathbf{X}_i \Delta_1 \mathbf{X}_i^T)$$
$$= \sigma^2 \sum_i \tilde{\mathbf{X}}_i \tilde{\mathbf{X}}_i^T \tilde{\mathbf{M}}^- \tilde{\mathbf{T}}_i. \tag{7.37}$$

Similarly, we have

$$\mathbb{E}(\Delta_1 \mathbf{X}^T \tilde{\mathbf{X}} \tilde{\mathbf{M}}^- \tilde{\mathbf{X}}^T \Delta_1 \mathbf{X}) = \sigma^2 \sum_i \tilde{\mathbf{X}}_i^T \tilde{\mathbf{M}}^- \tilde{\mathbf{X}}_i \tilde{\mathbf{T}}_i. \tag{7.38}$$

Thus the entire bias is given by

$$\mathbb{E}(\Delta_2 \mathbf{A}) = \sigma^2 \tilde{\mathbf{M}}^- \left[\frac{\tilde{\mathbf{A}}^T \tilde{\mathbf{H}}^* \tilde{\mathbf{A}}}{\tilde{\mathbf{A}}^T \tilde{\mathbf{N}} \tilde{\mathbf{A}}} \tilde{\mathbf{N}} \tilde{\mathbf{A}} - \tilde{\mathbf{H}}^* \tilde{\mathbf{A}} \right] + \mathscr{O}(\sigma^4), \qquad (7.39)$$

where

$$\mathbf{H}^* = 2\mathbf{T} - \mathbf{P} + \frac{1}{n^2} \sum_i \left[\mathbf{X}_i \mathbf{X}_i^T \mathbf{M}^- \mathbf{T}_i + \mathbf{T}_i \mathbf{M}^- \mathbf{X}_i \mathbf{X}_i^T + \mathbf{X}_i^T \mathbf{M}^- \mathbf{X}_i \mathbf{T}_i \right]. \quad (7.40)$$

We added the term $\mathbf{T}_i \mathbf{M}^- \mathbf{X}_i \mathbf{X}_i^T$ for the sole purpose of keeping $\mathbf{H}^*$ a symmetric matrix; its symmetry will be essential in the future. The added term does not affect (7.39), because $\tilde{\mathbf{X}}_i^T \tilde{\mathbf{A}} = 0$ for every $i = 1, \dots, n$.

A simpler formula for the essential bias. Next we simplify the formula (7.36) for the essential bias. Note that

$$\tilde{\mathbf{T}}_i \tilde{\mathbf{A}} = \mathbf{P} \tilde{\mathbf{A}} + 2 \tilde{\mathbf{A}} \tilde{\mathbf{X}}_i$$

for every $i = 1, \dots, n$, hence

$$\tilde{\mathbf{T}} \tilde{\mathbf{A}} = \mathbf{P} \tilde{\mathbf{A}} + 2 \tilde{\mathbf{A}} n^{-1} \sum_i \tilde{\mathbf{X}}_i$$

and therefore $\tilde{\mathbf{A}}^T \tilde{\mathbf{T}} \tilde{\mathbf{A}} = \tilde{\mathbf{A}}^T \mathbf{P} \tilde{\mathbf{A}}$. We also have

$$(2\tilde{\mathbf{T}} - \mathbf{P}) \tilde{\mathbf{A}} = \mathbf{P} \tilde{\mathbf{A}} + 4 \tilde{\mathbf{A}} n^{-1} \sum_i \tilde{\mathbf{X}}_i.$$

Also note that the vector $n^{-1} \sum_i \tilde{\mathbf{X}}_i$ coincides with the last column of the matrix $\tilde{\mathbf{M}}$, hence

$$n^{-1} \tilde{\mathbf{M}}^- \sum_i \tilde{\mathbf{X}}_i = [0, 0, 0, 1]^T.$$

Summarizing the above facts we reduce (7.36) to

$$\mathbb{E}(\Delta_2 \mathbf{A}) \stackrel{\text{ess}}{=} \sigma^2 \tilde{\mathbf{M}}^- \left[\frac{\tilde{\mathbf{A}}^T \mathbf{P} \tilde{\mathbf{A}}}{\tilde{\mathbf{A}}^T \tilde{\mathbf{N}} \tilde{\mathbf{A}}} \tilde{\mathbf{N}} \tilde{\mathbf{A}} - \mathbf{P} \tilde{\mathbf{A}} \right] - 4\sigma^2 \tilde{\mathbf{A}} [0, 0, 0, 1]^T. \qquad (7.41)$$

7.5   Comparison of algebraic circle fits

Bias of the Pratt and Taubin fits. We have seen that all the algebraic fits have the same asymptotic variance (7.29). We will see below that their asymptotic variance coincides with that of the geometric circle fit. Thus the difference between all our circle fits should be traced to the higher order terms, especially to their essential biases.

First we compare the Pratt and Taubin fits. For the Pratt fit, the constraint matrix is $N = \tilde{N} = P$, hence its essential bias (7.36) becomes

$$\mathbb{E}(\Delta_2 A_{\text{Pratt}}) \overset{\text{ess}}{=} -4\sigma^2 \tilde{A} [0,0,0,1]^T. \tag{7.42}$$

In other words, the Pratt constraint $N = P$ cancels the first term in (7.41); it leaves the second term intact.

For the Taubin fit, the constraint matrix is $N = T$ and its "true" value is $\tilde{N} = \tilde{T} = \frac{1}{n}\sum \tilde{T}_i$. Hence the Taubin's bias is

$$\mathbb{E}(\Delta_2 A_{\text{Taubin}}) \overset{\text{ess}}{=} -2\sigma^2 \tilde{A} [0,0,0,1]^T. \tag{7.43}$$

Thus, the Taubin constraint $N = T$ cancels the first term in (7.41) *and* a half of the second term; it leaves only a half of the second term in place.

As a result, the Taubin fit's essential bias is twice as small as that of the Pratt fit. Given that their main terms (variances) are equal, we see that the Taubin fit is statistically more accurate than that of Pratt; this difference might explain a slightly better performance of the Taubin fit, compared to Pratt, observed in our experiments in Chapter 5.

In his original paper [176] in 1991, Taubin expressed the intention to compare his method to that of Pratt, but no such comparison was published. We believe our analysis presents such a comparison.

Eliminating the essential bias. Interestingly, one can design an algebraic fit that has no essential bias at all. Let us set the constraint matrix to

$$N = H = 2T - P = \begin{bmatrix} 8\bar{z} & 4\bar{x} & 4\bar{y} & 2 \\ 4\bar{x} & 1 & 0 & 0 \\ 4\bar{y} & 0 & 1 & 0 \\ 2 & 0 & 0 & 0 \end{bmatrix}. \tag{7.44}$$

Then one can see that all the terms in (7.36) cancel out! The resulting essential bias vanishes:

$$\mathbb{E}(\Delta_2 A_{\text{Hyper}}) \overset{\text{ess}}{=} 0. \tag{7.45}$$

We call this fit *hyperaccurate*, or "Hyper" for short. The term *hyperaccuracy* was introduced by Kanatani [102, 103, 104] who was first to employ Taylor expansion up to the terms of order σ^4 for the purpose of comparing various algebraic fits and designing better fits (he treated more general quadratic models than circles).

The Hyper fit for circles is proposed in [4].

We note that the matrix H is not singular, three of its eigenvalues are positive and one is negative (these facts can be easily derived from the following simple observations: $\det H = -4$, trace $H = 8\bar{z} + 2 > 1$, and $\lambda = 1$ is one of its eigenvalues). If M is positive definite, then by Sylvester's law of inertia the matrix $H^{-1}M$ has the same signature as H does, i.e., the eigenvalues η of $H^{-1}M$

are all real, exactly three of them are positive and one is negative. In this sense the Hyper fit is similar to the Pratt fit, as the constraint matrix $\mathbf{P}$ also has three positive and one negative eigenvalues.

Invariance of the Hyper fit. We note that the Hyper fit is invariant under translations and rotations. This follows from the fact that the constraint matrix $\mathbf{H}$ is a linear combination of two others, $\mathbf{T}$ and $\mathbf{P}$, that satisfy the invariance requirements, as established in Sections 5.7 and 5.10.

The Hyper fit is also invariant under similarities, which can be verified as in Section 5.10. Indeed, whenever $(x,y) \mapsto (cx,cy)$, we have $(A,B,C,D) \mapsto (A/c^2, B/c, C/c, D)$, and hence $\mathbf{A}^T \mathbf{H} \mathbf{A} \mapsto c^{-2} \mathbf{A}^T \mathbf{H} \mathbf{A}$. Thus the constraint $\mathbf{A}^T \mathbf{H} \mathbf{A} = 1$ will be transformed into $\mathbf{A}^T \mathbf{H} \mathbf{A} = c^2$, which is irrelevant as our parameter vector $\mathbf{A}$ only needs to be determined up to a scalar multiple.

Hyper fit versus Pratt fit. Curiously, the Hyper fit and the Pratt fit return the same center of the circle[1]. This can be seen as follows. Recall that the Pratt fit is mathematically equivalent to the Chernov-Ososkov fit, and both minimize the objective function (5.16), which is

$$\mathscr{F}_{\text{Pratt}}(a,b,R) = \frac{\sum_{i=1}^{n} \left[(x_i - a)^2 + (y_i - b)^2 - R^2 \right]^2}{4R^2}.$$

Now expressing the Hyper fit in the geometric parameters (a,b,R) we can see that it minimizes the function

$$\mathscr{F}_{\text{Hyper}}(a,b,R) = \frac{\sum_{i=1}^{n} \left[(x_i - a)^2 + (y_i - b)^2 - R^2 \right]^2}{4n^{-1} \sum_{i=1}^{n} \left[2(x_i - a)^2 + 2(y_i - b)^2 - R^2 \right]}.$$

One can differentiate both functions with respect to R^2 and solve the equations

$$\partial \mathscr{F}_{\text{Pratt}} / \partial R^2 = 0 \qquad \text{and} \qquad \partial \mathscr{F}_{\text{Hyper}} / \partial R^2 = 0,$$

thus eliminating R^2. Then we arrive at an objective function in terms of a and b, and it happens to be the same for both Pratt and Hyper fits:

$$\mathscr{F}(a,b) = \left(\frac{1}{n} \sum_{i=1}^{n} \left[(x_i - a)^2 + (y_i - b)^2 \right]^2 \right)^{\frac{1}{2}} - \frac{1}{n} \sum_{i=1}^{n} \left[(x_i - a)^2 + (y_i - b)^2 \right],$$

up to irrelevant constant factors. Then its minimum gives the estimates of a and b, and it is the same for both fits. (The Taubin fit give a different estimate of the center.)

Since the Hyper fit returns the same circle center $(\hat{a}, \hat{b})$ as the Pratt fit does, the Hyper fit estimates $(\hat{a}, \hat{b}, \hat{R})$ also have infinite moments (Chapter 6.4).

---

[1]This interesting fact was discovered by A. Al-Sharadqah.

SVD-based Hyper fit. A numerically stable implementation of the Hyper fit is very similar to the SVD-based Pratt fit developed in Section 5.5. First, one computes the (short) SVD, $\mathbf{X} = \mathbf{U}\Sigma\mathbf{V}^T$, of the matrix $\mathbf{X}$. If its smallest singular value, σ_4, is less than a predefined tolerance ε (say, $\varepsilon = 10^{-12}$), then the solution $\mathbf{A}$ is simply the corresponding right singular vector (i.e., the fourth column of the $\mathbf{V}$ matrix). In the regular case ($\sigma_4 \geq \varepsilon$), one forms $\mathbf{Y} = \mathbf{V}\Sigma\mathbf{V}^T$ and finds the eigenpairs of the symmetric matrix $\mathbf{Y}\mathbf{H}^{-1}\mathbf{Y}$. Selecting the eigenpair $(\eta, \mathbf{A}_*)$ with the smallest positive eigenvalue and computing $\mathbf{A} = \mathbf{Y}^{-1}\mathbf{A}_*$ completes the solution. The corresponding MATLAB® code is available from [84].

An improved Hyper fit. Rangarajan and Kanatani [152] note that one can eliminate the entire $\mathscr{O}(\sigma^2)$ bias by adopting the constraint matrix $\mathbf{N} = \mathbf{H}^*$ given by (7.40); this fact is clear from (7.39). The matrix $\mathbf{H}^*$ is computationally more complex than our $\mathbf{H}$ (it requires finding the pseudo-inverse of $\mathbf{M}$), but numerical experiments [152] show that the accuracy of the resulting fit is higher than that of our Hyper fit for small n (such as $n = 10$ or $n = 20$); for larger n's there is apparently no difference in accuracy.

7.6 Algebraic circle fits in natural parameters

Our next goal is to express the covariance and the essential bias of algebraic circle fits in terms of the more natural geometric parameters (a, b, R).

Transition between parameter schemes. The conversion formulas between the algebraic circle parameters (A, B, C, D) and its geometric characteristics (a, b, R), see (3.11), are

$$a = -\frac{B}{2A}, \qquad b = -\frac{C}{2A}, \qquad R^2 = \frac{B^2 + C^2 - 4AD}{4A^2}. \tag{7.46}$$

Taking partial derivatives gives a 3×4 "Jacobian" matrix

$$\mathbf{J} = \begin{bmatrix} \frac{B}{2A^2} & -\frac{1}{2A} & 0 & 0 \\ \frac{C}{2A^2} & 0 & -\frac{1}{2A} & 0 \\ -\frac{R}{A} - \frac{D}{2A^2 R} & \frac{B}{4A^2 R} & \frac{C}{4A^2 R} & -\frac{1}{2AR} \end{bmatrix}.$$

Thus we have

$$\Delta_1 \Theta = \tilde{\mathbf{J}}\Delta_1 \mathbf{A} \qquad \text{and} \qquad \Delta_2 \Theta = \tilde{\mathbf{J}}\Delta_2 \mathbf{A} + \mathscr{O}_P(\sigma^2/n), \tag{7.47}$$

where $\tilde{\mathbf{J}}$ denotes the matrix $\mathbf{J}$ at the true parameters $(\tilde{A}, \tilde{B}, \tilde{C}, \tilde{D})$. The remainder term $\mathscr{O}_P(\sigma^2/n)$ comes from the second order partial derivatives of (a, b, R) with respect to (A, B, C, D). For example

$$\Delta_2 a = (\nabla a)^T (\Delta_2 \mathbf{A}) + \tfrac{1}{2}(\Delta_1 \mathbf{A})^T (\nabla^2 a)(\Delta_1 \mathbf{A}), \tag{7.48}$$

where $\nabla^2 a$ is the Hessian matrix of the second order partial derivatives of a with respect to (A, B, C, D). The last term in (7.48) can be actually discarded, as it is of order $\mathcal{O}_P(\sigma^2/n)$ because $\Delta_1 \mathbf{A} = \mathcal{O}_P(\sigma/\sqrt{n})$. We collect all such terms in the remainder term $\mathcal{O}_P(\sigma^2/n)$ in (7.47).

An auxiliary formula. Next we need a useful fact. Suppose a point (x_0, y_0) lies on the true circle $(\tilde{a}, \tilde{b}, \tilde{R})$, i.e.

$$(x_0 - \tilde{a})^2 + (y_0 - \tilde{b})^2 = \tilde{R}^2.$$

In accordance with our early notation we denote $z_0 = x_0^2 + y_0^2$ and $\mathbf{X}_0 = (z_0, x_0, y_0, 1)^T$. We also put $u_0 = (x_0 - \tilde{a})/\tilde{R}$ and $v_0 = (y_0 - \tilde{b})/\tilde{R}$, and consider the vector $\mathbf{Y}_0 = (u_0, v_0, 1)^T$. The following formula will be useful:

$$2\tilde{a}\tilde{R}\tilde{\mathbf{J}}\tilde{\mathbf{M}}^- \mathbf{X}_0 = -n(\mathbf{W}^T\mathbf{W})^{-1}\mathbf{Y}_0, \tag{7.49}$$

where the matrix $(\mathbf{W}^T\mathbf{W})^{-1}$ appears in (7.9) and the matrix $\tilde{\mathbf{M}}^-$ appears in (7.29). The identity (7.49) is easy to verify directly for the unit circle $\tilde{a} = \tilde{b} = 0$ and $\tilde{R} = 1$, and then one can check that it remains valid under translations and similarities.

Equation (7.49) implies that for every true point $(\tilde{x}_i, \tilde{y}_i)$

$$4\tilde{a}^2\tilde{R}^2\tilde{\mathbf{J}}\tilde{\mathbf{M}}^-\tilde{\mathbf{X}}_i\tilde{\mathbf{X}}_i^T\tilde{\mathbf{M}}^-\tilde{\mathbf{J}}^T = n^2(\mathbf{W}^T\mathbf{W})^{-1}\mathbf{Y}_i\mathbf{Y}_i^T(\mathbf{W}^T\mathbf{W})^{-1},$$

where $\mathbf{Y}_i = (\tilde{u}_i, \tilde{v}_i, 1)^T$ denote the columns of the matrix $\mathbf{W}$, cf. (7.4). Summing up over i gives

$$4\tilde{a}^2\tilde{R}^2\tilde{\mathbf{J}}\tilde{\mathbf{M}}^-\tilde{\mathbf{J}}^T = n(\mathbf{W}^T\mathbf{W})^{-1}. \tag{7.50}$$

Variance and bias of algebraic circle fits in the natural parameters. Now we can compute the variance (to the leading order) of the algebraic fits in the natural geometric parameters:

$$\mathbb{E}\left[(\Delta_1\Theta)(\Delta_1\Theta)^T\right] = \mathbb{E}\left[\tilde{\mathbf{J}}(\Delta_1\mathbf{A})(\Delta_1\mathbf{A})^T\tilde{\mathbf{J}}^T\right]$$
$$= n^{-1}\sigma^2(\tilde{\mathbf{A}}^T\mathbf{P}\tilde{\mathbf{A}})\tilde{\mathbf{J}}\tilde{\mathbf{M}}^-\tilde{\mathbf{J}}^T.$$

Using (7.50) and noting that

$$\tilde{\mathbf{A}}^T\mathbf{P}\tilde{\mathbf{A}} = \tilde{B}^2 + \tilde{C}^2 - 4\tilde{A}\tilde{D} = 4\tilde{A}^2\tilde{R}^2,$$

due to the third relation in (7.46), gives

$$\mathbb{E}\left[(\Delta_1\Theta)(\Delta_1\Theta)^T\right] = \sigma^2(\mathbf{W}^T\mathbf{W})^{-1}. \tag{7.51}$$

Thus the asymptotic variance of all the algebraic circle fits coincides with that of the geometric circle fit, cf. (7.6). Therefore all the circle fits achieve the minimal possible variance, to the leading order; the difference between them should be then characterized in terms of their biases, to which we turn next.

The essential bias of the Pratt fit is, due to (7.42),

$$\mathbb{E}(\Delta_2 \hat{\Theta}_{\text{Pratt}}) \overset{\text{ess}}{=} \frac{2\sigma^2}{\tilde{R}} \begin{bmatrix} 0 \\ 0 \\ 1 \end{bmatrix}. \tag{7.52}$$

Observe that the estimates of the circle center are essentially unbiased, and the essential bias of the radius estimate is $2\sigma^2/\tilde{R}$, which is independent of the number and location of the true points. We know (Section 7.5) that the essential bias of the Taubin fit is twice as small, hence

$$\mathbb{E}(\Delta_2 \hat{\Theta}_{\text{Taubin}}) \overset{\text{ess}}{=} \frac{\sigma^2}{\tilde{R}} \begin{bmatrix} 0 \\ 0 \\ 1 \end{bmatrix}, \tag{7.53}$$

Comparing to (7.10) shows that the geometric fit has an essential bias that is twice as small as that of Taubin and four times smaller than that of Pratt.

Therefore, the geometric fit has the smallest bias among all the popular circle fits, i.e., it is statistically most accurate. We observed that fact experimentally in Chapter 5.

Bias of the Kåsa fit. The formulas for the bias of the Kåsa fit can be derived, too, but in general they are complicated. However recall that all our fits, including Kåsa, are independent of the choice of the coordinate system, hence we can choose it so that the true circle has center at $(0,0)$ and radius $\tilde{R} = 1$; this will simplify the formulas. For this circle $\tilde{\mathbf{A}} = \frac{1}{\sqrt{2}}[1,0,0,-1]^T$, hence $\mathbf{P}\tilde{\mathbf{A}} = 2\tilde{\mathbf{A}}$ and so $\tilde{\mathbf{M}}^-\mathbf{P}\tilde{\mathbf{A}} = \mathbf{0}$, i.e., the middle term in (7.36) is gone. Also note that $\tilde{\mathbf{A}}^T\mathbf{P}\tilde{\mathbf{A}} = 2$, hence the last term in parentheses in (7.36) is $2\sqrt{2}[1,0,0,0]^T$. Then observe that $\tilde{z}_i = 1$ for every true point, thus

$$\mathbf{M} = \begin{bmatrix} 1 & \bar{x} & \bar{y} & 1 \\ \bar{x} & \overline{xx} & \overline{xy} & \bar{x} \\ \bar{y} & \overline{xy} & \overline{yy} & \bar{y} \\ 1 & \bar{x} & \bar{y} & 1 \end{bmatrix}.$$

Next, assume for simplicity that the true points are equally spaced on an arc of size θ (which is a typical arrangement in many studies). Choosing the coordinate system so that the east pole $(1,0)$ of the circle is at the center of that arc ensures $\bar{y} = \overline{xy} = 0$. It is not hard to see now that

$$\tilde{\mathbf{M}}^-[1,0,0,0]^T = \tfrac{1}{4}(\overline{xx} - \bar{x}^2)^{-1}[\overline{xx}, -2\bar{x}, 0, \overline{xx}]^T.$$

Using the formula (7.47) we obtain (omitting details as they are not so relevant)

total MSE = variance + (ess. bias)^2 + rest of MSE			
Pratt	5.6301	5.3541	0.2500	0.0260
Taubin	5.4945	5.3541	0.0625	0.0779
Geom.	5.4540	5.3541	0.0156	0.0843
Hyper.	5.4555	5.3541	0.0000	0.1014

Table 7.1 *Mean squared error (and its components) for four circle fits ($10^4 \times$ values are shown). In this test $n = 20$ points are placed (equally spaced) along a semicircle of radius $R = 1$ and the noise level is $\sigma = 0.05$.*

the essential bias of the Kåsa fit in the natural parameters (a, b, R):

$$\mathbb{E}(\Delta_2 \hat{\Theta}_{\text{Kåsa}}) \overset{\text{ess}}{=} 2\sigma^2 \begin{bmatrix} 0 \\ 0 \\ 1 \end{bmatrix} - \frac{\sigma^2}{\overline{xx} - \bar{x}^2} \begin{bmatrix} -\bar{x} \\ 0 \\ \overline{xx} \end{bmatrix}.$$

The first term here is the same as in (7.52) (recall that $\tilde{R} = 1$), but it is the second term above that causes serious trouble: it grows to infinity because $\overline{xx} - \bar{x}^2 \to 0$ as $\theta \to 0$. This explains why the Kåsa fit develops a heavy bias toward smaller circles when data points are sampled from a small arc (this phenomenon was discussed at length in Chapter 5). We will not dwell on the Kåsa fit anymore.

Some more experimental tests. To illustrate our analysis of various circle fits we have run two experiments. In the first one we set $n = 20$ true points equally spaced along a semicircle of radius $R = 1$. Then we generated random samples by adding a Gaussian noise at level $\sigma = 0.05$ to each true point. In the second experiment, we changed $n = 20$ to $n = 100$.

Table 7.1 summarizes the results of the first test, with $n = 20$ points; it shows the mean squared error (MSE) of the radius estimate $\hat{R}$ for each circle fit (obtained by averaging over 10^7 randomly generated samples). The table also gives the breakdown of the MSE into three components. The first two are the variance (to the leading order) and the square of the essential bias, both computed according to our theoretical formulas. These two components do not account for the entire mean squared error, due to higher order terms which our analysis discarded. The remaining part of the MSE is shown in the last column, it is relatively small.

We see that all the circle fits have the same (leading) variance, which accounts for the bulk of the MSE. Their essential bias is different, it is highest for the Pratt fit and smallest (zero) for the Hyper fit. Algorithms with smaller essential biases perform overall better, i.e., have smaller mean squared error.

	total MSE =	variance +	(ess. bias)^2 +	rest of MSE
Pratt	1.5164	1.2647	0.2500	0.0017
Taubin	1.3451	1.2647	0.0625	0.0117
Geom.	1.2952	1.2647	0.0156	0.0149
Hyper.	1.2892	1.2647	0.0000	0.0244

Table 7.2 *Mean squared error (and its components) for four circle fits ($10^4 \times$ values are shown). In this test $n = 100$ points are placed (equally spaced) along a semicircle of radius $R = 1$ and the noise level is $\sigma = 0.05$.*

However, note that the geometric fit is still slightly more accurate than the Hyper fit, despite its larger essential bias; this happens due to the contribution from the higher order terms, $\mathcal{O}(\sigma^4/n)$ and $\mathcal{O}(\sigma^6)$, which are not controlled by our analysis.

To suppress the contribution from terms $\mathcal{O}(\sigma^4/n)$ we increased the number of points to $n = 100$, and ran another test whose results are shown in Table 7.2. We see a similar picture here, but now the Hyper fit outperforms the (usually unbeatable) geometric fit. We see that the Hyper fit becomes the most accurate of all our circle fits when the number of points grows.

Overall, however, the geometric fit is the best for smaller samples ($n < 100$) and remains nearly equal to the Hyper fit for larger samples. Our experimental observations here agree with the popular belief in the statistical community that there is nothing better than minimizing the orthogonal distances. However, our analysis does not fully explain this phenomenon; apparently, the contribution from higher order terms (which our analysis ignores) accounts for the remarkable stability of the geometric fit. Perhaps one should derive explicitly some higher order terms, if one wants to understand why the geometric fit performs so well in practice.

On the practical side, our tests show that the best algebraic circle fit—the Hyper—is nearly equal to the geometric fit; the gain in accuracy (if any) is very small, barely noticeable. Kanatani [102, 103] also admits that "hyperaccurate" methods developed by the higher order analysis, which include $\mathcal{O}(\sigma^4)$ terms, only provide marginal improvement. Perhaps such studies, like Kanatani's and ours, may reveal interesting theoretical features of various estimators, but they fail to produce significantly better practical algorithms.

A real data example. We have also tested the Hyper circle fit on the real data example described in Section 5.12. Our new fit returned a circle with

$$\text{center} = (7.3871, 22.6674), \quad \text{radius} = 13.8144,$$

which is almost identical to the circles returned by the Pratt and Taubin fits; see Section 5.12.

More experimental tests involving the Hyper fit are reported in Section 8.10.

Summary. All the known circle fits (geometric and algebraic) have the same variance, to the leading order. The relative difference between them can be traced to higher order terms in the expansion for the mean squared error. The second leading term in that expansion is the essential bias, for which we have derived explicit expressions. Circle fits with smaller essential bias perform better overall. This explains a poor performance of the Kåsa fit, a moderate performance of the Pratt fit, and a better performance of the Taubin and geometric fits (in this order). We showed that there is no natural lower bound on the essential bias. In fact we constructed an algebraic fit with zero essential bias (the Hyper fit), which is nearly as accurate as the geometric fit, and sometimes exceeds it in accuracy.

7.7 Inconsistency of circular fits

We have seen in Section 6.10 that curve fitting methods are, generally, biased and inconsistent. Here we examine these features closely for our circle fits.

Inconsistency of circular fits. In the late 1980s Berman and Culpin [17, 18] investigated the inconsistency of two popular circle fitting algorithms: the geometric fit and the algebraic Kåsa fit. We present their results here and add other circle fits for comparison.

We use the notation and results of Section 6.10. According to Theorem 11, the limit point (a^*, b^*, R^*) of the geometric circle fit satisfies

$$(a_g^*, b_g^*, R_g^*) = \operatorname{argmin} \int \left[\sqrt{(x-a)^2 + (y-b)^2} - R \right]^2 d\mathbb{P}_\sigma \qquad (7.54)$$

(here the subscript "g" stands for "geometric"). One can easily minimize (7.54) with respect to R and get

$$R_g^* = \int \sqrt{(x - a_g^*)^2 + (y - b_g^*)^2} \, d\mathbb{P}_\sigma.$$

For the Kåsa fit, we have

$$(a_K^*, b_K^*, R_K^*) = \operatorname{argmin} \int \left[(x-a)^2 + (y-b)^2 - R^2 \right]^2 d\mathbb{P}_\sigma.$$

Again, minimization with respect to R^2 gives

$$(R_K^*)^2 = \int (x - a_K^*)^2 + (y - b_K^*)^2 \, d\mathbb{P}_\sigma.$$

The Pratt and Taubin fits are more easily expressed in terms of the algebraic circle parameters A, B, C, D, in which the circle is defined by equation

$$Az + Bx + Cy + D = 0, \quad \text{where} \quad z = x^2 + y^2.$$

Now the limit point of the Pratt fit satisfies

$$(A^*, B^*, C^*, D^*) = \text{argmin} \int \left[Az + Bx + Cy + D \right]^2 d\mathbb{P}_\sigma \qquad (7.55)$$

subject to the constraint

$$B^2 + C^2 - 4AD = 1. \qquad (7.56)$$

The limit point of the Taubin fit, see (5.49)–(5.51), is obtained by minimizing the same integral (7.55), but subject to *two* constraints: one is (7.56) and the other is

$$\int (Az + Bx + Cy + D) \, d\mathbb{P} = 0,$$

see Section 5.9. That additional constraint makes the limit point $(A_T^*, B_T^*, C_T^*, D_T^*)$ of the Taubin fit different from that of Pratt, $(A_P^*, B_P^*, C_P^*, D_P^*)$.

After determining the limit points (A^*, B^*, C^*, D^*) for the Pratt and Taubin fits one can apply the conversion formulas (3.11) to find the corresponding limit values (a^*, b^*, R^*). We omit analytical derivation and present the final results below.

The "hyperaccurate" circle fit developed in Section 7.5 cannot be expressed in the form (6.44), so we leave it out of our analysis.

Approximate limit points for circular fits. We give the limit points (a^*, b^*, R^*) for all our circle fits approximately, to the leading order in σ. We assume, as Berman [17] did, that the measure $\mathbb{P}_0$ describing the distribution of the true points is either uniform on the entire circle or uniform on an arc of length $\theta < 2\pi$.

Now in both cases the geometric radius estimate satisfies

$$R_g^* = R + \frac{\sigma^2}{2R} + \mathcal{O}(\sigma^3)$$

(note that it is independent of θ). This formula remarkably agrees with our early expression (7.11) for the essential bias of the geometric fit. The Taubin estimate is

$$R_T^* = R + \frac{\sigma^2}{R} + \mathcal{O}(\sigma^3),$$

i.e., its asymptotic bias is twice as large, to the leading order. The Pratt limit radius estimate is

$$R_P^* = R + \frac{2\sigma^2}{R} + \mathcal{O}(\sigma^3),$$

i.e., its asymptotic bias is four times larger than that of the geometric fit. These relations between the three major fits resemble those established in Section 7.5. Actually the Taubin and Pratt limits perfectly agree with the corresponding expressions (7.53) and (7.52) for their essential biases.

The asymptotic radius estimate of the Kåsa fit depends on the size of the arc θ where the true points are located. It is given by

$$R_K^* = R + \left[1 - \frac{\frac{2\sin^2(\theta/2)}{(\theta/2)^2}}{1 + \frac{\sin\theta}{\theta} - \frac{2\sin^2(\theta/2)}{(\theta/2)^2}} \right] \frac{\sigma^2}{R} + \mathcal{O}(\sigma^3). \qquad (7.57)$$

If $\theta = 2\pi$, i.e., if the true points cover the full circle, then

$$R_K^* = R + \frac{\sigma^2}{R} + \mathcal{O}(\sigma^3),$$

i.e., Kåsa's asymptotic bias is as small as Taubin's (and twice as small as Pratt's). In fact, the Kåsa fit is known to perform very well when the points are sampled along the entire circle, so our results are quite consistent with practical observations.

However, for small θ the fraction in (7.57) is large (it actually grows to infinity as $\theta \to 0$), and this is where the Kåsa fit develops a heavy bias toward smaller circles, the phenomenon already discussed at length in Chapter 5.

Illustration. Fig. 7.1 shows the asymptotic radius estimate R^* versus the size θ of the arc containing the true points for all the four circle fits (here we plot the exact R^*, rather than the approximations given above). We set $\tilde{R} = 1$ and $\sigma = 0.05$. Note again a heavy bias of the Kåsa fit for all $\theta < \pi$.

For the limit points of the center estimators $\hat{a}$ and $\hat{b}$ we have

$$a^* = a + \mathcal{O}(\sigma^3) \qquad \text{and} \qquad b^* = b + \mathcal{O}(\sigma^3) \qquad (7.58)$$

for the geometric fit, the Taubin fit, and the Pratt fit, independently of the arc size θ. Note that the second order terms $\mathcal{O}(\sigma^2)$ are missing from (7.58) altogether; this fact agrees with the absence of the essential bias of the corresponding estimates of the circle center established earlier in (7.10), (7.52), and (7.53).

We see again that the estimates of the circle center are substantially less biased than those of its radius. In fact, for $\theta = 2\pi$, due to the obvious rotational symmetry, we have precise identities: $a^* = a$ and $b^* = b$ (this fact was pointed out in [18]).

For the Kåsa fit, the formulas for a^* and b^* involve nonvanishing terms of order σ^2, which are given by complicated expressions similar to (7.57), and we omit them for brevity.

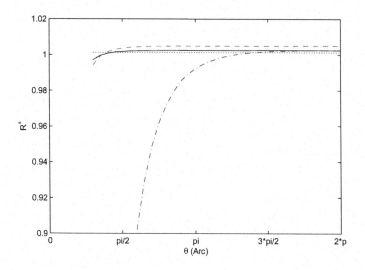

Figure 7.1 *The limit of the radius estimate R^*, as $n \to \infty$, versus the arc θ (the true value is $\tilde{R} = 1$). The geometric fit is marked by dots, the Pratt fit by dashes, the Taubin fit by a solid line, and the Kåsa fit by dash-dot (note how sharply the Kåsa curve plummets when $\theta < \pi$).*

7.8 Bias reduction and consistent fits via Huber

In Section 6.11 we mentioned some tricks that can be used to reduce the bias of the circle radius estimator. We also described a general procedure (called adjusted least squares) that produced asymptotically unbiased and consistent estimators, in the large sample limit $n \to \infty$. That procedure works for the polynomial model only, where one fits a polynomial $y = p(x) = \beta_0 + \beta_1 x + \cdots + \beta_k x^k$ to data points.

Here we describe similar procedures that work for implicit nonlinear models, in particular for circles and ellipses. These are based on fundamental ideas of Huber [85, 86, 87].

Unbiasing M-equations. Note that the maximum likelihood estimate (MLE) of a parameter Θ is a solution to the maximum likelihood (ML) equation

$$\nabla_\Theta \big(\ln L(\Theta) \big) = 0, \qquad (7.59)$$

where $L(\Theta)$ denotes the likelihood function and ∇_Θ the gradient. In the case of n independent observations $\mathbf{x}_i = (x_i, y_i)$, we have $L(\Theta) = \prod_{i=1}^n f(\mathbf{x}_i; \Theta)$, the product of the density functions; hence the equation (7.59) takes form

$$\sum_{i=1}^n \Psi_{\mathrm{MLE}}(\mathbf{x}_i; \Theta) = \mathbf{0} \qquad (7.60)$$

where $\Psi_{MLE}(\mathbf{x}; \Theta) = \nabla_\Theta \ln f(\mathbf{x}; \Theta)$.

Huber [85, 86, 87] considers more general estimators defined by

$$\sum_{i=1}^{n} \Psi(\mathbf{x}_i; \Theta) = \mathbf{0}. \tag{7.61}$$

This equation is similar to (7.61), but here Ψ is not necessarily related to the likelihood function. He calls (7.61) *M-equation* and its solution *M-estimate*; here "M" stands for "maximum likelihood type." Huber's main purpose is constructing robust estimators (i.e., estimators insensitive to outliers), but as it turns out, Huber's approach can be also used for construction of asymptotically unbiased (rather than robust) estimators. We describe it next.

The M-equation (7.61) is said to be *unbiased* if for every observed point $\mathbf{x}$

$$\mathbb{E}\left[\Psi(\mathbf{x}; \tilde{\Theta})\right] = \mathbf{0}, \tag{7.62}$$

where $\tilde{\Theta}$ again denotes the true parameter. (Note that we define the unbiasedness of *equation*, not *estimator*.)

Theorem 12 (Huber [86]) *If the M-equation is unbiased, then under certain regularity conditions the corresponding M-estimator $\hat{\Theta}$ has an asymptotically normal distribution. Furthermore, $n^{1/2}(\hat{\Theta} - \tilde{\Theta})$ converges to a normal law with mean zero. In particular, $\hat{\Theta}$ is asymptotically unbiased and consistent, in the limit $n \to \infty$.*

We note that the M-estimator obtained from an unbiased M-equation is *not* truly unbiased, but its bias vanishes as $n \to \infty$.

Thus in order to obtain an asymptotically unbiased and consistent estimators, one needs to modify the M-equation (7.61) defining a given estimator to "unbias" it, i.e., to make it satisfy (7.62). This approach has been used in the EIV regression analysis. In the case of fitting polynomials $y = \sum_{i=0}^{k} \beta_i x^i$ to data, this was done by Chan and Mak [30]. In the case of fitting conics (ellipses, hyperbolas), it was done by Kukush, Markovsky, and van Huffel [114, 115, 130, 167]. Here we apply this approach to circles, which will give novel methods for fitting circles.

Unbiasing algebraic circle fits. We have seen that algebraic circle fits are defined by a matrix equation

$$\mathbf{M}\mathbf{A} = \eta \mathbf{N}\mathbf{A}, \tag{7.63}$$

see (7.22), where the components of the matrices $\mathbf{M}$ and $\mathbf{N}$ are either constants or functions of the data $\mathbf{x}_i$, and $\mathbf{A} = (A, B, C, D)^T$ denotes the (algebraic) parameter vector. This equation is slightly different from Huber's M-equation (7.61), so we have to extend Huber's theory a bit. We say that a matrix $\mathbf{M}$ is unbiased if $\mathbb{E}(\mathbf{M}) = \tilde{\mathbf{M}}$, where $\tilde{\mathbf{M}}$ is the "true" value of $\mathbf{M}$, i.e., the one obtained by the substitution of the true points $(\tilde{x}_i, \tilde{y}_i)$ for the observed data points (x_i, y_i); one may call it the noise-free ($\sigma = 0$) version of $\mathbf{M}$. In the same way we construct $\tilde{\mathbf{N}}$, if the matrix $\mathbf{N}$ is also data-dependent. Then we have an analogue of Huber's theorem:

Theorem 13 *Suppose an estimator $\hat{\mathbf{A}}$ is a solution of a matrix equation (7.63), in which $\mathbf{M}$ and $\mathbf{N}$ are unbiased, i.e., $\mathbb{E}(\mathbf{M}) = \tilde{\mathbf{M}}$ and $\mathbb{E}(\mathbf{N}) = \tilde{\mathbf{N}}$. Then $n^{1/2}(\hat{\mathbf{A}} - \tilde{\mathbf{A}})$ converges to a normal law with mean zero; in particular, $\hat{\mathbf{A}}$ is asymptotically unbiased and consistent, in the limit $n \to \infty$.*

Thus in order to obtain an asymptotically unbiased and consistent estimator, one needs to modify the matrices $\mathbf{M}$ and $\mathbf{N}$ so that they become unbiased.

Interestingly, there is another benefit of making such a modification: it eliminates the essential bias of the estimator. Indeed, we have seen in (7.33) that the essential bias is proportional to $\mathbb{E}(\Delta_2 \mathbf{M})$, thus if the matrix $\mathbf{M}$ is unbiased, i.e., $\mathbb{E}(\Delta_2 \mathbf{M}) = \mathbf{0}$, then the estimator $\hat{\mathbf{A}}$ becomes *essentially unbiased* (its essential bias is zero). The unbiasedness of $\mathbf{N}$ does not affect the essential bias of $\hat{\mathbf{A}}$.

7.9   Asymptotically unbiased and consistent circle fits

Our construction of the unbiased versions of $\mathbf{M}$ and $\mathbf{N}$ for the algebraic circle fits is similar to that developed for more general quadratic models by Kanatani [93, 94, 95] and by Kukush, Markovsky, and van Huffel [114, 115, 130, 167].

First recall the structure of the matrix $\mathbf{M}$:

$$\mathbf{M} = \begin{bmatrix} \overline{zz} & \overline{zx} & \overline{zy} & \bar{z} \\ \overline{zx} & \overline{xx} & \overline{xy} & \bar{x} \\ \overline{zy} & \overline{xy} & \overline{yy} & \bar{y} \\ \bar{z} & \bar{x} & \bar{y} & 1 \end{bmatrix}.$$

A straightforward calculation shows that the mean values of its components are

$$\mathbb{E}(\bar{x}) = \bar{\tilde{x}}, \qquad \mathbb{E}(\bar{y}) = \bar{\tilde{y}}, \qquad \mathbb{E}(\overline{xy}) = \overline{\tilde{x}\tilde{y}},$$

$$\mathbb{E}(\overline{xx}) = \overline{\tilde{x}\tilde{x}} + V, \qquad \mathbb{E}(\overline{yy}) = \overline{\tilde{y}\tilde{y}} + V, \qquad \mathbb{E}(\bar{z}) = \bar{\tilde{z}} + 2V,$$

$$\mathbb{E}(\overline{zx}) = \overline{\tilde{z}\tilde{x}} + 4\bar{\tilde{x}}V, \qquad \mathbb{E}(\overline{zy}) = \overline{\tilde{z}\tilde{y}} + 4\bar{\tilde{y}}V, \qquad \mathbb{E}(\overline{zz}) = \overline{\tilde{z}\tilde{z}} + 8\bar{\tilde{z}}V + 8V^2,$$

where we denote, for convenience, $V = \sigma^2$, and our "sample mean" notation is applied to the true points, i.e., $\bar{\tilde{x}} = \frac{1}{n}\sum \tilde{x}_i$, etc.

Now it follows immediately that

$$\mathbb{E}(\overline{xx} - V) = \overline{\tilde{x}\tilde{x}}, \qquad \mathbb{E}(\overline{yy} - V) = \overline{\tilde{y}\tilde{y}}, \qquad \mathbb{E}(\bar{z} - 2V) = \bar{\tilde{z}},$$

$$\mathbb{E}(\overline{zx} - 4\bar{x}V) = \overline{\tilde{z}\tilde{x}}, \qquad \mathbb{E}(\overline{zy} - 4\bar{y}V) = \overline{\tilde{z}\tilde{y}}, \qquad \mathbb{E}(\overline{zz} - 8\bar{z}V + 8V^2) = \overline{\tilde{z}\tilde{z}}.$$

Thus the unbiased version of $\mathbf{M}$ is constructed as

$$\mathbf{M}_{\text{unbiased}} = \mathbf{M} - V\mathbf{H} + 8V^2\mathbf{I}_1, \qquad (7.64)$$

where

$$\mathbf{H} = \begin{bmatrix} 8\bar{z} & 4\bar{x} & 4\bar{y} & 2 \\ 4\bar{x} & 1 & 0 & 0 \\ 4\bar{y} & 0 & 1 & 0 \\ 2 & 0 & 0 & 0 \end{bmatrix} \quad \text{and} \quad \mathbf{I}_1 = \begin{bmatrix} 1 & 0 & 0 & 0 \\ 0 & 0 & 0 & 0 \\ 0 & 0 & 0 & 0 \\ 0 & 0 & 0 & 0 \end{bmatrix}.$$

We see now that $\mathbb{E}(\mathbf{M}_{\text{unbiased}}) = \tilde{\mathbf{M}}$.

Remarkably, the matrix $\mathbf{H}$ is our old friend: it is the constraint matrix (7.44) for the hyperaccurate circle fit. Actually, there is little wonder that we see it again in this section: indeed, our basic goal is the same as in Section 7.5, i.e., the removal of the essential bias. But now our goals extend further: we want an asymptotically unbiased and consistent estimator, in the limit $n \to \infty$, and these new features are ensured by the extra term $8V^2\mathbf{I}_1$.

Remark. Some authors [94, 198] discard the fourth order term $8\sigma^4\mathbf{I}_1 = 8V^2\mathbf{I}_1$ in an attempt to simplify computations. This simplification still allows them to eliminate the essential bias, but the resulting estimator is no longer consistent.

In the case of the Kåsa and Pratt circle fits, the matrix $\mathbf{N}$ is constant and needs no unbiasing. For the Taubin fit, $\mathbf{N}$ is data-dependent, cf. (7.21), but it has a structure similar to that of $\mathbf{M}$ (in fact, even simpler), and it can be unbiased similarly, we leave the details to the reader.

Actually, at this point the difference between our algebraic circle fits is no longer significant, as the unbiasing procedure guarantees the same statistical properties—the lack of essential bias and consistency—whichever fit we employ. So we will restrict our analysis to the simplest algebraic fit: the Kåsa fit.

Thus we arrive at the following circle fitting algorithm: assuming that the noise level $V = \sigma^2$ is known, find $\mathbf{A}$ satisfying

$$(\mathbf{M} - V\mathbf{H} + 8V^2\mathbf{I}_1)\mathbf{A} = \eta\mathbf{N}\mathbf{A}$$

and corresponding to the smallest nonnegative η. The resulting estimate has essentially bias zero, and in addition it is asymptotically unbiased and consistent in the limit $n \to \infty$.

Unknown noise level. In the above procedure, the parameter $V = \sigma^2$ is supposed to be known. In more realistic situations, however, it is unknown, and then one has to estimate it. The simplest way to do this is fit a circle by the above method with some (arbitrarily selected) $\sigma^2 > 0$, then estimate σ^2 by a standard formula

$$\hat{\sigma}^2 = \frac{1}{n-k}\sum_{i=1}^{n} d_i^2,$$

where d_i's are the distances from the data points to the fitted circle and k is the number of unknown parameters ($k = 3$ for circles); then one can reduce the bias using $\hat{\sigma}^2$ in place of σ^2.

This does not seem to be a perfect solution, though, as the estimate $\hat{\sigma}^2$ is computed from some initial fit, which may be essentially biased and inconsistent, thus $\hat{\sigma}^2$ may not be very accurate. Then, after one refits the circle by using an essentially unbiased and consistent fitting algorithm, it is tempting to reestimate $\hat{\sigma}^2$ (using the distances from the data points to the new circle) and then recompute the fit accordingly. Obviously, one can continue these adjustments recursively, hoping that the process would converge, and the limit circle would be indeed essentially unbiased and consistent.

The above iterative procedure is rather time consuming, and its convergence cannot be guaranteed. A better alternative would be to incorporate the estimation of σ^2 into the construction of the essentially unbiased and consistent fit. Such schemes have been designed by several authors independently, and we describe them next.

7.10 Kukush-Markovsky-van Huffel method

KMvH consistent estimator. In the early 2000s Kukush, Markovsky, and van Huffel proposed a consistent estimator for the parameters of quadratic curves (and surfaces) that are fitted to observed data. They started with ellipses whose center was known [115], then went on to ellipsoids [130] and general quadratic surfaces [114]; lastly, they provided complete and detailed mathematical proofs in [167]. Though they never treated the problem of fitting circles, we will adapt their approach to this task here. We abbreviate their algorithm as KMvH (after Kukush, Markovsky, and van Huffel).

Their basic observation is that, as $n \to \infty$, the matrix $\mathbf{M}_{\text{unbiased}}$ converges to its true (noise-free) value $\tilde{\mathbf{M}}$; in fact,

$$\mathbf{M}_{\text{unbiased}} = \tilde{\mathbf{M}} + \mathscr{O}_P(n^{-1/2}),$$

according to the classical central limit theorem. The true matrix $\tilde{\mathbf{M}}$ is positive semi-definite and singular; its only zero eigenvalue corresponds to the true parameter vector $\tilde{\mathbf{A}}$, because $\tilde{\mathbf{M}}\tilde{\mathbf{A}} = \mathbf{0}$, cf. Section 7.3.

Thus Kukush, Markovsky, and van Huffel estimate the unknown variance $V = \sigma^2$ by a value $\hat{V}$ that makes the variable matrix

$$\mathbf{M}_{\text{unbiased}}(V) = \mathbf{M} - V\mathbf{H} + 8V^2\mathbf{I}_1 \qquad (7.65)$$

positive semi-definite and singular, i.e., they find $\hat{V}$ such that

$$\mathbf{M}_{\text{unbiased}}(\hat{V}) \geq \mathbf{0} \qquad \text{and} \qquad \det \mathbf{M}_{\text{unbiased}}(\hat{V}) = 0. \qquad (7.66)$$

Then the eigenvector of $\mathbf{M}_{\text{unbiased}}(\hat{V})$ corresponding to its zero eigenvalue would be the desired estimate $\hat{\mathbf{A}}$ of the parameter vector, i.e., $\hat{\mathbf{A}}$ satisfies

$$\mathbf{M}_{\text{unbiased}}(\hat{V})\,\hat{\mathbf{A}} = \mathbf{0}. \tag{7.67}$$

Note that in this way one estimates σ^2 and $\mathbf{A}$ simultaneously, thus eliminating the need for further adjustments that were mentioned in Section 7.9.

Of course it is not immediately clear whether the value $\hat{V}$ satisfying (7.66) exists, or if it is unique. To this end Shklyar, Kukush, Markovsky, and van Huffel [167] prove the following:

Theorem 14 ([167]) *There is a unique value $\hat{V} \geq 0$ that satisfies (7.66). For all $V < \hat{V}$ the matrix $\mathbf{M}_{\text{unbiased}}(V)$ is positive definite, and for all $V > \hat{V}$ that matrix is indefinite, i.e., it has at least one negative eigenvalue.*

They actually proved that the minimal eigenvalue of the variable matrix $\mathbf{M}_{\text{unbiased}}(V)$, let us call it $\lambda_{\text{min}}(V)$, is a continuous function of V which is strictly positive for $V < \hat{V}$ and strictly negative for $V > \hat{V}$. Based on this theorem, Shklyar, Kukush, Markovsky, and van Huffel [167] propose to simultaneously estimate $\mathbf{A}$ and V via the bisection method as follows:

Step 1. Set $V_{\text{L}} = 0$ so that $\lambda_{\text{min}}(V_{\text{L}}) \geq 0$. If $\lambda_{\text{min}}(V_{\text{L}}) = 0$, go to Step 4. Set V_{R} to a large enough value such that $\lambda_{\text{min}}(V_{\text{R}}) < 0$. (It is shown in [167] that V_{R} may always be set to the minimum eigenvalue of the scatter matrix $\mathbf{S}$ defined by (1.5).)

Step 2. Compute $V_{\text{M}} = \frac{1}{2}(V_{\text{L}} + V_{\text{R}})$.

Step 3. If $\lambda_{\text{min}}(V_{\text{M}}) \geq 0$, reset $V_{\text{L}} = V_{\text{M}}$, otherwise reset $V_{\text{R}} = V_{\text{M}}$. If $\lambda_{\text{min}}(V_{\text{L}}) = 0$ or $V_{\text{R}} - V_{\text{L}} < \text{tolerance}$, go to Step 4. Otherwise return to Step 2.

Step 4. Set $\hat{V} = V_{\text{L}}$ and compute $\hat{\mathbf{A}}$ as an eigenvector of $\mathbf{M}_{\text{unbiased}}(\hat{V})$ corresponding to its smallest eigenvalue.

The advantage of this bisection scheme is that it is guaranteed to converge, though its convergence is slow (linear). We will see in Section 7.12 that one can accelerate the procedure by using (approximate) derivatives of $\lambda_{\text{min}}(V)$ and Newton's iterations; this was done by Kanatani. For the task of fitting circles, a much faster noniterative solution will be presented in Section 7.12.

Advantages of the KMvH estimator. Kukush, Markovsky, and van Huffel [114, 167] derive many statistical and geometric properties of their estimators $\hat{\mathbf{A}}$ and $\hat{V}$. First, both are strongly consistent, i.e.

$$\hat{\mathbf{A}} \to \tilde{\mathbf{A}} \qquad \text{and} \qquad \hat{V} \to \tilde{\sigma}^2 \qquad \text{as} \quad n \to \infty,$$

with probability one. In addition, $\hat{\mathbf{A}} = \tilde{\mathbf{A}} + \mathcal{O}_P(n^{-1/2})$, in accordance with the central limit theorem (see Remark 12 in [167]).

Second, they proved that their estimators $\hat{\mathbf{A}}$ and $\hat{V}$ are invariant under translations and rotations, thus the fitted curve and $\hat{\sigma}^2$ do not depend on the choice of the coordinate system; see Theorems 30 and 31 in [167]. Of course, if one rescales x and y by $x \mapsto cx$ and $y \mapsto cy$ with some constant $c > 0$, then $\hat{\sigma}^2$ is replaced by $c^2 \hat{\sigma}^2$, as required.

A real data example. We have tested the KMvH estimator on the real data example described in Section 5.12. That fit returned a circle with

$$\text{center} = (7.3857, 22.6688), \qquad \text{radius} = 13.8164,$$

which is very close to the circles returned by the Pratt and Taubin fits (Section 5.12) and the Hyper fit (Section 7.6). Actually, it is closer to the ideal circle (5.71) than circles returned by other fits.

Back to the known σ^2 case. We note that choosing $\hat{\mathbf{A}}$ as an eigenvector of $\mathbf{M}_{\text{unbiased}}(\hat{V})$ corresponding to its minimal (i.e., zero) eigenvalue is equivalent to minimizing $\mathbf{A}^T \mathbf{M}_{\text{unbiased}}(\hat{V}) \mathbf{A}$ subject to $\|\mathbf{A}\| = 1$. This may suggest that in the case of *known* $V = \sigma^2$, one could also minimize $\mathbf{A}^T \mathbf{M}_{\text{unbiased}}(V) \mathbf{A}$ subject to $\|\mathbf{A}\| = 1$, where the known V is used, instead of its estimate $\hat{V}$. One might expect that using the exact true value of V, rather than statistically imprecise estimate $\hat{V}$, will improve the accuracy of the resulting fit.

Surprisingly, the substitution of the known V for its estimate $\hat{V}$ can only lead to complications, as Shklyar and others [167] discover. First, the resulting estimate $\hat{\mathbf{A}}$ will *not* be translation invariant (though it remains rotation invariant); see [167]. To remedy the situation, Shklyar and others [167] propose, if one wants to use the known V, another constraint that ensures translation invariance. In the case of circles, it coincides with the Kåsa constraint, i.e., we need to minimize $\mathbf{A}^T \mathbf{M}_{\text{unbiased}}(V) \mathbf{A}$ subject to $\mathbf{A}^T \mathbf{K} \mathbf{A} = 1$, where $\mathbf{K}$ was introduced in (7.19) and V is the known variance. This method guarantees translation invariance.

Furthermore, even with the right constraint, the use of the known value of $V = \sigma^2$, instead of its estimate $\hat{V}$, may not improve the accuracy. Kukush, Markovsky, and van Huffel [114, 115] experimentally compared two estimation procedures for fitting ellipses. One uses the known value of $V = \sigma^2$ and computes a translation invariant ellipse fit, as mentioned above. The other ignores the known V and estimates it, according to Steps 1–4 described earlier in this section. Rather surprisingly, the second method consistently demonstrated the better performance, i.e., produced more accurate ellipse fits.

This may appear rather illogical. The authors of [114, 115] do not offer any explanation for their paradoxical observations. One may speculate that estimating σ^2 allows the algorithm to "adapt" to the data better, while trying to use the fixed value of σ^2 for every sample may be distractive. Perhaps, this phenomenon requires further investigation.

7.11 Renormalization method of Kanatani: 1st order

The renormalization method of Kanatani is one of the most interesting (and perhaps controversial) tools for designed fitting various quadratic models to data, in particular for fitting ellipses to observed points. Even though it has never been applied to circles yet, our exposition would be incomplete if we did not include Kanatani's method; besides, it happens to be very closely related to the KMvH algorithm of Section 7.10.

Historical remarks. Kanatani developed his method in the early 1990s (see [93, 94, 95]) aiming at the reduction of bias in two standard image processing tasks, one of them was fitting ellipses to data (the other, fundamental matrix computation, is not covered here). His method instantly captured attention of the computer vision community due to several well presented examples, in which an apparent bias was removed entirely. The effect of Kanatani's first publications was spectacular. But the theoretical construction of his method was quite obscure, and for several years it remained surrounded by mystery [101, 182, 181].

It took the community almost a decade to fully understand Kanatani's method. In 2001, Chojnacki et. al. [47] published a lengthy article aptly titled "Rationalising the renormalisation method of Kanatani," where they thoroughly compared Kanatani's method to other popular fitting algorithms and placed all of them within a unified framework. By now technical aspects of Kanatani's method are well understood, but its statistical properties and its relation to other statistical procedures are not yet fully determined.

Here we reveal a surprising fact: the renormalization method of Kanatani and the consistent estimator by Kukush, Markovsky, and van Huffel (KMvH, Section 7.10) are identical, they are twins! This is all the more surprising because Kanatani never aimed at consistency, his only stated goal was bias reduction. Nonetheless his algorithm produces exactly the same results as the consistent KMvH; in fact its practical implementation is perhaps superior to that of the KMvH.

Kanatani's method versus the KMvH algorithm. Kanatani studied parameter estimators which were based on the solution of a generalized eigenvalue problem

$$\mathbf{M}\mathbf{A} = \eta\mathbf{N}\mathbf{A}, \tag{7.68}$$

where $\mathbf{A}$ denotes the vector of unknown parameters, and $\mathbf{M}$ and $\mathbf{N}$ are some matrices that may depend on both data and unknown parameters. Many ellipse fitting methods and fundamental matrix computation fall into this category, and so do our algebraic circle fits; see (7.63).

Just like Huber (see our Section 7.8), Kanatani realized that the main source of bias for the estimate $\hat{\mathbf{A}}$, which solves (7.68), was the biasedness of the matrix $\mathbf{M}$. Then he constructed an unbiased version of the matrix $\mathbf{M}$, which in the circle fitting case is the matrix $\mathbf{M}_{\text{unbiased}}(V)$ given by (7.64).

Next, just like Kukush et al., Kanatani proposed to find $V = \hat{V}$ such that $\mathbf{M}_{\text{unbiased}}(V)$ would be positive semi-definite and singular; i.e., he proposed to estimate V by the rules (7.66). Lastly, he proposed to compute $\hat{\mathbf{A}}$ as an eigenvector of $\mathbf{M}_{\text{unbiased}}(\hat{V})$ corresponding to its zero eigenvalue. In other words, Kanatani arrived at the estimator (7.65)–(7.67); thus his method (theoretically) coincides with that of Kukush, Markovsky, and van Huffel.

Kanatani did not prove that the solution $\hat{V}$ of the problem (7.65)–(7.66) existed or was unique, but he proposed to solve the problem (7.65)–(7.66) by a relatively fast iterative scheme that he called *renormalization method*. In fact, he designed two versions of that method, we describe them next.

First order renormalization. In one (simpler) version, Kanatani drops the second order term $8V^2\mathbf{I}_1$ from (7.65) and solves a reduced problem: find V such that the matrix

$$\mathbf{M}_{\text{unbiased},1}(V) = \mathbf{M} - V\mathbf{H} \tag{7.69}$$

is positive semi-definite and singular. Then he computes $\hat{\mathbf{A}}$ as the eigenvector of this matrix corresponding to its zero eigenvalue.

Such an approximation still leads to an estimator $\hat{\mathbf{A}}$ with no essential bias, but $\hat{\mathbf{A}}$ is no longer consistent, cf. Remark in Section 7.9. (As consistency was not Kanatani's concern, the approximation (7.69) was legitimate for his purpose of reducing bias.)

Then Kanatani rewrites (7.69) as

$$\mathbf{M}_{\text{unbiased},1}(V)\mathbf{A}_V = \mathbf{M}\mathbf{A}_V - V\mathbf{H}\mathbf{A}_V = \lambda_{\min}(V)\mathbf{A}_V,$$

where $\lambda_{\min}(V)$ is the minimal eigenvalue of $\mathbf{M}_{\text{unbiased},1}(V)$ and $\mathbf{A}_V$ the corresponding unit eigenvector. Premultiplying by $\mathbf{A}_V^T$ gives

$$\mathbf{A}_V^T\mathbf{M}_{\text{unbiased},1}(V)\mathbf{A}_V = \mathbf{A}_V^T\mathbf{M}\mathbf{A}_V - V\mathbf{A}_V^T\mathbf{H}\mathbf{A}_V - \lambda_{\min}(V). \tag{7.70}$$

Now our goal is to solve equation $\lambda_{\min}(V) = 0$, i.e., find a zero of the above function. It is a nonlinear function of V, and Kanatani proposes to approximate its derivative by fixing the vector $\mathbf{A}_V$, which gives

$$\lambda'_{\min}(V) \approx -\mathbf{A}_V^T\mathbf{H}\mathbf{A}_V.$$

Then Kanatani applies the standard Newton method: given a current approximation V_k to $\hat{V}$, he finds the next one by

$$V_{k+1} = V_k + \frac{\lambda_{\min}(V_k)}{\mathbf{A}_{V_k}^T\mathbf{H}\mathbf{A}_{V_k}}, \tag{7.71}$$

then he computes $\mathbf{A}_{V_{k+1}}$ as the corresponding unit eigenvector of $\mathbf{M}_{\text{unbiased},1}(V_{k+1})$. These iterations continue until they converge. The initial guess is chosen as $V_0 = 0$. Kanatani calls this scheme *first order renormalization method*.

Kanatani did not prove the convergence of his iterative scheme, but it was tested in many experiments by various researchers and turned out to be very reliable.

Adaptation to circles. Kanatani actually developed his method for the more complicated problem of fitting ellipses, where the matrices $\mathbf{M}$ and $\mathbf{H}$ depended not only on the data, but also on the parameter vector $\mathbf{A}$, hence they needed to be recomputed at every iteration. In our case $\mathbf{M}$ and $\mathbf{H}$ are independent of $\mathbf{A}$, and then Kanatani's first order problem (7.69) easily reduces to the generalized eigenvalue problem

$$\mathbf{MA} = V\mathbf{HA}.$$

In other words, $\hat{V}$ is the smallest eigenvalue and $\hat{\mathbf{A}}$ is the corresponding eigenvector of the matrix pencil $(\mathbf{M}, \mathbf{H})$; this problem can be solved in one step (by using matrix algebra software) and needs no iterations. In fact, its solution is nothing but our familiar Hyper fit, see Section 7.5. Thus in the case of fitting circles, Kanatani's first order scheme coincides with the Hyperaccurate algebraic fit.

7.12 Renormalization method of Kanatani: 2nd order

Second order renormalization. This version keeps the second order term $8V^2\mathbf{I}_1$ in place. The equation (7.70) now takes its full form

$$\lambda_{\min}(V) = \mathbf{A}_V^T \mathbf{M} \mathbf{A}_V - V\mathbf{A}_V^T \mathbf{H} \mathbf{A}_V + 8V^2\mathbf{A}_V^T \mathbf{I}_1 \mathbf{A}_V = 0. \tag{7.72}$$

Kanatani proposes to solve (7.72) by the following iterative scheme. Given a current approximation V_k to $\hat{V}$, he fixes the vector $\mathbf{A}_{V_k}$ and then approximates the function $\lambda_{\min}(V)$ by a quadratic polynomial,

$$\lambda_{\min}(V) \approx \beta_0 + \beta_1 V + \beta_2 V^2, \tag{7.73}$$

where

$$\beta_0 = \mathbf{A}_{V_k}^T \mathbf{M} \mathbf{A}_{V_k}, \qquad \beta_1 = -\mathbf{A}_{V_k}^T \mathbf{H} \mathbf{A}_{V_k}, \qquad \beta_2 = 8\mathbf{A}_{V_k}^T \mathbf{I}_1 \mathbf{A}_{V_k}.$$

The next approximation to $\hat{V}$ is computed as the smaller root of (7.73), i.e.

$$V_{k+1} = \frac{-\beta_1 - \sqrt{\beta_1^2 - 4\beta_0\beta_2}}{2\beta_2}. \tag{7.74}$$

Kanatani takes the *smaller* root because he is aiming at the minimal eigenvalue of $\mathbf{M}_{\mathrm{unbiased}}(V)$. Normally, the polynomial (7.73) has two real roots, and the smaller one is computed by (7.74).

It may happen that the polynomial (7.73) has no real roots, which occurs whenever $\beta_1^2 < 4\beta_0\beta_2$. Then the quadratic approximation (7.73) is fruitless. In that case Kanatani goes back to the first order scheme and applies (7.71). Thus V_{k+1} is found by

$$V_{k+1} = \begin{cases} \frac{-\beta_1 - \sqrt{\beta_1^2 - 4\beta_0\beta_2}}{2\beta_2} & \text{if } \beta_1^2 \geq 4\beta_0\beta_2 \\ V_k + \frac{\lambda_{\min}(V_k)}{A_{V_k}^T H A_{V_k}} & \text{otherwise} \end{cases}.$$

Once V_{k+1} is found, Kanatani computes $A_{V_{k+1}}$ as the corresponding unit eigenvector of the matrix $M_{\text{unbiased}}(V_{k+1})$. These iterations continue until they converge. The initial guess is chosen again as $V_0 = 0$. Kanatani called this scheme the *second order renormalization method*.

Kanatani did not prove the convergence of this iterative scheme, but just like his first order method it was tested in many experiments by various researchers and turned out to be very reliable.

Adaptation to circles. Kanatani's original work [93, 94, 95] treats general quadratic curves (ellipses and hyperbolas). His matrices **M** and **H** depend on the parameter vector, hence they needed to be recomputed at every iteration.

In our case **M** and **H** are independent of **A**, and then Kanatani's second order method can be greatly simplified. First, it reduces to the *quadratic* generalized eigenvalue problem

$$MA - VHA + 8V^2 I_1 A = 0. \tag{7.75}$$

In other words, $\hat{V}$ is the smallest eigenvalue and $\hat{A}$ is the corresponding eigenvector of the *quadratic pencil* $(M, H, -8I_1)$.

The quadratic generalized eigenvalue problem (7.75) can be further reduced to a linear generalized eigenvalue problem by the following standard scheme, see e.g., [46, 47]:

$$\begin{bmatrix} M & -H \\ 0 & I \end{bmatrix} \begin{bmatrix} A \\ B \end{bmatrix} = V \begin{bmatrix} 0 & -8I_1 \\ I & 0 \end{bmatrix} \begin{bmatrix} A \\ B \end{bmatrix}, \tag{7.76}$$

where **I** is the 4×4 identity matrix. Note that the second line enforces $B = VA$. (This scheme is similar to a standard reduction of a second order differential equation $y'' = f(x, y, y')$ to a first order system of equations $z' = f(x, y, z)$ and $y' = z$.)

It is now tempting to solve the linear generalized eigenvalue problem (7.76) by using a standard matrix algebra software, for example by calling the function "eig" in MATLAB. We found, however, that the "eig" function tends to fail on (7.76) quite frequently (in 10 to 20% of simulated cases). The cause of failure is that the linear generalized eigenvalue problem (7.76) is *singular*; in fact, the double size (8×8) matrix on the right-hand side of (7.76) has rank 5.

A similar problem arises when one fits ellipses [46, 47]. To remedy the situation one should change variables and reduce the dimension of the matrices until all of them become nonsingular, cf. an example in [46].

In our case we can simply use the forth equation in (7.75), i.e.

$$A\bar{z} + B\bar{x} + C\bar{y} + D = 2VA,$$

as it allows us to replace $2VA$ in the last term of (7.75) by $A\bar{z} + B\bar{x} + C\bar{y} + D$. As a result, (7.75) transforms to

$$\mathbf{MA} = V\mathbf{H}_*\mathbf{A}, \tag{7.77}$$

where

$$\mathbf{H}_* = \begin{bmatrix} 4\bar{z} & 0 & 0 & -2 \\ 4\bar{x} & 1 & 0 & 0 \\ 4\bar{y} & 0 & 1 & 0 \\ 2 & 0 & 0 & 0 \end{bmatrix}.$$

Thus we reduce the quadratic generalized eigenvalue problem (7.75) to a *nonsingular* linear generalized eigenvalue problem, (7.77). Both matrices in (7.77) are nonsingular.

In fact, $\det \mathbf{H}_* = -4$, which guarantees that the matrix $\mathbf{H}_*$ is well conditioned (unless the coordinates of the data points assume abnormally large values, which can be easily prevented by a proper choice of the coordinate scale). Therefore, the generalized eigenvalue problem (7.77) can be safely transformed to an ordinary eigenvalue problem

$$\mathbf{H}_*^{-1}\mathbf{MA} = V\mathbf{A}, \tag{7.78}$$

which can be solved by standard matrix methods. Such a solution will be numerically stable; see Section 7.7 in [78]. In our numerical tests, we just called the MATLAB function "eig" to solve (7.77), and the obtained solution appeared to be stable and reliable.

On the other hand, the matrix $\mathbf{H}_*$ is not symmetric, hence one cannot apply Sylvester's law of inertia to determine the signs of the eigenvalues of (7.77). We found experimentally that in about 50% of our simulated samples all the eigenvalues were real and positive, and in the other 50% there were two complex eigenvalues and two real positive eigenvalues. We plan to look into this issue further.

Some more experimental tests. We modified the experiment reported in Section 7.6 to test the performance of various circle fits in the large sample limit, i.e., as $n \to \infty$ and $\sigma > 0$ fixed. We increased the sample size from $n = 20$ to $n = 50,000$. As before, for each n we set n true points equally spaced along a semicircle of radius $R = 1$ and generated random samples by adding a Gaussian noise at level $\sigma = 0.05$ to each true point.

	Pratt	Taubin	Geometric	Hyper	Consistent
$n = 20$	5.6301	5.4945	5.4540	5.4555	5.4566
$n = 100$	1.5164	1.3451	1.2952	1.2892	1.2895
$n = 1,000$	0.3750	0.1947	0.1483	0.1341	0.1341
$n = 10,000$	0.2557	0.0745	0.0286	0.0135	0.0134
$n = 50,000$	0.2450	0.0637	0.0180	0.00272	0.00270

Table 7.3 *Mean squared error for five circle fits ($10^4 \times$ values are shown). In this test n points are placed (equally spaced) along a semicircle of radius $R = 1$ and the noise level is $\sigma = 0.05$.*

The results are summarized in Table 7.3. It shows the mean squared error (MSE) of the radius estimate $\hat{R}$ for each circle fit (obtained by averaging over 10^6 randomly generated samples). We see that for small samples, $n < 100$, all the circle fits perform nearly equally well. For larger samples, they have very different characteristics. The MSE of the Pratt fit is the biggest, due to its highest essential bias; see Section 7.6. The MSE of the Taubin fit is about 4 times smaller, again because its essential bias is half of that of Pratt (Section 7.6). The MSE of the geometric fit is about 4 times smaller than that of Taubin, again in perfect agreement with our theoretical analysis of Section 7.1.

The Hyper fit introduced in Section 7.5 has zero essential bias, hence its MSE becomes very small for very large samples; its mean error is several times smaller than that of the geometric fit! However, the Hyper fit is inconsistent, in view of our early analysis, hence its MSE does not completely vanish as $n \to \infty$.

Lastly, the Consistent fit included in Table 7.3 is the one computed by (7.77), which is nothing but our implementation of the two (theoretically identical) estimators: the Kukush-Markovsky-van Huffel (KMvH) method from Section 7.10 and the second order renormalization method of Kanatani from this section. We see that the Consistent circle fit becomes the most efficient for very large samples ($n > 1,000$).

In image processing applications, the sample size n typically ranges from 10 to 20 to several hundreds. It is not normal to have $n > 1,000$ data points, although occasionally digitized images of circular arcs consist of thousands of points (pixels), see below.

For example, in an archaeological research by Chernov and Sapirstein [44], circular arcs appear as fragments of wheelmade and molded circular architectural terracottas whose original diameters are unknown and need be determined. Using a profile gauge, the profiles are transferred to graph paper, scanned, and transformed into an array of black and white pixels with a threshold filter. The data points are the x and y coordinates of black pixels measured in centimeters. The arcs usually consist of 5,000 to 20,000 points and subtend angles ranging from 5° to 270°. A typical digitized arc is shown in Fig. 7.2; on that arc there are 6045 black pixels. In a few arcs examined in [44], the number of pixels was close to 40,000.

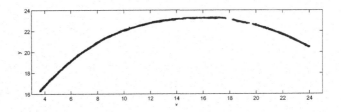

Figure 7.2 *A typical arc drawn by pencil with a profile gauge from a circular cover tile. This scanned image contains 6045 pixels.*

But even for samples that are very large by all practical standards, the advantage of the consistent circle fit over the Hyper fit or the geometric fit is minuscule and negligible. The authors of [44] reported that on samples consisting of 5,000 to 40,000 points there were no visible differences between circles obtained by the geometric fit or by algebraic fits.

We conclude therefore that, at least in the case of fitting circles to data in image processing, consistent fits appear to be interesting only from academic perspectives. On the other hand, in statistical applications, where both σ and n may be large, consistent estimators are of practical importance [114, 115, 130, 167].

Chapter 8

Various "exotic" circle fits

In this chapter we describe a few mathematically sophisticated circle fitting methods. Some involve complex numbers and conformal mappings of the complex plane [25, 82, 123, 159, 175, 173]. Others make use of trigonometric change of parameters [106, 107].

While the use of highly advanced mathematical tools seems promising and attractive, the resulting fits do not happen to perform better than the more basic fits we have covered already. Still these approaches are aesthetically appealing and may turn out productive in the future, so we review them in this chapter.

8.1   Riemann sphere

The idea of treating observed points (x_i, y_i) as complex numbers

$$\mathbf{z}_i = x_i + \mathbf{i} y_i, \qquad \mathbf{i} = \sqrt{-1}$$

and then using the geometry of the complex plane and elements of complex analysis is very attractive and has been explored by several authors in different ways. Here we present a general method that maps the observed points onto the Riemann sphere.

Extended complex plane. The complex plane is the regular xy plane where every point (x,y) becomes a complex number written as $\mathbf{z} = x + \mathbf{i}y$. Complex numbers are added and subtracted as 2D vectors

$$(a + \mathbf{i}b) \pm (c + \mathbf{i}d) = (a \pm b) + \mathbf{i}(c \pm d),$$

but their multiplication and division use the fact that $\mathbf{i}^2 = -1$:

$$(a + \mathbf{i}b)(c + \mathbf{i}d) = (ac - bd) + \mathbf{i}(bc + ad)$$

and

$$\frac{a + \mathbf{i}b}{c + \mathbf{i}d} = \frac{(a + \mathbf{i}b)(c - \mathbf{i}d)}{(c + \mathbf{i}d)(c - \mathbf{i}d)} = \frac{ac + bd}{c^2 + d^2} + \mathbf{i}\frac{bc - ad}{c^2 + d^2}.$$

The latter formula applies whenever $c^2 + d^2 \neq 0$, i.e., the divisor $c + \mathbf{i}d$ must differ from zero.

It is convenient to add a special point, "infinity" (denoted by ∞), to the complex plane, and allow formulas like $\frac{1}{0} = \infty$ and $\frac{1}{\infty} = 0$. That extra point makes the complex plane compact, in topological sense. More precisely, every sequence $\mathbf{z}_n = a_n + \mathbf{i}b_n$ of complex numbers whose absolute values grow, i.e., $|\mathbf{z}_n| = \sqrt{a_n^2 + b_n^2} \to \infty$, now converges to that special point ∞.

The complex plane is denoted by $\mathbb{C}$. The complex plane with the added point ∞ is denoted by $\bar{\mathbb{C}} = \mathbb{C} \cup \{\infty\}$, it is called the *extended complex plane*.

Stereographic projection. One can conveniently map the extended complex plane onto a sphere by using stereographic projection illustrated in Fig. 8.1.

Denote by

$$\mathbb{S} = \left\{ (x,y,z) : x^2 + y^2 + (z - \tfrac{1}{2})^2 = \tfrac{1}{4} \right\}$$

the sphere in the xyz space (whose xy coordinate plane is identified with the complex plane) of radius $r = \frac{1}{2}$ centered on the point $(0,0,\frac{1}{2})$. This is *Riemann sphere*. It "rests" on the complex plane: its south pole $(0,0,0)$ is right at the origin of the xy plane, and its north pole $(0,0,1)$ is right above it.

Now every point $(x,y,0)$ in the xy plane can be joined by a line with the north pole of the sphere. This line intersects the sphere in a unique point below the north pole. The coordinates of the intersection point are

$$\begin{aligned} x' &= x/(1 + x^2 + y^2) \\ y' &= y/(1 + x^2 + y^2) \\ z' &= (x^2 + y^2)/(1 + x^2 + y^2). \end{aligned} \tag{8.1}$$

This defines a map from the xy plane onto the sphere $\mathbb{S}$: it takes every point

$(x,y) \in \mathbb{C}$ to the point (x',y',z') on the sphere. This map can be visualized as the complex plane $\mathbb{C}$ "wrapped around" the sphere $\mathbb{S}$. The inverse map, taking the sphere back onto the plane, is called *stereographic projection*, it "unfolds" the sphere onto the plane under it.

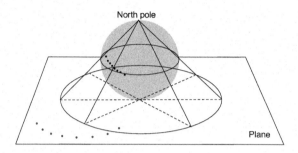

Figure 8.1 *Stereographic projection of the Riemann sphere onto the plane. Data points (hollow circles) are mapped onto the sphere (black dots).*

The above map is not defined at the north pole of the sphere. The map becomes complete if we transform the north pole to the special point $\infty \in \bar{\mathbb{C}}$. This makes sense as complex numbers with large absolute values are mapped to points near the north pole on the sphere. This establishes a one-to-one correspondence (bijection) between the extended complex plane $\bar{\mathbb{C}}$ and the Riemann sphere $\mathbb{S}$. This bijection is continuous in both directions; in mathematical terms, it is a homeomorphism.

Transformation of lines and circles. An important property of the Riemann sphere is that lines and circles in the complex plane are transformed into circles on the Riemann sphere.

Theorem 15 *If $\mathbb{L} \subset \mathbb{C}$ is a line in the complex plane, then its image on the Riemann sphere is a circle (passing through the north pole). If $\mathbb{O} \subset \mathbb{C}$ is a circle in the complex plane, then its image on the Riemann sphere is a circle (not passing through the north pole). Conversely, every circle on the Riemann sphere is mapped onto a line or a circle in $\mathbb{C}$.*

Thus, lines and circles in the xy plane uniquely correspond to circles on the Riemann sphere $\mathbb{S}$.

Recall that when one fits circles to observed data, then lines should be allowed as well, i.e., one should use the extended model consisting of lines and circles; see Chapter 3. On the Riemann sphere, both types of objects are conveniently represented by circles.

8.2 Simple Riemann fits

The previous section suggests the following circle fit, which was originally proposed in 1997 by Lillekjendlie [123] and further developed by him and his co-authors in [175, 173]. It consists of three steps:

Step 1. Map the data points (x_i, y_i) onto the Riemann sphere according to formulas (8.1). This produces n points (x_i', y_i', z_i') on the sphere.

Step 2. Fit a circle to the points (x_i', y_i', z_i') on the sphere. Since a circle on the sphere is obtained by intersection of the sphere with a plane, one just needs to fit a plane to the points (x_i', y_i', z_i').

Step 3. After finding the best fitting circle on the Riemann sphere, project it back onto the xy plane and get a circle (or a line) fitting the original data set (x_i, y_i).

The last step is a straightforward exercise in elementary geometry. Suppose the plane obtained at Step 2 is given by equation

$$\alpha x' + \beta y' + \gamma z' + c = 0. \tag{8.2}$$

Then the center (a, b) of the circle fitting the original data set is computed by

$$a = -\frac{\alpha}{2(c+\gamma)}, \qquad b = -\frac{\beta}{2(c+\gamma)},$$

and its radius R is computed by

$$R^2 = \frac{\alpha^2 + \beta^2 - 4c(c+\gamma)}{4(c+\gamma)^2}.$$

These formulas are derived in [123, 175]; we leave it to the reader to verify the details.

The above formulas fail if $c + \gamma = 0$, which happens exactly when the plane (8.2) passes through the north pole $(0, 0, 1)$. In that case the best fit to the original data set is achieved by a straight line (rather than a circle). The equation of the fitting line is

$$\alpha x + \beta y + c = 0.$$

The main issue is to fit a plane to the spacial points (x_i', y_i', z_i') in Step 2. We describe several approaches to this task.

Very simple plane fitting scheme. In his original paper [123] in 1997, Lillekjendlie followed the recipes of the classical linear regression. He describes the plane by an explicit equation

$$z = px + qy + r$$

and then minimizes

$$\mathscr{F}_{\text{classic}}(p,q,r) = \sum_{i=1}^{n} \left[z_i' - (px_i' + qy_i' + r) \right]^2.$$

This minimization problem easily reduces to a system of linear equations and has a simple explicit solution. Observe that this scheme actually minimizes the distances to the points (x_i', y_i', z_i') in the z direction.

Despite simplicity, this scheme has obvious disadvantages. Most notably, it does not account for vertical planes $px + qy + r = 0$. Omitting all vertical planes results in omitting all circles and lines passing through the origin in the x, y plane. This is a serious omission recognized by Lillekjendlie [123], and in the next paper he improved the fitting scheme, see below.

Geometric plane fitting. In 2000, Lillekjendlie and his co-authors [175] minimize geometric distances from the points (x_i', y_i', z_i') to the plane (8.2), i.e., they minimize the function

$$\mathscr{F}(\alpha, \beta, \gamma, c) = \sum_{i=1}^{n} \frac{[\alpha x_i' + \beta y_i' + \gamma z_i' + c]^2}{\alpha^2 + \beta^2 + \gamma^2}.$$

To resolve the issue of undeterminate parameters, they actually minimize

$$\mathscr{F}_{\text{geom}}(\alpha, \beta, \gamma, c) = \sum [\alpha x_i' + \beta y_i' + \gamma z_i' + c]^2 \qquad (8.3)$$

subject to the constraint $\alpha^2 + \beta^2 + \gamma^2 = 1$. This constraint means that $\mathbf{n} = (\alpha, \beta, \gamma)$ is a *unit normal* vector to the plane.

The geometric fit of a 2D plane to a set of points in 3D is very similar to the geometric fit of a line to a set of points in 2D thoroughly discussed in our Chapters 1 and 2. Without going into details (which can be found in [175]) we just say that the vector $\mathbf{n}$ must be the eigenvector of the 3×3 scatter matrix $\mathbf{S}$ corresponding to its smallest eigenvalue. The scatter matrix $\mathbf{S}$ is defined by

$$\mathbf{S} = \tfrac{1}{n} \sum_{i=1}^{n} (\mathbf{r}_i - \bar{\mathbf{r}})(\mathbf{r}_i - \bar{\mathbf{r}})^T, \qquad (8.4)$$

where $\mathbf{r}_i = (x_i', y_i', z_i')^T$ is the ith "data vector" and $\bar{\mathbf{r}} = \tfrac{1}{n} \sum \mathbf{r}_i$. This formula for $\mathbf{S}$ generalizes our earlier formula (1.5) for the 2D scatter matrix.

Lastly, c is determined from the equation

$$\alpha \sum x_i' + \beta \sum y_i' + \gamma \sum z_i' + nc = 0, \qquad (8.5)$$

which simply means that the best fitting plane passes through the centroid $\bar{\mathbf{r}}$ of the data set (x_i', y_i', z_i'); compare (8.5) to (1.14).

The geometric fitting scheme does not omit any planes, so all circles and lines in the xy plane can be obtained. Still, this scheme has a serious draw-back in that it minimizes the distances to the points (x_i', y_i', z_i'), which may have nothing to do with the distances from the original points (x_i, y_i) to the fitting circle. The authors of [175] recognized this fact and modified their scheme accordingly, see the next section.

The authors of [175] call circle fits involving the transformation to the Riemann sphere *Riemann fits*.

We call the above two versions *simple Riemann fits*, they are fast but statistically not sound.

8.3 Riemann fit: the SWFL version

To relate the distances in the xy plane to the distances on the Riemann sphere we invoke the following fact in complex analysis.

Conformality. Let Φ denote the map from the xy plane to the sphere $\mathbb{S}$ given by (8.1), i.e., we write $\Phi(x, y) = (x', y', z')$. Let $D\Phi$ denote the derivative of the map Φ, i.e., its linear part. Precisely, $D\Phi$ transforms tangent vectors (dx, dy) to the xy plane into tangent vectors (dx', dy', dz') to the sphere $\mathbb{S}$. Note that $D\Phi$ is a 3×2 matrix consisting of partial derivatives of expressions in (8.1) with respect to x and y.

Theorem 16 *The map Φ is conformal at every point (x, y). That is, its derivative $D\Phi$ satisfies two requirements:*
(i) it preserves angles between tangent vectors and
(ii) it contracts all tangent vectors uniformly.
More precisely, if $(dx', dy', dz')^T = D\Phi(dx, dy)^T$, then

$$\|(dx', dy', dz')\| = (1 + x^2 + y^2)^{-1} \|(dx, dy)\|.$$

Note that the contraction factor $(1 + x^2 + y^2)^{-1}$ does not depend on the direction of the tangent vector (dx, dy).

One can verify all these claims directly, by computing the matrix $D\Phi$ of partial derivatives of the formulas (8.1) with respect to x and y. This can be done a little easier if one rewrites (8.1) in polar coordinates (r, θ) on the xy plane:

$$x' = r\cos\theta / (1 + r^2)$$
$$y' = r\sin\theta / (1 + r^2)$$
$$z' = r^2 / (1 + r^2).$$

Now one can conveniently differentiate x', y', z' with respect to θ (in the tangent direction) and with respect to r (in the radial direction). We leave the details to the reader.

Relation between distances. Let P and Q be two nearby points on the xy plane and $d = \text{dist}(P,Q)$. Denote by d' the distance between their images $\Phi(P)$ and $\Phi(Q)$ on the sphere. According to the above theorem, d' and d are related by

$$d' = (1+x^2+y^2)^{-1}d + \mathcal{O}(d^2)$$

where (x,y) are the coordinates of P (or Q). Thus, to the first order, the distances between points are contracted by the factor of $(1+x^2+y^2)^{-1}$.

Furthermore, if a point $P = (x,y)$ is the distance d from a circle (or line) $\mathbb{O}$ on the plane, then its image $\Phi(P)$ is the distance

$$d' = (1+x^2+y^2)^{-1}d + \mathcal{O}(d^2) \tag{8.6}$$

from the circle $\Phi(\mathbb{O})$ on the sphere $\mathbb{S}$. Thus, to the first order, the distances to circles are contracted by the same factor of $(1+x^2+y^2)^{-1}$.

Next, let gain $\Phi(\mathbb{O})$ denote a circle on the sphere $\mathbb{S}$ and $\Phi(P)$ a nearby point a distance d' from the circle $\mathbb{O}$. Let $\mathbb{P}$ denote the plane containing the circle $\mathbb{O}$ and

$$\alpha x' + \beta y' + \gamma z' + c = 0$$

its equation in the spacial coordinates, as before. Let d'' denote the *orthogonal distance* from the point $\Phi(P)$ to the plane $\mathbb{P}$. Then d'' and d' are related by

$$d'' = \frac{\sqrt{\alpha^2+\beta^2-4c(\gamma+c)}}{\sqrt{\alpha^2+\beta^2+\gamma^2}}d' + \mathcal{O}([d']^2). \tag{8.7}$$

We note that whenever the fitting plane (8.2) intersects the Riemann sphere, the distance from its center $(0,0,\frac{1}{2})$ to the plane is $\leq \frac{1}{2}$, hence $\alpha^2+\beta^2-4c(\gamma+c) \geq 0$, which guarantees that the first square root in (8.7) is well defined. Combining the above formulas (8.6) and (8.7) gives

$$d'' = \frac{\sqrt{\alpha^2+\beta^2-4c(\gamma+c)}}{\sqrt{\alpha^2+\beta^2+\gamma^2}} \times \frac{d}{1+x^2+y^2} + \mathcal{O}(d^2).$$

If the point $\Phi(P)$ has coordinates (x',y',z'), then

$$d'' = \frac{|\alpha x' + \beta y' + \gamma z' + c|}{\sqrt{\alpha^2+\beta^2+\gamma^2}}$$

Combining the above two formulas gives

$$d = \frac{|(\alpha x' + \beta y' + \gamma z' + c)(1+x^2+y^2)|}{\sqrt{\alpha^2+\beta^2-4c(\gamma+c)}} + \mathcal{O}(d^2).$$

Objective function. Thus, if one wants to minimize distances d_i from the original points (x_i,y_i) to the fitting circle, then one should minimize

$$\mathscr{F}(\alpha,\beta,\gamma,c) = \sum_{i=1}^{n} \frac{(1+x_i^2+y_i^2)^2(\alpha x_i' + \beta y_i' + \gamma z_i' + c)^2}{\alpha^2+\beta^2-4c(\gamma+c)}, \tag{8.8}$$

which would be accurate to the first order. Again, to resolve the issue of undeterminate parameters, one can impose the constraint $\alpha^2 + \beta^2 + \gamma^2 = 1$.

The above method is similar to the gradient-weighted algebraic fit (GRAF) described in Section 6.5, both are based on the first order approximation to the geometric distances from data points to the fitting curve.

Riemann fit, the SWFL version. Before we continue the analysis of the objective function (8.8) we describe the fitting procedure proposed by Strandlie, Wroldsen, Frühwirth, and Lillekjendlie in [175].

They derive the formula (8.8), but then they made an improper allegation: *since the denominator $\alpha^2 + \beta^2 - 4c(\gamma + c)$ is constant, i.e., independent of the data point (x_i, y_i), then it does not affect the least squares procedure.* Based on this, they discard the denominator in (8.8). We will return to this issue later.

Thus the authors of [175] minimize a simpler objective function,

$$\mathscr{F}_{\text{SWFL}}(\alpha, \beta, \gamma, c) = \sum_{i=1}^{n} (1 + x_i^2 + y_i^2)^2 (\alpha x_i' + \beta y_i' + \gamma z_i' + c)^2 \qquad (8.9)$$

subject to the constraint $\alpha^2 + \beta^2 + \gamma^2 = 1$.

The numerical implementation of this procedure is quite straightforward. One precomputes the weights $w_i = (1 + x_i^2 + y_i^2)^2$ and then solves the weighted least squares problem by minimizing

$$\mathscr{F}_{\text{SWFL}}(\alpha, \beta, \gamma, c) = \sum_{i=1}^{n} w_i (\alpha x_i' + \beta y_i' + \gamma z_i' + c)^2 \qquad (8.10)$$

subject to the constraint $\alpha^2 + \beta^2 + \gamma^2 = 1$. The only difference between (8.10) and (8.3) is the presence of weights here. Thus the solution (α, β, γ) minimizing (8.10) must be the eigenvector of the 3×3 *weighted scatter matrix* $\mathbf{S}_{\text{w}}$ corresponding to its smallest eigenvalue. The weighted scatter matrix $\mathbf{S}_{\text{w}}$ is defined by

$$\mathbf{S}_{\text{w}} = \frac{1}{\sum w_i} \sum_{i=1}^{n} w_i (\mathbf{r}_i - \bar{\mathbf{r}})(\mathbf{r}_i - \bar{\mathbf{r}})^T, \qquad (8.11)$$

where $\mathbf{r}_i = (x_i', y_i', z_i')^T$ is the ith "data vector" and $\bar{\mathbf{r}} = (\sum w_i \mathbf{r}_i) / (\sum w_i)$. This formula generalizes our earlier (8.4). Note that the factor $1/(\sum w_i)$ in (8.11) is irrelevant and can be dropped. The last parameter c is determined from the equation

$$\alpha \sum w_i x_i' + \beta \sum w_i y_i' + \gamma \sum w_i z_i' + c \sum w_i = 0;$$

compare it to (8.5).

This is the Riemann fit proposed by Strandlie, Wroldsen, Frühwirth, and Lillekjendlie in [175], we abbreviate it as SWFL, by the names of the authors.

So we call this procedure the *Riemann fit, the SWFL version*. Computationally, it is only slightly more expensive than the second simple Riemann fit of

Section 8.2, as the present fit involves weights; those must be precomputed and used as extra factors in various formulas. The statistical analysis of the SWFL Riemann fit will be completed in the next section.

A real data example. We have tested the SWFL Riemann fit on the real data example described in Section 5.12. That fit returned a circle with

$$\text{center} = (7.4071, 22.6472), \qquad \text{radius} = 13.7863,$$

which is good, but slightly worse than the circles returned by simpler and faster algebraic circle fits (see Chapters 5 and 7).

8.4 Properties of the Riemann fit

Algebraic formulation of the SWFL Riemann fit. We can relate the SWFL Riemann fit to the algebraic circle fits discussed in Chapter 5 by expressing the objective function (8.9) in terms of the original data points (x_i, y_i). By using (8.1) we obtain

$$\mathcal{F}_{\text{SWFL}}(\alpha, \beta, \gamma, c) = \sum_{i=1}^{n} \left[\alpha x_i + \beta y_i + \gamma(x_i^2 + y_i^2) + c(1 + x_i^2 + y_i^2)\right]^2 \quad (8.12)$$

subject to the constraint $\alpha^2 + \beta^2 + \gamma^2 = 1$.

Now let us use our old notation $z_i = x_i^2 + y_i^2$. Then the above fit is equivalent to minimizing

$$\mathcal{F}_{\text{SWFL}}(\alpha, \beta, \gamma, c) = \sum_{i=1}^{n} \frac{\left[\alpha x_i + \beta y_i + \gamma z_i + c(1 + z_i)\right]^2}{\alpha^2 + \beta^2 + \gamma^2}, \quad (8.13)$$

without constraints. Now let us change parameters as

$$A = \gamma + c, \quad B = \alpha, \quad C = \beta, \quad D = c. \quad (8.14)$$

Then we can rewrite the objective function as

$$\mathcal{F}_{\text{SWFL}}(A, B, C, D) = \sum_{i=1}^{n} \frac{[Az_i + Bx_i + Cy_i + D]^2}{(A - D)^2 + B^2 + C^2}. \quad (8.15)$$

Equivalently, one can minimize

$$\mathcal{F}(A, B, C, D) = \sum_{i=1}^{n} [Az_i + Bx_i + Cy_i + D]^2 \quad (8.16)$$

subject to the constraint

$$(A - D)^2 + B^2 + C^2 = 1. \quad (8.17)$$

The objective function (8.16) is used by every algebraic circle fit in Chapter 5, those fits only differ by different constraints. The one given here by (8.17) is new, we have not seen it yet. Thus, the SWFL Riemann fit is nothing but yet another algebraic circle fit characterized by the unusual constraint (8.17).

Analysis of the SWFL Riemann fit. The constraint of an algebraic circle fit can be written in the matrix form as $\mathbf{A}^T \mathbf{N} \mathbf{A} = 1$, where $\mathbf{A} = (A, B, C, D)^T$. The constraint (8.17) corresponds to the matrix

$$\mathbf{N} = \begin{bmatrix} 1 & 0 & 0 & -1 \\ 0 & 1 & 0 & 0 \\ 0 & 0 & 1 & 0 \\ -1 & 0 & 0 & 1 \end{bmatrix}. \tag{8.18}$$

Recall that all the algebraic circle fits invariant under translations must have a constraint matrix given by (5.40), and the above matrix (8.18) is not of that type. Thus the SWFL Riemann fit is *not invariant* under translations. Note, on the other hand, that it is invariant under rotations, as the matrix $\mathbf{N}$ is of the form (5.41). At the same time, the SWFL Riemann fit is not invariant under similarities; see also Section 5.7.

In other words, the circle returned by the SWFL Riemann fit depends on the choice of the origin (pole) of the complex plane and on the unit of length. The noninvariance is clearly a disadvantage of the fit.

Generally, this fit is similar to the Gander-Golub-Strebel (GGS) fit and the Nievergelt fit mentioned in Section 5.8, which just use an arbitrarily chosen constraint matrix regardless of statistical efficiency. None of these three fits is invariant under translations or similarities. The statistical analysis of such fits, along the lines of Chapter 7, is impossible as they are not well defined — the returned circle depends on the choice of the coordinate system and on the unit of length.

Enforcing invariance. There is a standard trick enforcing the invariance of a fit under translation — a prior centering of the data set, as mentioned in Section 5.8. This means, precisely, placing the origin of the coordinate system at the centroid $(\bar{x}, \bar{y})$. This trick guarantees that $\bar{x} = \bar{y} = 0$. Many authors (see e.g., [53, 176]) claim that centering the data prior to the fit helps reduce round-off errors.

A similar trick can be applied to enforce invariance under similarities. Assuming that the origin is already placed at the centroid, one simply needs to choose the unit of length so that $\bar{z} = b^2$, where b^2 is a preselected constant. More precisely, given $b > 0$, one scales the data by the rule

$$(x_i, y_i) \mapsto (c x_i, c y_i), \qquad c = \frac{b}{\sqrt{\frac{1}{n}\left[\sum x_i^2 + \sum y_i^2\right]}}.$$

While any choice of b guarantees invariance under similarities, in practice one can adjust b empirically, depending on the type of data. We found that $b = 0.5$ works quite well in most cases, see the last section in this chapter.

Proper Riemann fit. On the other hand, the SWFL Riemann fit can be reorganized if one corrects a little mistake inadvertently made by its authors in [175]. Indeed, the denominator $\alpha^2 + \beta^2 - 4c(\gamma + c)$ in the formula (8.10) is independent of the data points, but it depends on the parameters, hence it is improper to discard it.

Now let us keep the denominator and again change variables according to (8.14). This gives us the objective function

$$\mathscr{F}_{\text{Riemann}}(A, B, C, D) = \sum_{i=1}^{n} \frac{[Az_i + Bx_i + Cy_i + D]^2}{B^2 + C^2 - 4AD}. \tag{8.19}$$

Equivalently, one can minimize

$$\mathscr{F}(A, B, C, D) = \sum_{i=1}^{n} [Az_i + Bx_i + Cy_i + D]^2$$

subject to the constraint

$$B^2 + C^2 - 4AD = 1.$$

Perhaps, this should be called the *proper Riemann fit.*

However, a closer look reveals that it is nothing but our familiar Pratt fit, see Section 5.5. Thus the proper version of the Riemann fit is just identical to the Pratt fit. It is one of the best circle fits, it is invariant under translations and rotations, and it has good statistical properties, in particular small bias. We have reported its properties in our extensive error analysis in Chapter 7 and numerous simulated experiments there and in Chapter 5. Everything we have said about the Pratt fit applies to the proper Riemann fit as well.

8.5   Inversion-based fits

In this and the next sections we describe some more specialized circle fitting schemes based on conformal mappings in the complex plane.

Inversion. The transformation of the complex plane that takes $\mathbf{z} = x + iy$ to

$$\mathscr{I}(\mathbf{z}) = \frac{1}{\mathbf{z}} = \frac{x - iy}{x^2 + y^2}$$

is called *inversion.* One can define it on the extended complex plane $\bar{\mathbb{C}}$ by setting $\mathscr{I}(0) = \infty$ and $\mathscr{I}(\infty) = 0$. It is a one-to-one map of $\bar{\mathbb{C}}$ onto itself (a bijection), which is continuous in both directions (a homeomorphism). It coincides with its inverse, i.e., $\mathscr{I}^{-1} = \mathscr{I}$.

In the xy coordinates, the map $\mathscr{I}$ is computed by

$$x' = \frac{x}{x^2 + y^2}, \qquad y' = -\frac{y}{x^2 + y^2}.$$

Since the negative sign before the second fraction is not essential for our purpose of fitting circles, many researchers just drop it and use the slightly simplified rules

$$x' = \frac{x}{x^2 + y^2}, \qquad y' = \frac{y}{x^2 + y^2}. \tag{8.20}$$

From now on we will call this map the inversion $\mathscr{I}$. It has all the properties of the actual inversion mentioned above.

The map $\mathscr{I}$ has many remarkable properties summarized next.

Theorem 17 *The map $\mathscr{I}$ is conformal at every point (x,y), except the origin $(0,0)$. This means that its derivative $D\mathscr{I}$ (the linear part of $\mathscr{I}$) preserves angles between tangent vectors and it expands (or contracts) all tangent vectors uniformly. Precisely, if (dx,dy) is a tangent vector at $(x,y) \neq (0,0)$, then its image $(dx',dy') = D\mathscr{I}(dx,dy)$ has length*

$$\|(dx',dy')\| = (x^2 + y^2)^{-1}\|(dx,dy)\|.$$

Note that the factor $(x^2 + y^2)^{-1}$ does not depend on the direction of the tangent vector (dx,dy).

We have seen conformal maps in Section 8.3. This theorem can be verified directly, by computing the matrix $D\mathscr{I}$ of partial derivatives of the formulas (8.20) with respect to x and y, we leave this exercise to the reader.

Theorem 18 *The map $\mathscr{I}$ transforms every line into a line or a circle. It transforms every circle into a line or a circle. In particular, every circle passing through the origin $(0,0)$ is transformed into a line not passing through the origin $(0,0)$.*

This theorem is illustrated in Fig. 8.2.

Inversion-based circle fit. The key fact in the last theorem is that $\mathscr{I}$ transforms *some* circles (precisely, those passing through the origin) to lines and back.

Now suppose one looks for a circle passing through the origin. Then the above theorem suggests the following fit:

Step 1. Map the data points (x_i, y_i) into (x'_i, y'_i) by formulas (8.20).

Step 2. Fit a line to the points (x'_i, y'_i).

Step 3. Transform the line obtained at Step 2 into a circle passing through the origin by the map $\mathscr{I}$ (recall that $\mathscr{I}^{-1} = \mathscr{I}$).

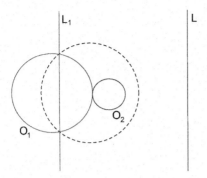

Figure 8.2 *Inversion map $\mathscr{I}$ transforms the circle O_1 into the line L_1 and the circle O_2 into the line L_2 The dashed circle is the unit circle $x^2 + y^2 = 1$.*

This procedure was originally proposed by Brandon and Cowley [25] in 1983, but they only described it in vague terms. An explicit description was published by Rusu, Tico, Kuosmanen, and Delp in [159], and we present it here.

The authors of [159] propose to fit a line at Step 2 by the orthogonal least squares, i.e., employ the geometric fit. It was thoroughly described in Chapters 1 and 2, so we need not repeat details here.

Lastly, Step 3 is a straightforward exercise in elementary geometry. Suppose the line obtained at Step 2 is given by equation

$$\alpha x' + \beta y' + c = 0. \tag{8.21}$$

Then the parameters (a, b, R) of the circle fitting the original data set is computed by

$$a = -\frac{\alpha}{2c}, \qquad b = -\frac{\beta}{2c}, \qquad R^2 = a^2 + b^2.$$

These formulas are derived in [159], we leave it to the reader to verify the details.

Limitation and generalization. An obvious limitation of the above fit is that it can only produce circles passing through the origin $(0, 0)$.

It can be used in a more general situation when one looks for a circle passing through a certain point (x_0, y_0). In that case one simply translates the coordinate system to the point (x_0, y_0), i.e., replaces the data points (x_i, y_i) with $(x_i - x_0, y_i - y_0)$. Then one applies the above fit, finds a circle with center (a, b) and radius R passing through the (new) origin, and lastly translates the coordinate system back, i.e., replaces (a, b) with $(a + x_0, b + y_0)$.

To summarize, the inversion-based fit works whenever a point is known on the desired circle. Then one uses that point as the pole of the inversion transformation. The big question is how to find such a point.

The choice of a pole. Assuming that all the data points are close to the desired circle, one can use any one of them as the pole, i.e., select $i = 1, \ldots, n$ and set $(x_0, y_0) = (x_i, y_i)$. In most cases this would be a good approximation to a point lying on the desired circle. Such approximations are employed in some realizations [159]. Then one can find a better point on the desired circle by using an iterative scheme; see Section 8.7.

Choosing one of the data points as a pole has its drawbacks. First, that point will be transformed by the inversion to ∞, thus its image cannot be used at Step 2, i.e., the sample is effectively reduced. Moreover, nearby data points will be mapped far away from the origin, their new coordinates (x', y') may be very inaccurate (to the extent that they may cause more trouble than good).

In fact, a more stable version of the fit (described in Section 8.6) requires assigning weights to data points and applying a weighted least squares fit at Step 2. The weights must be very small (close to zero) for points near the pole, hence their influence on the fitting circle is strongly suppressed. This effectively reduces the sample size even further.

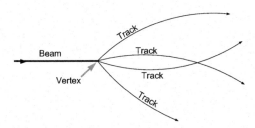

Figure 8.3 *Tracks coming out of a vertex after a collision that occurs on the beam line.*

In some applications a point on the desired circle may be known from the nature of the data. In high energy physics, experimenters commonly deal with elementary particles created (born) in a collision of an energetic particle with a target. The energetic (primary) particle moves in a beam driven by magnetic fields in an accelerator; the new (secondary) particles move along circular arcs (called *tracks*) emanating from the point of collision (called *the vertex*); see a typical picture in Fig. 8.3. The location of the vertex is often known very precisely, then one can fit circular arcs to individual tracks using the coordinates of the vertex.

A modification of the inversion-based fit. In some high energy physics experiments the tracks do not exactly pass through the known vertex, and their deviation from the vertex is an important parameter ε, called the *impact of the track*.

The following modification of the inversion-based fit was proposed in 1988 by Hansroul, Jeremie, and Savard [82] to account for the impact parameter. Let (a,b,R) denote, as usual, the center and radius of the circle (track) and

$$\delta = R^2 - a^2 - b^2$$

be small, $|\delta| \ll R^2$, which guarantees that the circle passes close to the origin (the vertex is at the origin). Assume that the track is rotated so that it is almost parallel to the x axis. Then the inversion (8.20) maps the circle (a,b,R) to another, bigger circle, and the images (x_i', y_i') of the data points are located along a nearly horizontal arc of that bigger circle. The authors of [82] approximate that arc by a parabola

$$y' = \alpha + \beta x' + \gamma [x']^2.$$

Practically, they propose to fit a parabola to the points (x_i', y_i') by the classical regression techniques; this is a linear problem that has an exact and simple solution. Then the track parameters can be computed by

$$a = -\frac{\beta}{2\alpha}, \quad b = \frac{1}{2\alpha}, \quad \varepsilon = -\frac{\gamma}{(1+\beta^2)^{3/2}}, \quad R = \varepsilon + \sqrt{a^2 + b^2},$$

where ε is the impact parameter. Thus one determines all the track characteristics.

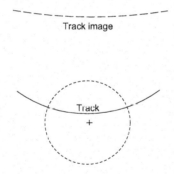

Figure 8.4 *A track passing near the pole. Its image under the inversion map is an arc that can be approximated by a parabola.*

Conclusions. We see that the inversion-based fit, in its original form, is useful only in rather special applications where one knows a point that lies *on the desired circle* or *very close to it*. However, such applications are rare; in most cases one does not have the luxury of knowing a point near the desired circle. In the next sections we present inversion-based circle fits that can handle more general situations.

8.6   The RTKD inversion-based fit

In 2003, an interesting inversion-based circle fit was developed by Rusu, Tico, Kuosmanen, and Delp [159]. First, it is a fairly accurate procedure because it properly accounts for the distance change under the inversion map. Secondly, it can work iteratively without a prior knowledge of a point on the desired circle (this will be presented in the next section).

Relation between distances. First, the authors of [159] introduced proper weights to stabilize the inversion-based fit and achieve a good accuracy. Their analysis is actually similar to the one presented in Section 8.3 for the Riemann fit.

Let P and Q be two nearby points on the xy plane and $d = \text{dist}(P,Q)$. Denote by d' the distance between their images $\mathscr{I}(P)$ and $\mathscr{I}(Q)$. According to Theorem 17,

$$d' = (x^2+y^2)^{-1}d + \mathcal{O}(d^2)$$

where (x,y) are the coordinates of P (or Q). Furthermore, if a point $P = (x,y)$ is the distance d from a circle $\mathbb{O}$ passing through the pole, then its image $\mathscr{I}(P)$ is the distance

$$d' = (x^2+y^2)^{-1}d + \mathcal{O}(d^2) \tag{8.22}$$

from the line $\mathbb{L} = \mathscr{I}(\mathbb{O})$. If the line $\mathbb{L}$ is given by equation $\alpha x' + \beta y' + c = 0$, then we can solve (8.22) for d and get

$$d = \frac{\left|(\alpha x' + \beta y' + c)(x^2+y^2)\right|}{[\alpha^2 + \beta^2]^{1/2}} + \mathcal{O}(d^2). \tag{8.23}$$

Thus, if one wants to minimize distances d_i from the original points (x_i, y_i) to the fitting circle, then one should minimize

$$\mathscr{F}_{\text{RTKD}}(\alpha, \beta, c) = \sum_{i=1}^{n} \frac{(x_i^2 + y_i^2)^2(\alpha x_i' + \beta y_i' + c)^2}{\alpha^2 + \beta^2}, \tag{8.24}$$

which would be accurate to the first order. We call this procedure the Rusu-Tico-Kuosmanen-Delp inversion-based fit, by the names of the authors of [159], and abbreviate it by RTKD.

We call the above method the *RTKD inversion-based fit*. This is perhaps the most accurate variant of the inversion-based fit.

This method is similar to the gradient-weighted algebraic fit (GRAF) described in Section 6.5; it is also similar to the proper Riemann fit of Section 8.4, which minimizes (8.19). All these fits are based on the first order approximation to the geometric distances from data points to the fitting curve, which provides the best accuracy possible for noniterative (algebraic) fits.

The RTKD inversion-based fit. The numerical implementation of the RTKD fit is quite straightforward. One precomputes the weights

$$w_i = (x_i^2 + y_i^2)^2 \tag{8.25}$$

and then solves the linear weighted least squares problem by minimizing

$$\mathscr{F}_{\text{RTKD}}(\alpha, \beta, c) = \sum_{i=1}^{n} w_i (\alpha x_i' + \beta y_i' + c)^2 \tag{8.26}$$

subject to the constraint $\alpha^2 + \beta^2 = 1$. The solution (α, β) minimizing (8.26) must be the eigenvector of the 2×2 *weighted scatter matrix* $\mathbf{S}_w$ corresponding to its smallest eigenvalue. The weighted scatter matrix $\mathbf{S}_w$ is defined by

$$\mathbf{S}_w = \frac{1}{\sum w_i} \sum_{i=1}^{n} w_i (\mathbf{r}_i - \bar{\mathbf{r}})(\mathbf{r}_i - \bar{\mathbf{r}})^T, \tag{8.27}$$

where $\mathbf{r}_i = (x_i', y_i')^T$ is the ith "data vector" and $\bar{\mathbf{r}} = \left(\sum w_i \mathbf{r}_i\right) / \left(\sum w_i\right)$; compare this to (1.7). Note that the factor $1/\left(\sum w_i\right)$ in (8.27) is irrelevant and can be dropped. The last parameter c is determined from the equation

$$\alpha \sum w_i x_i' + \beta \sum w_i y_i' + c \sum w_i = 0.$$

This concludes our description of the RTKD circle fit.

We remark that the authors of [159] implement the above procedure slightly differently. They describe the line by equation $y' = a + bx'$, instead of our (8.21), and then propose to compute a and b by the classical formulas (1.14)–(1.16) (with extra weights). This approach, however, excludes vertical lines and becomes numerically unstable when the line is close to vertical. The above eigenvalue method, using standard matrix software, ensures numerical stability and handles all lines without exception.

Analysis of the RTKD inversion-based fit. The authors of [159] do not provide a statistical analysis of their fit. We just indicate its relation to another fit here.

First, we use (8.20) to express the objective function (8.24) in the original coordinates (x_i, y_i):

$$\mathscr{F}_{\text{RTKD}}(\alpha, \beta, c) = \sum_{i=1}^{n} \frac{\left[\alpha x_i + \beta y_i + c(x_i^2 + y_i^2)\right]^2}{\alpha^2 + \beta^2}.$$

Second, we use our old notation $z_i = x_i^2 + y_i^2$ and change parameters as

$$A = c, \qquad B = \alpha, \qquad C = \beta,$$

which gives us the objective function

$$\mathscr{F}_{\text{RTKD}}(A,B,C) = \sum_{i=1}^{n} \frac{\left[Az_i + Bx_i + Cy_i\right]^2}{B^2 + C^2}. \tag{8.28}$$

Comparing this to (8.19) we see that the RTKD inversion fit is obtained from the proper Riemann fit (which is identical to the Pratt fit) by setting $D = 0$. The constraint $D = 0$ simply forces the fitting circle to pass through the origin. Thus, the RTKD inversion fit is nothing but a constrained version of the classical Pratt fit, when the latter is reduced to circles passing through the origin.

8.7   The iterative RTKD fit

Here we describe an iterative procedure developed by Rusu, Tico, Kuosmanen, and Delp [159] that is based on their inversion method introduced in the previous section, but is capable of fitting circles without prior knowledge of a point on the circumference.

First we observe that the weights $w_i = (x_i^2 + y_i^2)^2$ for the RTKD fit, cf. (8.25), are small for data points close to the origin (the pole); moreover, they vanish at the pole $(0,0)$. This means that the influence of data points near the pole is severely suppressed. If there are several such points, then the sample size is effectively reduced, leading to an undesirable loss of information. This observation prompts the following considerations.

Position of the pole. The authors of [159] examined how the accuracy of the fit depends on the location of the pole on the circle.

Of course if the data points are samples from the entire circle uniformly, then any location is as good as another. So suppose the data points are sampled from a small arc C of the true circle. Then the experiments reported in [159] show that

- If the pole is placed on the arc C (i.e., close to the data points), then the accuracy of the inversion fit is low.
- If the pole is placed outside the arc C (i.e., far from the data points), then the accuracy of the inversion fit is good.
- If the pole is placed diametrically opposite to the arc C (i.e., the farthest from the data points), then the accuracy of the inversion fit is the best.

Fig. 8.5 illustrates these conclusions; we refer to [159] for a more detailed account on their experimental results.

Iterative scheme. To avoid unfortunate locations of the pole close to the data points, the authors of [159] propose the following iterative scheme.

Step 1. Set $k = 0$. Suppose an initial pole P_0 is chosen.

Figure 8.5 *Possible locations for the pole on the circle. The observed points (marked by crosses) are clustered on the right.*

Step 2. Apply the RTKD inversion fit (Section 8.6) with the pole P_k, obtain a circle $\mathbb{O}_k$.

Step 3. Place the new pole P_{k+1} at point on the circle $\mathbb{O}_k$ diametrically opposite to P_k (recall that $P_k \in \mathbb{O}_k$ by the nature of the fit).

Step 4. If $\text{dist}(P_{k+1}, P_{k-1})$ is small enough, stop. Otherwise increment k and return to Step 2.

The underlying idea is that if the initial pole happens to be in a bad part of the circle (close to the data points), then the next location will be in a very good part of it (diametrically opposite to the data points), and then the locations will alternate between bad and very good; the latter will stabilize the fit. If the initial pole is in a good part of the circle, then its locations will remain in good parts.

Relaxing the constraint. Next, Rusu, Tico, Kuosmanen, and Delp [159] realized that their RTKD iterative scheme can be applied without a prior knowledge of a point on the circumference.

Indeed, suppose the initial pole P_0 at Step 1 is chosen away from the circle. Then the procedure will change the pole at every step, and the poles may get closer and closer to the circle. Thus the procedure may converge to a limit where pole will be on circle.

Practically, the authors of [159] place the initial pole at one of the observed points, which is a very reasonable strategy (Section 8.5). In their numerical tests, the iterative procedure always converged to a circle that was quite close to the one returned by the geometric fit. The number of iterations was between 5 and 15 in all reported cases.

We call this algorithm the *iterative RTKD fit*.

The authors of [159] also investigated the sensitivity of their iterative scheme to the choice of the initial pole in general, when the latter is selected anywhere in the plane. They discovered that when the pole was chosen inside the true circle or relatively close to it, then the scheme almost certainly converged. If the pole is selected far from the true circle, the picture is mixed.

The authors of [159] found large areas in the plane from which the procedure converged, but also nearly equally large areas from which it diverged. Roughly speaking, if a pole is selected randomly from a very big disk around the observed points, then about 50–60% of the times the procedure converges, and about 40–50% it does not. More details may be found in [159].

Such a statistics of convergence, by the way, is not unusual. Recall that the geometric fit, implemented by standard Gauss-Newton or Levenberg-Marquardt numerical schemes, also converges 50-60% of the times, if started from a point randomly chosen in a very large area; we discovered (and rigorously proved!) this fact in Chapter 3.

Analysis of the iterative RTKD fit. The authors of [159] do not investigate the convergence of their fit or its statistical properties. We do that here.

Recall that if the pole coincides with the origin $(0,0)$, then the RTKD inversion fit minimizes the function (8.28), i.e.,

$$\mathscr{F}_{\text{RTKD}}(A,B,C) = \sum_{i=1}^{n} \frac{[Az_i + Bx_i + Cy_i]^2}{B^2 + C^2}.$$

If the pole is at an arbitrary point $P = (p,q)$, then the RTKD inversion fit will minimize the function

$$\mathscr{F}(A,B,C) = \sum_{i=1}^{n} \frac{\left[A\left((x_i - p)^2 + (y_i - q)^2\right) + B(x_i - p) + C(y_i - q)\right]^2}{B^2 + C^2}$$

$$= \sum_{i=1}^{n} \frac{\left[Az_i + (B - 2Ap)x_i + (C - 2Aq)y_i + A(p^2 + q^2) - Bp - Cq\right]^2}{B^2 + C^2}.$$

Let us change parameters as follows:

$$\check{A} = A, \quad \check{B} = B - 2Ap, \quad \check{C} = C - 2Aq, \quad \check{D} = A(p^2 + q^2) - Bp - Cq. \quad (8.29)$$

Now the RTKD inversion fit minimizes the function

$$\mathscr{F}(\check{A}, \check{B}, \check{C}, \check{D}) = \sum_{i=1}^{n} \frac{\left[\check{A}z_i + \check{B}x_i + \check{C}y_i + \check{D}\right]^2}{\check{B}^2 + \check{C}^2 - 4\check{A}\check{D}}, \quad (8.30)$$

the formula in the denominator follows from

$$B^2 + C^2 = \check{B}^2 + \check{C}^2 + 4A\left[A(p^2 + q^2) + \check{B}p + \check{C}q\right]$$
$$= \check{B}^2 + \check{C}^2 + 4\check{A}\left[-A(p^2 + q^2) + Bp + Cq\right].$$

We see that the RTKD inversion fit with the pole at $P = (p,q)$ minimizes the objective function (8.30). The coordinates of the pole do not explicitly enter the formula (8.30), but they impose a constraint on the parameters $\check{A}, \check{B}, \check{C}, \check{D}$, which is dictated by (8.29):

$$\check{D} = \check{B}p + \check{C}q - \check{A}(p^2 + q^2).$$

This constraint precisely means that the circle must pass through the pole (p, q).

Now let P_k and P_{k+1} be two successive poles obtained by the iterative RTKD scheme. Recall that both P_k and P_{k+1} belong to the circle $\mathbb{O}_k$ found by the RTKD inversion fit with the pole P_k. At the next step, the RTKD inversion fit with pole P_{k+1} find a new circle, $\mathbb{O}_{k+1}$. Observe that

$$\mathscr{F}(\mathbb{O}_{k+1}) = \min_{\mathbb{O} \ni P_{k+1}} \mathscr{F}(\mathbb{O}) \leq \mathscr{F}(\mathbb{O}_k),$$

where $\mathscr{F}(\mathbb{O})$ denotes the value of the objective function (8.30) at parameters corresponding to the circle $\mathbb{O}$. We see that the function $\mathscr{F}$ monotonically decreases at every step of the iterative scheme. This is a good feature. In a sense, it guarantees convergence: the iterations simply cannot "wander aimlessly" or loop periodically.

It is reasonable to assume that the RTKD iterative scheme should converge to the global minimum of the function (8.30), i.e., it should find a circle for which $\mathscr{F}$ takes its minimum value. Our experimental tests show that this is indeed the case: the RTKD algorithm either converges to the global minimum of (8.30), or diverges altogether. We have observed this in millions of simulated examples and have not seen a single exception.

Recall that the objective function (8.30) is also minimized by the Pratt fit, see (5.21), as well as by the proper Riemann fit, see (8.19); those fits always (!) find the global minimum of $\mathscr{F}$. Thus the iterative RTKD fit and the Pratt fit and the proper Riemann fit are theoretically identical, hence everything we have said about the Pratt fit in Chapters 5 and 7 applies to the iterative RTKD fit as well. Practically, though, the Pratt fit is preferable because it is fast, non-iterative, and 100% reliable.

A real data example. We have tested the Rusu-Tico-Kuosmanen-Delp (RTKD) circle fits on the real data example described in Section 5.12. As a pole, we used the first data point in the set.

The original version (Section 8.6) returned a circle with

$$\text{center} = (5.2265, 24.7238), \quad \text{radius} = 16.7916,$$

so it grossly overestimates the radius. The improved (iterative) version of the RTKD fit returned a circle with

$$\text{center} = (7.3871, 22.6674), \quad \text{radius} = 13.8146,$$

which is identical to that found by the Pratt fit (Section 5.12).

8.8 Karimäki fit

An interesting fitting scheme was developed in 1991 by Karimäki [106, 107] for applications in high energy physics. His scheme is similar, in spirit, to inversion-based fits by Brandon and Cowley [25] and Hansroul, Jeremie, and Savard [82] discussed in Section 8.5. But instead of conformal maps, Karimäki uses a trigonometric change of parameters.

Karimäki's parameters. Recall (Section 8.5) that in nuclear physics experiments, one fits circular arcs to tracks of elementary particles; tracks commonly have a large radius (i.e., a small curvature) and pass near a fixed region in the detector (the vertex). If the origin of the coordinate system is close to the vertex, then the tracks pass near the origin.

In these experiments, the traditional circle parameters (a, b, R) are not convenient as they frequently take large values leading to numerical instability. In the limit of straight line tracks (which are characteristic for particles with very high energy), these parameters turn infinite and the computations fail. This problem was discussed at length in Section 3.2.

Karimäki [106] proposes an alternative set of parameters defined geometrically (by using trigonometric functions). He replaces the radius R with a signed curvature $\rho = \pm R^{-1}$, see below, and the center (a, b) with two other parameters, d and φ, illustrated in Fig. 8.6. He calls d the distance of the closest approach to the origin (i.e., the distance from $(0, 0)$ to the circle), and φ the direction of propagation at the point of closest approach.

Karimäki's choice of signs for ρ and d is the following: $\rho > 0$ for tracks turning clockwise and $\rho < 0$ for those turning counterclockwise (at the point closest to the origin); the sign of d is determined from the equation

$$d = x_d \sin \varphi - y_d \cos \varphi,$$

where (x_d, y_d) is the point of the circle closest to the origin. Our Fig. 8.6 shows a configuration where both ρ and d are positive.

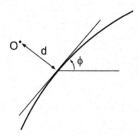

Figure 8.6: *Karimäki's parameters d and ϕ.*

The three parameters (ρ, d, ϕ) completely (and uniquely) describe the circular arc. They are convenient in the high energy physics experiments, as they never have to take dangerously large values, and they represent exactly the quantities of practical interest.

Conversion formulas. We refer to the illustration in Fig. 8.6, where both ρ and d are positive (other cases are treated similarly). Observe that the line joining the origin with the nearest point on the arc also passes through the circle's center.

Now given (ρ, d, φ) one can compute the traditional parameters by $R = 1/|\rho|$ and

$$a = \left(d + \tfrac{1}{\rho}\right) \sin \varphi \qquad \text{and} \qquad b = -\left(d + \tfrac{1}{\rho}\right) \cos \varphi. \qquad (8.31)$$

Conversely, given (a, b, R) one can compute the new parameters by $\rho = \pm 1/R$ and by solving the following two equations for d and φ, respectively:

$$\left(d + \tfrac{1}{\rho}\right)^2 = a^2 + b^2, \qquad \tan \varphi = -a/b.$$

Karimäki's objective function. To construct his objective function, Karimäki approximates distances between data points (x_i, y_i) and the circle (a, b, R) following an earlier work by Chernov and Ososkov [45]. Namely, he approximates the geometric distance

$$\varepsilon_i = \sqrt{(x_i - a)^2 + (y_i - b)^2} - R$$

by

$$\varepsilon_i \approx \varepsilon_i = (2R)^{-1} \left[(x_i - a)^2 + (y_i - b)^2 - R^2\right].$$

He calls this a "highly precise approximation" [107]; we refer to Section 5.4 for the motivation and related discussion.

Then Karimäki expresses the above approximate distance in terms of his circle parameters (ρ, d, φ):

$$\varepsilon_i = \tfrac{1}{2} \rho r_i - (1 + \rho d) r_i \sin(\varphi - \theta_i) + \tfrac{1}{2} \rho d^2 + d, \qquad (8.32)$$

where r_i and θ_i are the polar coordinates of the ith data point, i.e.

$$x_i = r_i \cos \theta_i \qquad \text{and} \qquad y_i = r_i \sin \theta_i.$$

Next Karimäki rewrites (8.32) as

$$\varepsilon_i = (1 + \rho d)\left(\kappa r_i^2 - r_i \sin(\varphi - \theta_i) + \mu\right), \qquad (8.33)$$

where

$$\kappa = \frac{\rho}{2(1 + \rho d)} \qquad \text{and} \qquad \mu = \frac{1 + \tfrac{1}{2}\rho d}{1 + \rho d} d, \qquad (8.34)$$

are shorthand notation. Now Karimäki arrives at the objective function

$$\mathscr{F}(\rho, d, \varphi) = (1 + \rho d)^2 \sum_{i=1}^{n} \left(\kappa r_i^2 - r_i \sin(\varphi - \theta_i) + \mu \right)^2. \tag{8.35}$$

This is, of course, the same objective function (5.16) used by Chernov and Ososkov [45], but expressed in terms of the new parameters ρ, d, φ.

Karimäki further simplifies the objective function (8.35) by discarding the factor $(1 + \rho d)^2$, so he actually minimizes

$$\mathscr{F}_{\text{Kar}}(\rho, d, \varphi) = \sum_{i=1}^{n} \left(\kappa r_i^2 - r_i \sin(\varphi - \theta_i) + \mu \right)^2$$

$$= \sum_{i=1}^{n} \left(z_i \kappa - x_i \sin \varphi + y_i \cos \varphi + \mu \right)^2, \tag{8.36}$$

where we use our standard notation $z_i = x_i^2 + y_i^2$.

Karimäki's rational for dropping the factor $(1 + \rho d)^2$ from (8.35) is that in high energy physics experiments

$$|d| \ll R = 1/|\rho|, \qquad \text{hence} \qquad 1 + \rho d \approx 1.$$

Furthermore, Karimäki argues that the factor $(1 + \rho d)^2$ is almost constant, in the sense that it varies slowly (as a function of ρ and d), while the function $\mathscr{F}_{\text{Kar}}$ in (8.36) has a sharp minimum, thus the location of that minimum should not be affected much by an extra slowly varying factor.

He supports his reasoning by experimental evidence: he shows that in typical cases the parameters minimizing (8.36) also provide the minimum of (8.35), with a relative error $< 10^{-3}$; see [106]. In another paper [107] Karimäki describes the range of parameter values for which his reduction of (8.35) to (8.36) remains reasonably accurate.

We will return to this issue in Section 8.9.

Karimäki's minimization scheme. The formula (8.36) suggests that κ and μ can be treated as new parameters that will temporary replace ρ and d. The minimization of (8.36) with respect to κ, μ, and φ turns out to be a relatively simple problem that has an explicit closed form solution. It was found by Karimäki [106], we present it next.

First, one differentiates (8.36) with respect to κ and μ and finds

$$\frac{1}{2n} \frac{\partial \mathscr{F}_{\text{Kar}}}{\partial \kappa} = \overline{zz} \kappa - \overline{xz} \sin \varphi + \overline{yz} \cos \varphi + \bar{z} \mu = 0,$$

$$\frac{1}{2n} \frac{\partial \mathscr{F}_{\text{Kar}}}{\partial \mu} = \bar{z} \kappa - \bar{x} \sin \varphi + \bar{y} \cos \varphi + \mu = 0, \tag{8.37}$$

where we again employ our standard "sample mean" notation

$$\bar{x} = \frac{1}{n}\sum x_i, \quad \overline{xz} = \frac{1}{n}\sum x_i z_i, \quad \overline{zz} = \frac{1}{n}\sum z_i^2, \tag{8.38}$$

etc. The equations (8.37) are linear in κ and μ, hence these two parameters can be eliminated as follows:

$$\kappa = \frac{C_{xz}\sin\varphi - C_{yz}\cos\varphi}{C_{zz}} \tag{8.39}$$

$$\mu = -\bar{z}\kappa + \bar{x}\sin\varphi - \bar{y}\cos\varphi.$$

Now one substitutes (8.39) into (8.36) and arrives at the minimization of a single-variable function

$$\mathcal{F}_{\text{Kar}}(\varphi) = (C_{zz}C_{xx} - C_{xz}^2)\sin^2\varphi + (C_{zz}C_{yy} - C_{yz}^2)\cos^2\varphi$$
$$- 2(C_{zz}C_{xy} - C_{xz}C_{yz})\sin\varphi\cos\varphi. \tag{8.40}$$

Here C_{zz}, C_{xy}, etc. denote "sample covariances:"

$$\begin{aligned}
C_{xx} &= \overline{xx} - \bar{x}^2, & C_{yy} &= \overline{yy} - \bar{y}^2, \\
C_{xy} &= \overline{xy} - \bar{x}\bar{y}, & C_{xz} &= \overline{xz} - \bar{x}\bar{z}, \\
C_{yz} &= \overline{yz} - \bar{y}\bar{z}, & C_{zz} &= \overline{zz} - \bar{z}^2.
\end{aligned} \tag{8.41}$$

By differentiating (8.40) one arrives at the following equation for φ:

$$\tan 2\varphi = \frac{2(C_{zz}C_{xy} - C_{xz}C_{yz})}{C_{zz}(C_{xx} - C_{yy}) - C_{xz}^2 + C_{yz}^2}. \tag{8.42}$$

We note that this is an analogue of the classical Pearson formula (1.17).

After solving (8.42) for φ one computes κ and μ by (8.39). Next one recovers ρ and d from (8.34):

$$\rho = \frac{2\kappa}{\sqrt{1 - 4\kappa\mu}} \quad \text{and} \quad d = \frac{2\mu}{1 + \sqrt{1 - 4\kappa\mu}}. \tag{8.43}$$

Lastly, if $\rho \neq 0$, then one can compute the radius of the fitting circle by $R = 1/|\rho|$ and its center (a,b) by (8.31). This completes the Karimäki fitting procedure.

We call it simply the *Karimäki fit*.

8.9   Analysis of Karimäki fit

Karimäki's fit has been used in several high energy physics laboratories around the world. Its fast code (in FORTRAN 77, optimized for speed) is included in the CPC Program Library, Queen's University of Belfast; see [107]. Due to its popularity we devote this section to its analysis.

Algebraic description of the Karimäki fit. Karimäki employs a trigonometric change of parameters, but his scheme can be expressed, equivalently, in pure algebraic terms. Indeed, we can rewrite his objective function (8.36) as

$$\mathscr{F}_{Kar} = \sum_{i=1}^{n} [Az_i + Bx_i + Cy_i + D]^2 \qquad (8.44)$$

subject to the constraint

$$B^2 + C^2 = 1. \qquad (8.45)$$

This places his fit in the context of algebraic circle fits discussed in Chapter 5. Recall that the objective function (8.44) is used by every algebraic circle fit, those fits only differ by different constraints. The one given here by (8.45) is new, we have not seen it yet.

The constraint of an algebraic circle fit can be written in the matrix form as $\mathbf{A}^T \mathbf{N} \mathbf{A} = 1$, where $\mathbf{A} = (A, B, C, D)^T$, and the constraint (8.45) corresponds to the matrix

$$\mathbf{N} = \begin{bmatrix} 0 & 0 & 0 & 0 \\ 0 & 1 & 0 & 0 \\ 0 & 0 & 1 & 0 \\ 0 & 0 & 0 & 0 \end{bmatrix}. \qquad (8.46)$$

Recall that all the algebraic circle fits invariant under translations must have a constraint matrix given by (5.40), and the above matrix (8.46) is not of that type. Thus the Karimäki fit is *not invariant* under translations. Note, on the other hand, that it is invariant under rotations, as the matrix $\mathbf{N}$ is of the form (5.41). It is also invariant under similarities (see Section 5.7).

We also note that all the algebraic circle fits can be reduced to a generalized eigenvalue problem:

$$\mathbf{M} \mathbf{A} = \eta \mathbf{N} \mathbf{A}, \qquad (8.47)$$

see (7.63). In many cases this problem can be efficiently solved by standard matrix algebra software (for example, by calling the "eig" function in MAT-LAB), but not in the present case. The trouble is that the matrix $\mathbf{N}$ in (8.46) is singular (its rank is 2), hence standard matrix library functions may run into exceptions and break down or become unstable.

To remove the singularity from the problem (8.47), one can eliminate two variables and reduce it to a 2×2 nonsingular generalized eigenvalue problem. It is most convenient to remove A and D, but then the resulting solution will be practically identical to the original Karimäki's formulas given in Section 8.8. Thus the algebraic description of the Karimäki fit does not lead to any better implementation.

Similarity with inversion-based fits. Karimäki's fit, like all the algebraic circle fits, is designed as an approximation to the geometric fit. But his fit is a double approximation, as it employs two subsequent approximations to the geometric distances.

First, Karimäki uses the formula (8.32) introduced by Chernov and Ososkov [45] and later by Pratt [150] and Kanatani [96]; it works well when $\sigma \ll R$ and puts no restrictions on circles. If Karimäki's approximations were limited to (8.32), then his fit would be mathematically equivalent to those by Chernov-Ososkov (Section 5.4) and Pratt (Section 5.5).

But Karimäki further reduces (8.35) to (8.36), and this reduction is based on his assumption $|d| \ll R$, which puts explicit restrictions on circles. It precisely means that the circle passes close to the origin (compared to its size). This assumption makes the Karimäki fit similar to the inversion-based circle fits.

Recall that the standard inversion-based fit (Section 8.5) assumes that the circle passes right through the origin. A modification by Hansroul, Jeremie, and Savard [82] works also when the circle passes near the origin; see Section 8.5. Thus all these fits (including Karimäki's) are restricted to the same rather special type of applications. This type, however, is quite common in high energy physics (and it is hardly a coincidence that all these fits have been developed by nuclear physics experimenters).

Correction formulas. Karimäki recognizes that the removal of the factor $(1 + \rho d)^2$ from the objective function (8.35) affects the location of its minimum, i.e., the minimum of the new function (8.36) found by his algorithm is only an approximation to the minimum of the more appropriate function (8.35).

Karimäki proposes to correct his estimates of ρ, d, φ by making a Newton's step toward the true minimum of (8.35). Precisely, he increments his estimates by

$$
\begin{aligned}
\Delta \rho &= -\tfrac{n}{2}(dV_{\rho\rho} + \rho V_{\rho d})K, \\
\Delta d &= -\tfrac{n}{2}(dV_{d\rho} + \rho V_{dd})K, \qquad\qquad (8.48) \\
\Delta \varphi &= -\tfrac{n}{2}(dV_{\varphi\rho} + \rho V_{\varphi d})K,
\end{aligned}
$$

where

$$
\begin{aligned}
K = {}& 2(1 + \rho d)(\overline{xx} \sin^2 \varphi - 2\overline{xy} \sin \varphi \cos \varphi + \overline{yy} \cos^2 \varphi) \\
& - \rho(\overline{xz} \sin \varphi - \overline{yz} \cos \varphi) - d(2 + \rho d)(\bar{x} \sin \varphi - \bar{y} \cos \varphi),
\end{aligned}
$$

and $V_{\rho\rho}, V_{\rho d}, \ldots$ are the components of the matrix $\mathbf{V} = \mathbf{H}^{-1}$, where $\mathbf{H}$ is the 3×3 Hessian matrix consisting of the second order partial derivatives of the objective function (8.35) evaluated at its minimum. We refer to [106] for explicit formulas of the components of $\mathbf{H}$.

Karimäki remarks [106] that the computation of the Hessian matrix $\mathbf{H}$ and its inverse $\mathbf{V}$ is a part of the standard "error estimation" procedure (i.e., the

evaluation of the covariance matrix of the estimates), so his corrections (8.48)
would come as an additional benefit at almost no extra cost.

Karimäki's assumption revisited. Karimäki designed his fit under a
seemingly restrictive assumption, $|d| \ll R$, but in fact his fit works well un-
der a less restrictive condition:

$$1 + \rho d \qquad \text{should not be close to } 0. \qquad (8.49)$$

Indeed, let us expand the objective function (8.35) into Taylor series

$$\mathscr{F}(\Theta) = a_1 + \mathbf{D}_1^T(\Theta - \hat{\Theta}) + \tfrac{1}{2}(\Theta - \hat{\Theta})^T \mathbf{H}_1(\Theta - \hat{\Theta}) + \cdots \qquad (8.50)$$

where higher order terms are omitted; here Θ denotes the vector of parameters,
$\mathbf{D}_1$ is the gradient and $\mathbf{H}_1$ the Hessian of $\mathscr{F}$. Suppose that the expansion is
made at the point $\hat{\Theta}$ where (8.35) takes its minimum, then $\mathbf{D}_1 = \mathbf{0}$. Note that $\hat{\Theta}$
is in fact Pratt's estimate of the circle parameters (see Section 5.5).

Karimäki's objective function (8.36) can be written as $\mathscr{F}_{\text{Kar}} = g\mathscr{F}$, where
$g = (1 + \rho d)^{-2}$. Let us expand the extra factor g into Taylor series, too:

$$g(\Theta) = a_2 + \mathbf{D}_2^T(\Theta - \hat{\Theta}) + \tfrac{1}{2}(\Theta - \hat{\Theta})^T \mathbf{H}_2(\Theta - \hat{\Theta}) + \cdots$$

Multiplying the above expansions gives Karimäki's objective function (8.36):

$$\mathscr{F}_{\text{Kar}}(\Theta) = a_1 a_2 + a_1 \mathbf{D}_2^T(\Theta - \hat{\Theta}) + \tfrac{1}{2}(\Theta - \hat{\Theta})^T (a_1 \mathbf{H}_2 + a_2 \mathbf{H}_1)(\Theta - \hat{\Theta}) + \cdots$$

The minimum of $\mathscr{F}_{\text{Kar}}$ is taken at

$$\hat{\Theta}_{\text{Kar}} = \hat{\Theta} - a_1 (a_1 \mathbf{H}_2 + a_2 \mathbf{H}_1)^{-1} \mathbf{D}_2 + \cdots$$

where higher order terms are omitted. Let us examine how the minimum $\hat{\Theta}_{\text{Kar}}$
differs from the minimum $\hat{\Theta}$ of the function $\mathscr{F}$ given by (8.50).

As usual, suppose the noise is Gaussian at a level $\sigma \ll R$; then the observed
points are at distance $\mathscr{O}(\sigma)$ from the best fitting circle. Then $\mathscr{F} = \mathscr{O}(\sigma^2)$ at
its minimum, hence $a_1 = \mathscr{O}(\sigma^2)$. Now the difference between the two minima
$\hat{\Theta}_{\text{Kar}}$ and $\hat{\Theta}$ will be of order σ^2 provided the vector $(a_1 \mathbf{H}_2 + a_2 \mathbf{H}_1)^{-1} \mathbf{D}_2$ is
not too large, i.e., if its magnitude is of order one. This vector depends on
several factors which are hard to trace completely, but it appears that indeed its
magnitude is of order one provided $g(\hat{\Theta}) = a_2$ is not too large, i.e., $1 + \rho d$ is
not too close to zero, i.e., we arrive at the mild condition (8.49).

Recall that the statistical error of the estimate $\hat{\Theta}$ (i.e., its standard deviation) is of order σ, hence an additional error of order σ^2 would be indeed
relatively small. In particular, both fits will have the same covariance matrix,
to the leading order. We have seen in Chapter 7 that indeed all the algebraic
circle fits have the same covariance matrix. Moreover, the corrections (8.48)
may reduce it further, perhaps down to $\mathscr{O}(\sigma^3)$. This is, of course, a conjecture

to be verified. If this is true, i.e., if $\hat{\Theta}_{Kar} - \hat{\Theta} = \mathcal{O}(\sigma^3)$, then Karimäki's estimator $\hat{\Theta}_{Kar}$ would not only have the same covariance matrix as other algebraic fits (Chapter 7), but also the same essential bias as the Pratt fit. Our numerical tests support this hypothesis. We observed that the mean squared error of the Karimäki fit is usually larger than that of the Pratt fit, but after the corrections (8.48) are applied, the mean squared error becomes nearly equal to that of the Pratt fit.

We recall that the Pratt fit is one of the best circle fits, though its statistical accuracy is slightly lower than that of the Taubin fit or the geometric fit or the "Hyperaccurate" fit (see a detailed analysis in Chapter 7).

Possible extensions. There are several ways to relax even the mild restriction (8.49). They are not discussed in Karimäki's papers [106, 107], but we include them here in an attempt to extend the applicability of his fit.

First, if all the observed points are close to the circle (i.e., $\sigma \ll R$), one can choose any of them as the origin. In other words, one can choose $1 \leq j \leq n$ and then transform $(x_i, y_i) \mapsto (x_i - x_j, y_i - y_j)$ for all $i = 1, \ldots, n$. This places the origin at the jth observed point. In most cases such a shift would guarantee the validity Karimäki's main assumption (8.49). Furthermore, such a choice of the coordinate system does not reduce the size of the sample, as it did in the case of inversion-based fits, see Section 8.5. Even the jth point itself, where the origin is placed, effectively participates in the fitting procedure.

Next, one can apply an iterative scheme analogous to that of Rusu, Tico, Kuosmanen, and Delp [159]; see Section 8.6. Namely, one can adjust the origin (pole) $P = (0,0)$ at every iteration. Precisely, suppose for the current origin P_k (where k is the iteration number) the Karimäki fit returns a circle $\mathbb{O}_k$. Then one can set the next origin (pole) P_{k+1} at the point on $\mathbb{O}_k$ closest to P_k; then one refits the circle $\mathbb{O}_{k+1}$, etc.

This iterative scheme may converge much faster than the RTKD algorithm of Section 8.6, because Karimäki's circle is more flexible, so it is not rigidly constrained to pass through the current pole. On the other hand, repeating the Karimäki fit iteratively would instantly deprive it of its best asset—fast performance—because its computational time would double or triple.

A real data example. We have tested the Karimäki circle fits on the real data example described in Section 5.12. As a pole, we used the first data point in the set.

The original Karimäki fit (Section 8.8) returned a circle with

$$\text{center} = (7.4408, 22.6155), \qquad \text{radius} = 13.7401,$$

so it underestimates the radius. The corrected version described in this section returned a circle with

$$\text{center} = (7.3909, 22.6636), \qquad \text{radius} = 13.8092,$$

which is closer to the ideal circle (5.71) but still slightly worse than circles found by simpler and faster algebraic fits (Section 5.12).

8.10 Numerical tests and conclusions

We have tested all the algorithms described in this chapter in a set of numerical experiments with simulated data. Our goal was to "try them on" under various conditions and compare them to the geometric and algebraic circle fits described in the previous chapters.

Numerical tests. First, we generated $N = 10^7$ random samples of $n = 20$ points located (equally spaced) on the entire unit circle $x^2 + y^2 = 1$ and corrupted by Gaussian noise at level $\sigma = 0.05$. For each circle fit we have empirically determined the mean squared error of the parameter vector, i.e.,

$$\text{MSE} = \frac{1}{N} \sum_{i=1}^{N} \left[(\hat{a}_i - a)^2 + (\hat{b}_i - b)^2 + (\hat{R}_i - R)^2 \right],$$

where $\hat{a}_i, \hat{b}_i, \hat{R}_i$ denote the estimates of the circle parameters computed from the ith sample, and $a = b = 0$, $R = 1$ are their true values. The resulting MSE's are given in the first column of Table 8.1.

We note that the inversion-based fit and Karimäki fit require a pole (i.e., a point presumably lying on the circle) to be supplied. We did not give them such a luxury as a point on the true circle, but we used a randomly selected data point as a pole (which is a standard strategy; see Section 8.5).

The first column of Table 8.1 shows that the exotic circle fits described in this chapter are generally less accurate than the standard geometric and algebraic fits. The plain RTKD inversion-based fit (the second line) is especially poor; apparently, it heavily depends on the choice of the pole, and a randomly selected data point is just not good enough. The subsequent iterations improve the performance of the inversion-based fit, making it similar to that of the Pratt fit.

The plain Karimäki fit is not very accurate either, but after the subsequent corrective step it performs similarly to the Pratt fit. This fact confirms our theoretical predictions that the iterative RTKD fit and the corrected Karimäki fit would produce nothing but good approximations to the Pratt fit.

The SWFL Riemann fit was implemented with data centering and rescaling, as described in Section 8.4, to enforce its invariance under translations and similarities. These steps were added to the original version published in [175, 173], as without such steps the performance of the Riemann fit was unreliable. Surprisingly, with these additional steps, the Riemann fit becomes very stable, it even slightly outperforms the Pratt fit! But the best fitting schemes are still our "old friends"—Taubin, Hyper, and geometric fits.

Next we simulated more a difficult situation, where the $n = 20$ data points

$n = 20$ data points	360° $(\sigma = 0.05)$	180°	90° $(\sigma = 0.01)$	45°
Riemann fit (the SWFL version)	6.439	19.32	9.970	166.7
Inversion-based fit (before iterations)	18.42	52.33	27.95	661.3
Inversion-based fit (after iterations)	6.549	19.56	9.990	169.0
Karimäki fit (before correction)	6.601	20.16	10.01	169.3
Karimäki fit (after correction)	6.548	19.55	9.991	169.0
Pratt fit	6.549	19.56	9.990	169.0
Taubin fit	6.397	19.42	9.990	169.0
Hyper fit	6.344	19.38	9.990	169.0
Geometric fit	6.317	19.29	9.991	169.3

Table 8.1 *Mean squared error for several circle fits ($10^4 \times$ values are shown). In this test n = 20 points are placed (equally spaced) along an arc of specified size of a circle of radius R = 1. For arcs of 360° and 180°, the noise level is $\sigma = 0.05$. For arcs of 90° and 45°, the noise level is $\sigma = 0.01$.*

were equally spaced on a semicircle (180°), see the second column of Table 8.1. The picture has not changed much, except the SWFL Riemann fit now fares even better; it performs nearly as well as the geometric fit, and better than the Hyper fit.

Next we have simulated even more challenging data sets located on smaller arcs, such as a quarter of the circle (90°), and 1/8 of the circle (45°), see the last two columns of Table 8.1. Though in these cases we had to lower the Gaussian noise to $\sigma = 0.01$, to avoid erratic behavior demonstrated by several fits (notably, by the inversion-based RTKD fit). The picture remains the same, but the SWFL Riemann fit now demonstrates the best performance of all—it beats all the other fits in our test! We will address its phenomenal behavior below.

Then we have repeated our experiment with the number of points increased from $n = 20$ to $n = 100$; the results can be seen in Table 8.2. The overall picture is similar to the one observed in Table 8.1, except on the smallest arc (45°) the inversion-based RTKD fit has diverged all too frequently, making the MSE evaluation impossible; this fact is marked by the "infinity" in the corresponding lines of the last column. Note also that the SWFL Riemann fit is no longer the top performer for the 90° arc, it falls behind the algebraic and geometric fits.

$n = 100$ data points	360° 180° ($\sigma = 0.05$)	90° 45° ($\sigma = 0.01$)
Riemann fit (the SWFL version)	1.382 4.529	2.328 37.12
Inversion-based fit (before iterations)	15.03 46.96	25.98 ∞
Inversion-based fit (after iterations)	1.505 4.639	2.320 ∞
Karimäki fit (before correction)	1.567 5.465	2.342 37.81
Karimäki fit (after correction)	1.503 4.634	2.320 37.39
Pratt fit	1.505 4.639	2.320 37.38
Taubin fit	1.326 4.468	2.320 37.38
Hyper fit	1.266 4.412	2.319 37.38
Geometric fit	1.270 4.387	2.319 37.38

Table 8.2 *Mean squared error for several circle fits ($10^4 \times$ values are shown). In this test $n = 100$ points are placed (equally spaced) along an arc of specified size of a circle of radius $R = 1$. For arcs of 360° and 180°, the noise level is $\sigma = 0.05$. For arcs of 90° and 45°, the noise level is $\sigma = 0.01$.*

On the other hand, the Riemann fit still is still the winner for the smallest 45° arc.

The unexpectedly strong performance of the SWFL fit in some cases is a pleasant surprise. It would be tempting to investigate the accuracy of this fit theoretically, as we did in Chapter 7. Unfortunately, this task appears prohibitively difficult, as our version of this fit involves the centering and rescaling of the data described in Section 8.4 (recall that these steps were added for the sake of invariance under translations and similarities). It can be easily seen that these steps bring serious nonlinearities to the mathematical formulas, and we are currently unable to handle such complications in our theoretical analysis.

Final conclusions. In this last chapter, we have seen various sophisticated mathematical tools (conformal maps, stereographic projection, trigonometric change of parameters) applied to the circle fitting problem, in an attempt to simplify the original objective function (4.18), decouple its variables, and produce a fast and reliable minimization scheme. These attempts do achieve some success, they provide acceptable fits under certain conditions, especially when additional information is available (e.g., a point on the fitting circle is given).

But a detailed analysis shows that most of the resulting fits turn out to be just approximations to the algebraic circle fit proposed by Pratt (Section 5.5).

Actually one of the new fits—the proper Riemann fit (see Section 8.4)—is identical to the Pratt fit, while the RTKD inversion-based fit and the Karimäki fit only approximate it (to various degrees).

A special note must be made on the "improper" SWFL Riemann fit. Its original version published in [175, 173] is not invariant under translations and similarities, thus it is unreliable. To remedy the situation, we have added two steps to the procedure—centering and rescaling of the data points (see Section 8.4)—to enforce invariance. With these additional steps, the fit turns out to be fairly robust and occasionally beats all the other fits in our studies. Its remarkable performance remains to be investigated theoretically... In any case, it serves as a good evidence that mathematically sophisticated fits described in this chapter have great potential and perhaps should be developed further.

Bibliography

[1] R. J. Adcock. Note on the method of least squares. *Analyst, London*, 4:183–184, 1877.

[2] R. J. Adcock. A problem in least squares. *Analyst, London*, 5:53–54, 1878.

[3] S. J. Ahn, W. Rauh, and H. J. Warnecke. Least-squares orthogonal distances fitting of circle, sphere, ellipse, hyperbola, and parabola. *Pattern Recog.*, 34:2283–2303, 2001.

[4] A. Al-Sharadqah and N. Chernov. Error analysis for circle fitting algorithms. *Electr. J. Statist.*, 3:886–911, 2009.

[5] Y. Amemiya and W. A. Fuller. Estimation for the nonlinear functional relationship. *Annals Statist.*, 16:147–160, 1988.

[6] D. A. Anderson. The circular structural model. *J. R. Statist. Soc. B*, 27:131–141, 1981.

[7] T. W. Anderson. Estimation of linear functional relationships: Approximate distributions and connections with simultaneous equations in econometrics. *J. R. Statist. Soc. B*, 38:1–36, 1976.

[8] T. W. Anderson. Estimating linear statistical relationships. *Annals Statist.*, 12:1–45, 1984.

[9] T. W. Anderson and H. Rubin. Statistical inference in factor analysis. In *Proc. Third Berkeley Symp. Math. Statist. Prob.*, volume 5, pages 111–150. Univ. California Press, 1956.

[10] T. W. Anderson and T. Sawa. Exact and approximate distributions of the maximum likelihood estimator of a slope coefficient. *J. R. Statist. Soc. B*, 44:52–62, 1982.

[11] I. O. Angell and J. Barber. An algorithm for fitting circles and ellipses to megalithic stone rings. *Science and Archaeology*, 20:11–16, 1977.

[12] A. Atieg and G. A. Watson. Fitting circular arcs by orthogonal distance regression. *Appl. Numer. Anal. Comput. Math.*, 1:66–76, 2004.

[13] J. Bentz. Private communication, 2001.

[14] R. H. Berk. Sphericity and the normal law. *Annals Prob.*, pages 696–701, 1986.

[15] M. Berman. Estimating the parameters of a circle when angular differences are known. *Appl. Statist.*, 32:1–6, 1983.

[16] M. Berman. The asymptotic statistical behaviour of some estimators for a circular structural model. *Statist. Prob. Lett.*, 7:413–416, 1989.

[17] M. Berman. Large sample bias in least squares estimators of a circular arc center and its radius. *CVGIP: Image Understanding*, 45:126–128, 1989.

[18] M. Berman and D. Culpin. The statistical behavior of some least squares estimators of the center and radius of a circle. *J. R. Statist. Soc. B*, 48:183–196, 1986.

[19] M. Berman and D. Griffiths. Incorporating angular information into models for stone circle data. *Appl. Statist.*, 34:237–245, 1985.

[20] M. Berman and P. I. Somlo. Efficient procedures for fitting circles and ellipses with application to sliding termination measurements. *IEEE Trans. Instrum. Meas.*, 35:31–35, 1986.

[21] R. H. Biggerstaff. Three variations in dental arch form estimated by a quadratic equation. *J. Dental Res.*, 51:1509, 1972.

[22] P. T. Boggs, R. H. Byrd, J. E. Rogers, and R. B. Schnabel. Users reference guide for ODRPACK version 2.01. Technical Report NISTIR 4834, NIST, US Department of Commerce, 1992.

[23] P. T. Boggs, R. H. Byrd, and R. B. Schnabel. A stable and efficient algorithm for nonlinear orthogonal distance regression. *SIAM J. Sci. Stat. Comput.*, 8:1052–1078, 1987.

[24] F. L. Bookstein. Fitting conic sections to scattered data. *Computer Graphics and Image Processing*, 9:56–71, 1979.

[25] J. A. Brandon and A. Cowley. A weighted least squares method for circle fitting to frequency response data. *J. Sound and Vibration*, 89:419–424, 1983.

[26] D. Calvetti and L. Reichel. Gauss quadrature applied to trust region computations. *Numerical Algorithms*, 34:85–102, 2003.

[27] R. J. Carroll, D. Ruppert, and L. A. Stefansky. *Measurement Error in Nonlinear Models*. Chapman & Hall, London, 1st edition, 1995.

[28] R. J. Carroll, D. Ruppert, L. A. Stefansky, and C. M. Crainiceanu. *Measurement Error in Nonlinear Models: A Modern Perspective*. Chapman & Hall, London, 2nd edition, 2006.

[29] G. Casella and R. L. Berger. *Statistical Inference*. Brooks/Cole, 3rd edition, 2002.

[30] L. K. Chan and T. K. Mak. On the polynomial functional relationship. *J. R. Statist. Soc. B*, 47:510–518, 1985.

[31] N. N. Chan. On circular functional relationships. *J. R. Statist. Soc. B*, 27:45–56, 1965.

[32] Y. T. Chan, Y. Z. Elhalwagy, and S. M. Thomas. Estimation of circle parameters by centroiding. *J. Optimiz. Theory Applic.*, 114:363–371, 2002.

[33] Y. T. Chan, B. H. Lee, and S. M. Thomas. Unbiased estimates of circle parameters. *J. Optimiz. Theory Applic.*, 106:49–60, 2000.

[34] Y. T. Chan, B. H. Lee, and S. M. Thomas. Approximate maximum likehood estimation of circle parameters. *J. Optimiz. Theory Applic.*, 125:723–733, 2005.

[35] Y. T. Chan and S. M. Thomas. Cramer-Rao lower bounds for estimation of a circular arc center and its radius. *Graph. Models Image Proc.*, 57:527–532, 1995.

[36] B. B. Chaudhuri and P. Kundu. Optimum circular fit to weighted data in multi-dimensional space. *Patt. Recog. Lett.*, 14:1–6, 1993.

[37] C.-L. Cheng and A. Kukush. Non-existence of the first moment of the adjusted least squares estimator in multivariate errors-in-variables model. *Metrika*, 64:41–46, 2006.

[38] C.-L. Cheng and J. W. Van Ness. On the unreplicated ultrastructural model. *Biometrika*, 78:442–445, 1991.

[39] C.-L. Cheng and J. W. Van Ness. On estimating linear relationships when both variables are subject to errors. *J. R. Statist. Soc. B*, 56:167–183, 1994.

[40] C.-L. Cheng and J. W. Van Ness. *Statistical Regression with Measurement Error*. Arnold, London, 1999.

[41] N. Chernov. Fitting circles to scattered data: parameter estimates have no moments, *METRIKA, in press*, 2010.

[42] N. Chernov and C. Lesort. Statistical efficiency of curve fitting algorithms. *Comp. Stat. Data Anal.*, 47:713–728, 2004.

[43] N. Chernov and C. Lesort. Least squares fitting of circles. *J. Math. Imag. Vision*, 23:239–251, 2005.

[44] N. Chernov and P. Sapirstein. Fitting circles to data with correlated noise. *Comput. Statist. Data Anal.*, 52:5328–5337, 2008.

[45] N. I. Chernov and G. A. Ososkov. Effective algorithms for circle fitting. *Comp. Phys. Comm.*, 33:329–333, 1984.

[46] W. Chojnacki, M. J. Brooks, and A. van den Hengel. Fitting surfaces to data with covariance information: fundamental methods applicable to computer vision. Technical Report TR99-03, Department of Computer Science, University of Adelaide, 1999. Available at

http://www.cs.adelaide.edu.au/~mjb/mjb_ abstracts.html.

[47] W. Chojnacki, M. J. Brooks, and A. van den Hengel. Rationalising the renormalisation method of Kanatani. *J. Math. Imaging & Vision*, 14:21–38, 2001.

[48] W. Chojnacki, M. J. Brooks, A. van den Hengel, and D. Gawley. From FNS to HEIV: a link between two vision parameter estimation methods. *IEEE Trans. Pattern Analysis Machine Intelligence*, 26:264–268, 2004.

[49] W. Chojnacki, M. J. Brooks, A. van den Hengel, and D. Gawley. FNS, CFNS and HEIV: A unifying approach. *J. Math. Imaging Vision*, 23:175–183, 2005.

[50] W. R. Cook. On curve fitting by means of least squares. *Philos. Mag. Ser. 7*, 12:1025–1039, 1931.

[51] I. D. Coope. Circle fitting by linear and nonlinear least squares. *J. Optim. Theory Appl.*, 76:381–388, 1993.

[52] C. Corral and C. Lindquist. On implementing Kasa's circle fit procedure. *IEEE Trans. Instrument. Measur.*, 47:789–795, 1998.

[53] J. F. Crawford. A non-iterative method for fitting circular arcs to measured points. *Nucl. Instr. Meth.*, 211:223–225, 1983.

[54] P. Delogne. Computer optimization of Deschamps' method and error cancellation in reflectometry. In *Proc. IMEKO-Symp. Microwave Measurement (Budapest)*, pages 117–123, 1972.

[55] W. E. Deming. The application of least squares. *Philos. Mag. Ser. 7*, 11:146–158, 1931.

[56] G. R. Dolby. The ultrastructural relation: A synthesis of the functional and structural relations. *Biometrika*, 63:39–50, 1976.

[57] I. Fazekas and A. Kukush. Asymptotic properties of an estimator in nonlinear functional errors-in-variables models with dependent error terms. *Computers Math. Appl.*, 34:23–39, 1997.

[58] I. Fazekas and A. Kukush. Asymptotic properties of estimators in nonlinear functional errors-in-variables with dependent error terms. *J. Math. Sci.*, 92:3890–3895, 1998.

[59] I. Fazekas, A. Kukush, and S. Zwanzig. Correction of nonlinear orthogonal regression estimator. *Ukr. Math. J.*, 56:1101–1118, 2004.

[60] W. Feller. *An Introduction to Probability Theory and Its Applications, Vol. 2*. John Wiley & Sons, 2nd edition, 1971.

[61] A. Fitzgibbon, M. Pilu, and R. Fisher. Direct least-square fitting of ellipses. In *International Conference on Pattern Recognition*, volume 21, pages 476–480, Vienna, August 1996.

[62] A. W. Fitzgibbon and R. B. Fisher. A buyer's guide to conic fitting. In

British Machine Vision Conf., pages 513–522, 1995.

[63] A. W. Fitzgibbon, M. Pilu, and R. B. Fisher. Direct least squares fitting of ellipses. *IEEE Trans. Pattern Analysis and Machine Intelligence*, 21:476–480, 1999.

[64] J. L. Fleiss. The distribution of a general quadratic form in normal variables. *J. Amer. Statist. Assoc.*, 66:142–144, 1971.

[65] P. R. Freeman. Note: Thom's survey of the Avebury ring. *J. Hist. Astronom.*, 8:134–136, 1977.

[66] W. A. Fuller. *Measurement Error Models*. L. Wiley & Son, New York, 1987.

[67] W. Gander, G. H. Golub, and R. Strebel. Least squares fitting of circles and ellipses. *BIT*, 34:558–578, 1994.

[68] J. Gates. Testing for circularity of spacially located objects. *J. Applied Statist.*, 20:95–103, 1993.

[69] C.-F. Gauss. *Theoria Motus Corporum Coelestium in sectionibus conicis solem ambientium*. Book 2, 1809.

[70] D. M. Gay. Computing optimal locally constrained steps. *SIAM J. Scientif. Statist. Computing*, 2:186–197, 1981.

[71] P. E. Gill and W. Murray. Algorithms for the solution of the nonlinear least-squares problem. *SIAM J. Numer. Anal.*, 15:977–992, 1978.

[72] C. Gini. Sull'interpolazione di una retta quando i valori della variabile indipendente sono affetti da errori accidentali. *Metron*, 1:63–82, 1921.

[73] L. J. Gleser. Estimation in a multivariate 'errors in variables' regression model: Large sample results. *Annals Statist.*, 9:24–44, 1981.

[74] L. J. Gleser. Functional, structural and ultrastructural errors-in-variables models. In *Proc. Bus. Econ. Statist. Sect. Am. Statist. Ass.*, pages 57–66, 1983.

[75] L. J. Gleser. A note on G. R. Dolby's unreplicated ultrastructural model. *Biometrika*, 72:117–124, 1985.

[76] L. J. Gleser and G. S. Watson. Estimation of a linear transformation. *Biometrika*, 60:525–534, 1973.

[77] G. H. Golub and C. F. Van Loan. An analysis of the total least squares problem. *SIAM J. Numer. Anal.*, 17:883–893, 1980.

[78] G. H. Golub and C. F. van Loan. *Matrix computations*. J. Hopkins U. Press, 3rd edition, 1996.

[79] Z. Griliches and V. Ringstad. Error-in-the-variables bias in nonlinear contexts. *Econometrica*, 38:368–370, 1970.

[80] R. Haliř and J. Flusser. Numerically stable direct least squares fitting of

ellipses. In *Sixth Conf. Cent. Europe Comput. Graph. Vis., WSCG'98, Conf. Proc.*, volume 1, pages 125–132, Plzen, Czech Rep., 1998.

[81] R. Haliř and Ch. Menard. Diameter estimation for archaeological pottery using active vision. In Axel Pinz, editor, *Proc. 20th Workshop Austrian Assoc. Pattern Recognition (ÖAGM'96)*, pages 251–261, Schloss Seggau, Leibnitz, 1996.

[82] M. Hansroul, H. Jeremie, and D. Savard. Fast circle fit with the conformal mapping method. *Nucl. Instr. Methods Phys. Res. A*, 270:498–501, 1988.

[83] R. M. Haralick and L. G. Shapiro. *Computer and Robot Vision, Vol. I.* Addison-Wesley, Reading, MA, 1992.

[84] http://www.math.uab.edu/ chernov/cl.

[85] P. J. Huber. Robust estimation of a location parameter. *Annals Math. Statist.*, 35:73–101, 1964.

[86] P. J. Huber. The behavior of maximum likelihood estimates under nonstandard conditions. In *Proc. Fifth Berkeley Symp. Math. Stat. Prob.*, volume 1, pages 221–233, Univ. California Press, Berkeley, 1967.

[87] P. J. Huber. *Robust Statistics.* Wiley-Interscience, 1981.

[88] J. P. Imhof. Computing the distribution of quadratic forms in normal variables. *Biometrika*, 48:419–426, 1961.

[89] W. N. Jessop. One line or two? *Appl. Statist.*, 1:131–137, 1952.

[90] S. H. Joseph. Unbiased least-squares fitting of circular arcs. *Graph. Mod. Image Process.*, 56:424–432, 1994.

[91] J. B. Kadane. Testing overidentifying restrictions when the disturbances are small. *J. Amer. Statist. Assoc.*, 65:182–185, 1970.

[92] J. B. Kadane. Comparison of k-class estimators when the disturbances are small. *Econometrica*, 39:723–737, 1971.

[93] K. Kanatani. Renormalization for unbiased estimation. In *Proc. 4th Intern. Conf. Computer Vision (ICCV'93)*, pages 599–606, Berlin, Germany, 1993.

[94] K. Kanatani. Statistical bias of conic fitting and renormalization. *IEEE Trans. Pattern Analysis Machine Intelligence*, 16:320–326, 1994.

[95] K. Kanatani. *Statistical Optimization for Geometric Computation: Theory and Practice.* Elsevier Science, Amsterdam, Netherlands, 1996.

[96] K. Kanatani. Cramer-Rao lower bounds for curve fitting. *Graph. Mod. Image Process.*, 60:93–99, 1998.

[97] K. Kanatani. For geometric inference from images, what kind of statistical method is necessary? *Mem. Faculty Engin. Okayama Univ.*, 37:19–28, 2002.

[98] K. Kanatani. How are statistical methods for geometric inference justified? In *3rd Intern. Conf. Statist. Comput. Theories of Vision (SCTV 2003)*, 2003, Nice, France, 2003.

[99] K. Kanatani. For geometric inference from images, what kind of statistical model is necessary? *Syst. Comp. Japan*, 35:1–9, 2004.

[100] K. Kanatani. Optimality of maximum likelihood estimation for geometric fitting and the KCR lower bound. *Memoirs Fac. Engin. Okayama Univ.*, 39:63–70, 2005.

[101] K. Kanatani. Unraveling the mystery of renormalization. *IPSJ SIG Technical Reports*, 38:15–22, 2005.

[102] K. Kanatani. Ellipse fitting with hyperaccuracy. *IEICE Trans. Inform. Syst.*, E89-D:2653–2660, 2006.

[103] K. Kanatani. Ellipse fitting with hyperaccuracy. In *Proc. 9th European Conf. Computer Vision (ECCV 2006)*, volume 1, pages 484–495, 2006.

[104] K. Kanatani. Statistical optimization for geometric fitting: Theoretical accuracy bound and high order error analysis. *Int. J. Computer Vision*, 80:167–188, 2008.

[105] Y. Kanazawa and K. Kanatani. Optimal line fitting and reliability evaluation. *IEICE Trans. Inform. Syst.*, E79-D:1317–1322, 1996.

[106] V. Karimäki. Effective circle fitting for particle trajectories. *Nucl. Instr. Meth. Phys. Res. A*, 305:187–191, 1991.

[107] V. Karimäki. Fast code to fit circular arcs. *Computer Physics Commun.*, 69:133–141, 1992.

[108] I. Kåsa. A curve fitting procedure and its error analysis. *IEEE Trans. Inst. Meas.*, 25:8–14, 1976.

[109] M. G. Kendall. Regression, structure, and functional relationships, Part I. *Biometrica*, 38:11–25, 1951.

[110] M. G. Kendall. Regression, structure, and functional relationships, Part II. *Biometrica*, 39:96–108, 1952.

[111] M. G. Kendall and A. Stuart. *The Advanced Theory of Statistics, Volume 2, Inference and Relationship*. Griffin, London, 4th edition, 1979.

[112] D. Kincaid and W. Cheney. *Numerical Analysis*. Duxbury Press, 2nd edition, 2002.

[113] T. C. Koopmans. *Linear regression analysis of economic time series*. De erven F. Bohn n.v., Haarlem, 1937.

[114] A. Kukush, I. Markovsky, and S. Van Huffel. Consistent estimation in an implicit quadratic measurement error model. *Comput. Statist. Data Anal.*, 47:123–147, 2004.

[115] A. Kukush, I. Markovsky, and S. Van Huffel. Consistent estimation of

an ellipsoid with known center. In *COMPSTAT Proc. Comput. Statistics, 16th Symp., Prague, Czech Republic*, pages 1369–1376, Heidelberg, Germany, 2004. Physica-Verlag.

[116] A. Kukush and E. O. Maschke. The efficiency of adjusted least squares in the linear functional relationship. *J. Multivar. Anal.*, 87:261–274, 2003.

[117] C. H. Kummell. Reduction of observation equations which contain more than one observed quantity. *Analyst, London*, 6:97–105, 1879.

[118] N. Kunitomo. Asymptotic expansions of the distributions of estimators in a linear functional relationship and simultaneous equations. *J. Amer. Statist. Assoc.*, 75:693–700, 1980.

[119] U. M. Landau. Estimation of a circular arc center and its radius. *CVGIP: Image Understanding*, 38:317–326, 1987.

[120] Y. Leedan and P. Meer. Heteroscedastic regression in computer vision: Problems with bilinear constraint. *Intern. J. Comp. Vision*, 37:127–150, 2000.

[121] A.-M. Legendre. *Nouvelles méthodes pour la détermination des orbites des comètes*. Paris, 1806.

[122] K. Levenberg. A method for the solution of certain non-linear problems in least squares. *Quart. Appl. Math.*, 2:164–168, 1944.

[123] B. Lillekjendlie. Circular arcs fitted on a Riemann sphere. *Computer Vision Image Underst.*, 67:311–317, 1997.

[124] K. B. Lim, K. Xin, and G. S. Hong. Detection and estimation of circular arc segments. *Pattern Recogn. Letters*, 16:627–636, 1995.

[125] R. A. Liming. *Mathematics for Computer Graphics*. Aero Publishers, Fallbrook, CA, 1979.

[126] D. V. Lindley. Regression lines and the linear functional relationship. *Suppl. J. R. Statist. Soc.*, 9:218–244, 1947.

[127] A. Madansky. The fitting of straight lines when both variables are subject to error. *J. Amer. Statist. Ass.*, 54:173–205, 1959.

[128] E. Malinvaud. *Statistical Methods of Econometrics*. North Holland Publ. Co, Amsterdam, 3rd edition, 1980.

[129] K. V. Mardia and D. Holmes. A statistical analysis of megalithic data under elliptic pattern. *J. R. Statist. Soc. A*, 143:293–302, 1980.

[130] I. Markovsky, A. Kukush, and S. Van Huffel. Consistent least squares fitting of ellipsoids. *Numerische Mathem.*, 98:177–194, 2004.

[131] D. Marquardt. An algorithm for least squares estimation of nonlinear parameters. *SIAM J. Appl. Math.*, 11:431–441, 1963.

[132] P. A. P. Moran. Estimating structural and functional relationships. *J.*

Multivariate Anal., 1:232–255, 1971.

[133] J. J. Moré. The Levenberg-Marquardt algorithm: Implementation and theory. In *Lecture Notes in Mathematics, Vol. 630*, pages 105–116, Berlin, 1978. Springer-Verlag.

[134] J. J. Moré, B. B. Garbow, and K. E. Hillstrom. User guide for MINPACK-1. Technical Report ANL-80-74, Argonne National Lab., IL (USA), 1980.

[135] J. J. Moré and D. C. Sorensen. Computing a trust region step. *SIAM J. Scientif. Statist. Computing*, 4:553–572, 1983.

[136] L. Moura and R. I. Kitney. A direct method for least-squares circle fitting. *Comp. Phys. Commun.*, 64:57–63, 1991.

[137] Y. Nievergelt. Hyperspheres and hyperplanes fitted seamlessly by algebraic constrained total least-squares. *Linear Algebra Appl.*, 331:43–59, 2001.

[138] Y. Nievergelt. A finite algorithm to fit geometrically all midrange lines, circles, planes, spheres, hyperplanes, and hyperspheres. *J. Numerische Math.*, 91:257–303, 2002.

[139] N. Nussbaum. Asymptotic optimality of estimators of a linear functional relation if the ratio of the error variances is known. *Math. Op. Statist. Ser. Statist.*, 8:173–198, 1977.

[140] M. O'Neill, L. G. Sinclair, and F. J. Smith. Polynomial curve fitting when abscissas and ordinates are both subject to error. *Computer J.*, 12:52–56, 1969.

[141] G. A. Ososkov. P10-83-187 (in Russian). Technical report, JINR Dubna, 1983.

[142] W. M. Patefield. On the validity of approximate distributions arising in fitting linear functional relationship. *J. Statist. Comput. Simul.*, 5:43–60, 1976.

[143] W. M. Patefield. Multivariate linear relationships: Maximum likelihood estimation and regression bounds. *J. R. Statist. Soc. B*, 43:342–352, 1981.

[144] K. Pearson. On lines and planes of closest fit to systems of points in space. *Phil. Mag.*, 2:559–572, 1901.

[145] S.-C. Pei and J.-H. Horng. Optimum approximation of digital planar curves using circular arcs. *Pattern Recogn.*, 29:383–388, 1996.

[146] W. A. Perkins. A model-based vision system for industrial parts. *IEEE Trans. Comput.*, 27:126–143, 1978.

[147] M. Pilu, A. Fitzgibbon, and R.Fisher. Ellipse-specific direct least-square fitting. In *IEEE International Conference Image Processing*, Lausanne,

September 1996.

[148] M. J. D. Powell. A hybrid method for nonlinear equations. In P. Robinowitz, editor, *Numerical Methods for Nonlinear Programming Algebraic Equations*, pages 87–144, London, 1970. Gordon and Beach Science.

[149] M. J. D. Powell. A new algorithm for unconstrained optimization. In J. B. Rosen, O. L. Mangasarian, and K. Ritter, editors, *Nonlinear Programming, Proc. Sympos. Math. Res. Center*, pages 31–66, New York, 1970.

[150] V. Pratt. Direct least-squares fitting of algebraic surfaces. *Computer Graphics*, 21:145–152, 1987.

[151] W. H. Press, S. A. Teukolsky, W. T. Vetterling, and B. P. Flannery. *Numerical Recipes in C++*. Cambridge Univ. Press., 2nd edition, 2002.

[152] P. Rangarajan and K. Kanatani. Improved algebraic methods for circle fitting. *Electr. J. Statist.*, 3:1075–1082, 2009.

[153] H. Robbins and J. Pitman. Application of the method of mixtures to quadratic forms in normal variates. *Annals Math. Statist.*, 20:552–560, 1949.

[154] J. Robinson. The distribution of a general quadratic form in normal variables. *Austral. J. Statist.*, 7:110–114, 1965.

[155] S. M. Robinson. Fitting spheres by the method of least squares. *Commun. Assoc. Comput. Mach.*, 4:491, 1961.

[156] C. F. Roos. A general invariant criterion of fit for lines and planes where all variates are subject to error. *Metron*, 13:3–30, 1937.

[157] C. Rorres and D. G. Romano. Finding the center of a circular starting line in an ancient Greek stadium. *SIAM Rev.*, 39:745–754, 1997.

[158] P. L. Rosin and G. A. W. West. Segmentation of edges into lines and arcs. *Image Vision Comp.*, 7:109–114, 1989.

[159] C. Rusu, M. Tico, P. Kuosmanen, and E. J. Delp. Classical geometrical approach to circle fitting – review and new developments. *J. Electron. Imaging*, 12:179–193, 2003.

[160] Ed. S. van Huffel. *Recent advances in total least squares techniques and errors-in-variables modeling*. SIAM, 1997.

[161] Ed. S. van Huffel. *Total Least Squares and Errors-in-Variables Modeling*. Kluwer, Dordrecht, 2002.

[162] R. Safaee-Rad, I. Tchoukanov, B. Benhabib, and K. C. Smith. Accurate parameter estimation of quadratic curves from grey-level images. *CVGIP: Image Understanding*, 54:259–274, 1991.

[163] P. D. Sampson. Fitting conic sections to very scattered data: an iterative

refinement of the Bookstein algorithm. *Comp. Graphics Image Proc.*, 18:97–108, 1982.

[164] B. Sarkar, L. K. Singh, and D. Sarkar. Approximation of digital curves with line segments and circular arcs using genetic algorithms. *Pattern Recogn. Letters*, 24:2585–2595, 2003.

[165] E. Saund. Identifying salient circular arcs on curves. *CVGIP: Image Understanding*, 58:327–337, 1993.

[166] C. Shakarji. Least-squares fitting algorithms of the NIST algorithm testing system. *J. Res. Nat. Inst. Stand. Techn.*, 103:633–641, 1998.

[167] S. Shklyar, A. Kukush, I. Markovsky, and S. Van Huffel. On the conic section fitting problem. *J. Multivar. Anal.*, 98:588–624, 2007.

[168] M. E. Solari. 'Maximum Likelihood solution' of the problem of estimating a linear functional relationship. *J. R. Statist. Soc. B*, 31:372–375, 1969.

[169] P. I. Somlo and J. D. Hunter. A six-port reflectometer and its complete characterization by convenient calibration procedures. *IEEE Trans. Microwave Theory Tech.*, 30:186–192, 1982.

[170] H. Späth. Least-squares fitting by circles. *Computing*, 57:179–185, 1996.

[171] H. Späth. Orthogonal least squares fitting by conic sections. In *Recent Advances in Total Least Squares techniques and Errors-in-Variables Modeling*, pages 259–264. SIAM, 1997.

[172] T. Steihaug. Conjugate gradient method and trust regions in large scale optimization. *SIAM J. Numer. Anal.*, 20:626–637, 1983.

[173] A. Strandlie and R. Frühwirth. Error analysis of the track fit on the Riemann sphere. *Nucl. Instr. Methods Phys. Res. A*, 480:734–740, 2002.

[174] A. Strandlie, J. Wroldsen, and R. Frühwirth. Treatment of multiple scattering with generalized Riemann sphere track fit. *Nucl. Instr. Methods Phys. Res. A*, 488:332–341, 2002.

[175] A. Strandlie, J. Wroldsen, R. Frühwirth, and B. Lillekjendlie. Particle tracks fitted on the Riemann sphere. *Computer Physics Commun.*, 131:95–108, 2000.

[176] G. Taubin. Estimation of planar curves, surfaces and nonplanar space curves defined by implicit equations, with applications to edge and range image segmentation. *IEEE Trans. Pattern Analysis Machine Intelligence*, 13:1115–1138, 1991.

[177] A. Thom. A statistical examination of the megalithic sites in britain. *J. Royal Statist. Soc. A*, 118:275–295, 1955.

[178] A. Thom and A. S. Thom. A megalithic lunar observatory in Orkney:

the ring of Brogar and its cairns. *J. Hist. Astronom.*, 4:111–123, 1973.

[179] A. Thom, A. S. Thom, and T. R. Foord. Avebury (1): a new assessment of the geometry and metrology of the ring. *J. Hist. Astronom.*, 7:183–192, 1976.

[180] S. M. Thomas and Y. T. Chan. A simple approach for the estimation of circular arc center and its radius. *CVGIP: Image Understanding*, 45:362–370, 1989.

[181] B. Triggs. A new approach to geometric fitting, ICCV'98, Bobbay (unpublished), 1998.

[182] B. Triggs. Optimal estimation of matching constraints. In *3D Structure from Multiple Images of Large-Scale Environments*, volume 1506, pages 63–77, Berlin/Heidelberg, 1998. Springer.

[183] K. Turner. *Computer perception of curved objects using a television camera*. PhD thesis, University of Edinburgh, 1974. Dept. of Machine Intelligence.

[184] D. Umbach and K. N. Jones. A few methods for fitting circles to data. *IEEE Trans. Instrument. Measur.*, 52:1181–1885, 2003.

[185] S. van Huffel and J. Vandewalle. *The Total Least Squares Problem: Computational Aspects and Analysis*. SIAM, Philadelphia, 1991.

[186] G. Vosselman and R. M. Haralick. Performance analysis of line and circle fitting in digital images. In *Workshop on Performance Characteristics of Vision Algorithms*, April 1996.

[187] A. Wald. The fitting of straight lines if both variables are subject to error. *Ann. Math. Statist.*, 11:284–300, 1940.

[188] C. M. Wang and C. T. Lam. A mixed-effects model for the analysis of circular measurements. *Technometrics*, 39:119–126, 1997.

[189] M. Werman and D. Keren. A Bayesian method for fitting parametric and nonparametric models to noisy data. *IEEE Trans. Pattern Analysis and Machine Intelligence*, 23:528–534, 2001.

[190] M. E. Whalen. Ceramic vessel size estimation from sherds: An experiment and a case study. *J. Field Archaeology*, 25:219–227, 1998.

[191] J. Wolfowitz. Estimation of structural parameters when the number of incidental parameters is unbounded. *Annals Math. Statist.*, 25:811, 1954.

[192] K. M. Wolter and W. A. Fuller. Estimation of nonlinear errors-in-variables models. *Annals Statist.*, 10:539–548, 1982.

[193] Z. Wu, L. Wu, and A. Wu. The robust algorithms for finding the center of an arc. *Comp. Vision Image Under.*, 62:269–278, 1995.

[194] S. J. Yin and S. G. Wang. Estimating the parameters of a circle by

heteroscedastic regression models. *J. Statist. Planning Infer.*, 124:439–451, 2004.

[195] P. C. Yuen and G. C. Feng. A novel method for parameter estimation of digital arcs. *Pattern Recogn. Letters*, 17:929–938, 1996.

[196] E. Zelniker and V. Clarkson. A statistical analysis of the Delogne-Kåsa method for fitting circles. *Digital Signal Proc.*, 16:498–522, 2006.

[197] S. Zhang, L. Xie, and M. D. Adams. Feature extraction for outdoor mobile robot navigation based on a modified gaussnewton optimization approach. *Robotics Autonom. Syst.*, 54:277–287, 2006.

[198] Z. Zhang. Parameter estimation techniques: A tutorial with application to conic fitting. *Inter. J. Image & Vision Comp.*, 15:59–76, 1997.

Index